Environmental Biogeography

Environmental Biogeography

Paul Ganderton and Paddy Coker

Harlow, England • London • New York • Boston • San Francisco • Toronto • Sydney • Singapore • Hong Kong
Tokyo • Seoul • Taipei • New Delhi • Cape Town • Madrid • Mexico City • Amsterdam • Munich • Paris • Milan

Pearson Education Limited
Edinburgh Gate
Harlow
Essex CM20 2JE
England

and Associated Companies throughout the world

Visit us on the World Wide Web at:
www.pearsoned.co.uk

First published 2005

ISBN 0582 31829 7

British Library Cataloguing-in-Publication Data
A catalogue record for this book is available from the British Library

Library of Congress Cataloging-in-Publication Data

A catalog record for this book is available from the Library of Congress

10 9 8 7 6 5 4 3 2 1
10 09 08 07 06 05

Typeset in 10/12.5 Minion by 30
Printed by Ashford Colour Press Ltd, Gosport

The publisher's policy is to use paper manufactured from sustainable forests.

CONTENTS

PREFACE

When we set out to write this book, we wanted to produce something that would bring a fresh dimension to the subject and to this end, we decided upon a structure that would look at biogeography in theory, practice and in synthesis and make it even more readily accessible to ecologists, environmental scientists and geographers at undergraduate and senior high school level. We looked through the available texts, some of which had been around for a number of years, and felt that although many were soundly based, they lacked breadth of coverage, particularly on ecological and environmental issues and how biogeographical knowledge could be applied to management. In addition, we felt that while many books give a solid treatment the data were often old and new ideas could not fit easily into them. In *Environmental Biogeography*, we hope that we have achieved our aim of making a book and the associated website a framework which can help provide a greater understanding of the way in which ecology and biogeography (in combination) can contribute to a greater knowledge of what goes on in the natural world which, we hope, will prompt its readers to go and find out more for themselves – and exercise their natural curiosity.

Both authors would like to thank Andrew Taylor, our acquisitions editor at Pearson Education, for his patience and kindness and for giving us a truly memorable level of support. Matthew Smith (who commissioned the book some years ago) believed in us, helped us get under way and provided us with a great deal of support for which we are truly grateful. The anonymous reviewers for the original submission and sample chapters did a great job, providing excellent additional ideas, tempering constructive criticism with praise and urging us to carry on with the project (– we worked out who you are!). In the latter stages of production our thanks must also go to Emma Travis and Natasha Dupont, whose knowledge and unfailing help got this book from conceptual to actual paper pages. Philip Tye did a great job of copyediting the manuscript.

Many people have contributed ideas, material or help in various ways, sometimes without even realising it. In particular we would like to thank Dr Hilary Thomas, Dr Roy Alexander, Professor Kasue Fujiwara (for organising the trip to Yakushima), Professor Martin Kent, Professor John Matthews, Dr Peter Glaves, Dr Anne-Maria Brennan, Dr Charles Nelson, Neil Redgate, Maria Hartley and Mike Wood. Apart from colleagues we must also thank several generations of students who have learned from/put up with us in our quest to show the relevance and importance of this subject. Although we have learned a lot from all of you, the usual comment about errors being our own still applies!

PC wants particularly to thank his co-author for keeping the momentum going, raiding online resources and being there, cheerful and encouraging at the end of a 12,000-mile phone call or through email. Rosemary and Bryony gave their help and encouragement unstintingly. His thanks are also due to Dr Francis Rose and the late Professor Jan Barkman for introducing him to biogeography back in the late 1960s and to the late Professor John Pugh for giving him the chance to work in the Geography Department at King's College London and to begin to find out what *really* makes geographers and biogeographers tick!

PSG wants to thank Maggie for her patience and especially Tim, Beth and Tom for their constant enthusiasm for learning which is what got us started in the first place. Also, thanks to PC for starting him along the environmental road too many years ago!

Paul Ganderton, Sydney, Australia
Paddy Coker, Farnborough, Kent, UK
February 2005

Supporting resources

Visit **www.pearsoned.co.uk/ganderton** to find valuable online resources

For students and instructors

- Suggestions for projects and coursework
- Guest articles from international biogeographers
- Regular updates and book reviews
- Links to current issues, relevant journals, magazines and other sites
- An online glossary to explain key terms

For more information please contact your local Pearson Education sales representative or visit **www.pearsoned.co.uk/ganderton**

ACKNOWLEDGEMENTS

We are grateful to the following for permission to reproduce copyright material:

Figure for Box 3.3 courtesy of John Campbell, USDA Forest Service; Figure 7.3 courtesy of the Royal Scottish Geographical Society.

In some instances we have been unable to trace the owners of copyright material and we would appreciate any information that would enable us to do so.

PART ONE

Theoretical aspects of biogeography

CHAPTER 1

What is biogeography?

Key points

- Biogeography is an important and multidisciplinary area of contemporary science;
- It is primarily concerned with current and historical distribution patterns of plants and animals rather than species–environment interactions;
- Biogeography operates through observation and detection of pattern, both small and large scale;
- It has links with almost every aspect of human activity on the planet – from agriculture to reclamation.

1.1 Introduction

Biogeography, because of its multifaceted nature, encourages people to *share* experiences and expertise, and to see the wider value in the approach of a discipline with which they may be unfamiliar, or to take a sideways look at their own expertise to see how relevant it is for problem solving in another area. This is a far cry from the straitjacketed school and university curricula of the first half of the last century when far too many geographers and ecologists never talked to each other or had even a modicum of interest outside the narrow confines of 'geography' or 'biology'. One explanation for this lack of communication might be that specialists often felt uncomfortable when dealing with topics outside their speciality. There remain a number of specialists who still, even in the twenty-first century, work within narrowly defined areas of interest, but the advent of the Internet and ready access to a treasure trove of information from all over the world has made a tremendous difference. Even taking into account that not all Internet material is peer-reviewed or valid, there is still a wealth of information from reliable academic, industrial, non-governmental and governmental organisations, both text and video-based. It is as well to be a little sceptical and to, if possible, find independent confirmation of material before relying on its authenticity or veracity.

The study of biogeography has been part of the scientific landscape for almost 200 years but its importance as a link between the earth sciences (geology and geography) and the life sciences has only been realised fully in the past few decades. It was an intriguingly new topic in some university geography departments during the 1960s and was sometimes known as the 'ecological fifth column', bent on infusing conventional physical geography with 'dangerous' biological ideas. Its influence spawned a more environmental and holistic view of the world at an important time and contributed in a number of ways to a small but significant broadening of the traditional primary and secondary education curriculum in many parts of the world (Carson 1978).

In the UK and elsewhere, new school curricula included environmental courses that were often team-taught by teachers from conventionally disparate disciplines such as biology, sociology, economics, geology and geography. At the same time, the numbers of higher education courses at degree and diploma level in which environmental and biogeographical concerns were significant elements increased massively, paralleling growing environmental awareness throughout the 1980s and early 1990s (Coker 1978).

1.2 Biogeography in context

So, how could one define the term 'biogeography'? It is certainly a science that attempts to document and understand spatial patterns of biodiversity. It is also the study of distribution patterns, both past and present, of plants and animals and their variations in numbers and species on a global scale. Geographers tend to regard it as a component of physical geography that deals with organisms, and biologists regard it as relating more to ecology and environment. It deals with three fundamental, time-related processes – evolution, dispersal and extinction; evolution arises as a result of an irreversible genetic change in an individual or population, dispersal occurs when organisms move out from their origins and extinction occurs when a species is permanently eliminated by natural (and sometimes unnatural) actions which ensure that no living examples remain. A trawl through the literature reveals a definition (Christopherson 1994) which encompasses all these aspects:

> [Biogeography is] the study of the distribution of plants and animals and related ecosystems, the geographical relationships with related environments over time, essentially a spatial ecology. To better understand an ecosystem, biogeographers must look across the ages and re-assemble Earth's tectonic plates to recreate environmental relationships.

Biogeography is not generally regarded as an experimental science, neither is it entirely based upon theories. It is, above all, a science of observation and comparison, often dealing with spatial and temporal scales which are on a global or continental scale. It is, unjustly, regarded as a 'soft science' since the conventional ways of investigation, so often found in the physical and molecular sciences which often yield 'exact' answers, are rarely found in biogeography. Biogeography tends to deal with species and habitat distributions whereas ecology tends to concentrate upon species interactions with each other and the environment.

Biogeography functions through the development of theories derived from searching for patterns, testing assumptions and predictions, with new observations. Very often the testing and pattern searching is done with the aid of statistical and computational techniques. It utilises natural events such as volcanic eruptions or the movements of glacier ice as experiments to shed light on rates of change through long-term monitoring. The topics of plate tectonics and continental drift, first proposed 150 years ago, have, since their ultimate acceptance in the 1960s, been a major reason for the development of biogeography. Distribution patterns of both fossil and living species were now seen in perspective, not as isolated chances but as temporal changes of distribution over many millions of years, aided by the movement of continental plates. Even disjunct distributions began to make sense once it was understood that continental-scale movements of the earth's crust were happening over extended periods of time, and that populations of species might become isolated as a result – as shown by the remarkably similar fossil records of parts of South America and Southern Africa. Biogeography also helps to explain many aspects of dispersal and the effect that barriers may have on the movement of species. Barriers may be as simple as a watercourse which is too wide for a land mammal to cross, or a mountain chain which has no viable passes. Small-scale barriers might include such aspects as the variation in a habitat factor such as soil moisture or nutrient availability, or even the impact of a particular isotherm (Dahl 1951).

A biogeographer may specialise and become a phytogeographer (dealing with plants), a zoogeographer (studying animals), or a historical biogeographer, whose work focuses upon reconstruction of the evolution, origin, dispersal and extinction patterns of groups of animals and/or plants. Ecological biogeographers study the present-day patterns of distributions of plants and animals in relation to their physical environment. This last specialism is the authors' preferred way of tackling the topic.

1.3 Biogeography – an inquiring discipline

One of the best aspects of biogeography is the freedom it gives to ask seemingly simple yet profound questions such as:

- Why are there so many more species in tropical rainforest than in temperate woodland?
- What allows a species to live in one area and does not allow it to disperse and colonise other areas?
- Why do animals living on isolated islands often grow larger than their relatives on the mainland?
- What is (or was) distinctive about the plants and animals of large, isolated islands such as Madagascar, Mauritius, New Zealand and Australia?

Questions such as these frequently have no single, simple answer but the biogeographer is able to interpret and analyse what information is currently available, possibly from many diverse sources, and to come up with theories which may throw some light on the problem.

Biogeography is, above all, a science of synthesis, bringing in the best aspects of many other disciplines, not only of conventional sciences, but also those of history, mathematics and language. It relates critically to topics of immediate importance such as global climate change, the ecology of extinctions, as well as environmental management and world population growth and resource availability. See, for example, Brown and Lomolino (1996).

1.4 Biogeography in real-life situations

Very many aspects of biogeography have some relevance to our present planetary lifestyle because they are intimately linked with our use of natural resources and management of the environment, both biotic and abiotic. Thomas (2004) refers to four important aspects for biogeography as a management tool – agriculture, conservation, environmental tools and recreation/amenity. She goes on to show that management of, for example, agricultural systems focuses on the provision of optimal conditions for particular plants or animals of economic use through the direction and control of normal processes of succession in that particular biome. Control often amounts to elimination of competing species and provision of extra plant nutrients.

Very often, species of economic use are now grown in areas far removed from their origin (many of our cereal species such as wheat originated in western Asia and are grown throughout the world, and the ever-present orange has its natural home in South-East Asia), demonstrating the adaptability of species to new environmental conditions. Alien plant species such as these may require management of native competing species which would tend to exclude them – as happens when ornamental gardens are neglected and become overgrown with native weeds and scrub. Aims of agriculture and conservation frequently differ and the need to provide land for agriculture may have important implications for the conservation of rare or endemic species or habitats. While it is possible from the comfort of a developed world standpoint to applaud the designation of a conservation area in the less developed world to allow a species of bird or mammal to survive, the setting up of such areas often has a human cost. The profound effects on the lives and traditions of indigenous people who may be translocated to areas away from the original territory have all too often been ignored, and a reflective view of the ethics of conservation might lead one to seriously question the value of conserving a species which may be on the edge of extinction through natural causes such as plant succession or climatic change, compared with uprooting and destroying the traditions and ways of life of other human beings. One should also not overlook the progressive loss of biodiversity in many intensively farmed environments where management practices tend directly and indirectly to reduce the numbers of species through the use of pesticides or supplementary fertilisers, nor the environmental damage which can be caused to fragile environments through recreation and ecotourism and the need to develop 'facilities' which will encourage the tourists, often at the expense of the local environment and people. Similarly, the use of plants as environmental tools can have important economic and environmental implications. Stabilisation of slopes with plants (such as marram grass (*Ammophila*) on sand dunes or grass mixtures on roadside embankments) or the planting of resistant strains of plants in areas subject to pollution from heavy metals are cases where the plants perform a *specific* function, rather than being present as part of a natural ecosystem. Where appropriate in each of the following chapters, we have included some examples in an Applications section.

1.5 What happens when we ignore biogeography?

The biogeographical and ecological consequences of the introduction of non-native species for sport such as rabbits in Australia, or for their aesthetic or visual effect, as with European gorse (*Ulex europaeus*) in New Zealand or *Rhododendron ponticum* (from Turkey) in the British Isles, have been quite serious (see Figure 1.1). These species are highly competitive against native species and reproduce rapidly, colonising quickly because there are few, if any, natural enemies in their new habitats. Control measures – in particular the introduction of the myxomatosis virus about 50 years ago which only affects rabbits – have reduced the problem to more manageable proportions in Australia. As far as New Zealand gorse is concerned, insect and myco-herbicide (based on fungi) controls have been introduced with some recent success. Biological control methods are less environmentally damaging than use of chemical sprays, but sometimes the control agent goes on to become a pest in its own right as with the African mongoose which was introduced

Figure 1.1 Rhododendrons have beautiful flowers but little else to commend them in areas outside their natural occurrence. The seeds are extremely small and produced in huge numbers. After a slow start, rhododendrons grow very rapidly and aggressively shade out ground flora

into the sugar-producing areas of the Caribbean to control an economic pest, the cane rat. The problem was that once the cane rat had been controlled, the thriving mongoose population started predating on bird species, eating eggs and young, and has had to be controlled in turn.

Overall, there have been remarkably few cases worldwide where introduced species have integrated well into their new habitat. Something usually goes wrong – either the introduced species fails to thrive and dies out or more seriously, it outcompetes native species which are then lost as a result of excessive predation or habitat change brought about by the newcomer. The problem is that many governments and international organisations have not, until comparatively recently, listened to scientific advice based upon sound ecological and biogeographical principles and research.

One of the most striking examples of an alien species' power to spread is provided by a freshwater bivalve mollusc, the zebra mussel (*Dreissena polymorpha*) whose origins are in the area of the Black, Caspian and Aral seas in eastern Europe and western Asia. The mussel spreads from one water body to another, usually on the hulls of ships or in ballast water, and by this means spread through Europe (Hungary in 1790, France and Germany by 1830, and England and Denmark at about the same time). Interestingly, it did not reach Scandinavia until the 1940s and is now colonizing rivers in the western part of Russia. The zebra mussel is unable to tolerate normal salinity so any sea transits would have to be in ballast or as an accidental part of cargo.

Its arrival in North America in the mid to late 1980s and its rapid spread are documented in an animated map which clearly shows its spread throughout the river systems of the eastern United States from a presumed entry point in the Lake Erie area on the Canadian–US border where it was first noted in 1988. Within 6 years, it had spread down as far as the Mississippi delta and throughout the larger rivers such as the Ohio, Tennessee and Arkansas, probably on the hulls of river barges. Checks have shown that zebra mussels can survive for a time out of water since they have been detected on the hulls of trailered small craft as far west as California. Accidental transport from

CASE STUDY

A prickly problem

Occasionally, introduction of species to new areas goes very badly wrong, as in the case of the prickly pear cacti (*Opuntia* spp.), introduced from America to eastern Australia in the nineteenth century both as a decorative garden plant and as an impenetrable barrier to control livestock and act as an alternate food source in time of drought. *Opuntia* spreads rapidly since its segments can root individually if they fall on the ground and it has succulent and edible fruits which are eaten by birds and other animals. The seeds either pass through or stick to the animal and are deposited elsewhere. Having no natural predators in its new habitat, prickly pear rapidly assumed pest proportions by the early twentieth century and seriously affected the sheep and cattle ranching industry. Since 1930, it has been biologically controlled by the larvae of an introduced moth, *Cactoblastis*, which occurs as a natural predator of the plant in America. An alternative biological control agent for some prickly pear species is the cochineal insect (the source of a red dye). For further information, it is worth looking at the following fact sheet produced by the government of Queensland in Australia:

http://www.nrm.qld.gov.au/factsheets/pdf/pest/PP29.pdf

one water body to another is thus made extremely easy and the only way to prevent such happenings is to clean and disinfect boat hulls and fishing gear very thoroughly.

http://nationalatlas.gov/dynamic/dyn_zm.html

The problem with zebra mussels lies in their fecundity and ability to survive in a very wide range of freshwater environments where they compete with other filter-feeding molluscs and outcompetes them by sheer numbers. This will have serious effects for the equilibrium of the water body and may eventually cause loss of fish stocks and bird life.

1.6 How this book is organised

We have divided the book into three parts. Part One comprises Chapters 1–6 and is concerned with the basic theory underlying biogeography and ecology. Part Two puts theory into practice and deals with methods and data manipulation, while Part Three is a synthesis, bringing together human aspects, case studies and future trends.

Part one – Theoretical aspects of biogeography

1 What is biogeography?

An introduction to the subject, putting it in a global, temporal and cultural context.

2 Issues in biogeography

In this chapter, we see that biogeography is a constantly changing subject; and that the issues arising from this change in knowledge can impact at individual, species and community levels and in particular on the natural environment through loss of habitat and biodiversity on a global scale.

3 The physical environment

Here, we focus upon the interdependence of environmental cycles which make up the physical environment which operate at scales ranging from the individual organism, through populations to the community or biome, controlling existence, distribution and composition respectively. Changes in equilibrium between the physical environment and plants or animals occur through time and living organisms alter their environmental responses through evolution or location (succession and, possibly, zonation).

4 The biological environment

This chapter looks at the biological basis of species distribution at the individual, population and community level as in Chapter 3. Genetics is seen as playing a key role in the characteristics of species and their survival; the key issues for populations are those such as human impact or environmental change which control their range. Community-scale responses to the environment are seen to rely on rule-based species interactions.

5 Putting it together – classification, biodiversity and ecosystems

The main emphasis of this chapter is upon the usefulness of classification as a means of understanding a complex environment and the importance of biodiversity as a measure not only of the number of different species, but of genetic diversity and community differentiation.

6 Global patterns in biogeography

Here we examine the way in which biogeography can help make sense of plant and animal distribution patterns at a variety of scales, the way in which they are shaped and how they can be found, described and analysed.

Part two Biogeography in practice

7 Describing and studying vegetation and its distribution

In this chapter we look at the historical development of vegetation study and the way in which modern methods have enabled biogeographers to understand how vegetation develops in relation to its environment and management.

8 Studying animals and their distributions

Variations in distribution can be seen as an interrelated set of factors – long-, medium- and short-range, including adaptation, competition and food supply. Other medium-range factors include environmental conditions, while the longest-range factors have a temporal and global focus. As with plants, an increasingly wide range of techniques are used from trapping and counting to radio tagging and satellite tracking, as well as an increasing knowledge of the environmental factors which are often of great interest in determining global distribution patterns.

9 Studying environmental factors and gradients

Undoubtedly, environmental factors play a great part in determining the extent and success of species distributions, and in this chapter we will look at those factors which appear to be most significant, and their effects either singly or in combination. Changes in the impact of a particular environmental factor will produce an environmental gradient and this in turn will influence species distributions.

10 Simple statistics in ecology and biogeography

The purpose of this chapter is to guide readers towards a range of resources for statistical and other analyses, both printed and available through computer packages or via the Internet, and to show how the more quantitative aspects of a survey can be dealt with and how to choose the most appropriate statistical method for a particular application.

11 Modelling

In recent years, modelling has become a very important aspect of biogeography, not just through simulations but also through the aid of GIS. This chapter introduces the ideas behind modelling and information systems and shows their increasing relevance in biogeography.

Part three The human dimension in biogeography

12 Environments under threat

Ecosystems are dynamic and subject to constant change both in the short and longer term. Factors causing changes in ecosystems should be regarded as a challenge or threat whether from a human or natural agency. This chapter looks at how we should assess and evaluate threats using biogeographical methods in order to find and put solutions into effect.

13 Fragile environments – case studies

The concept of environmental fragility is not well defined ecologically, and it is better thought of in terms of species gain and loss which in themselves are seen to be complex interactions between chaos dynamics, genetics and system organisation and disruption by human or other agencies. We will see how it is appropriate to model fragility using these guidelines and to apply insights to our study of those factors which can cause catastrophic change to ecosystems. It is our contention that this will be one of the most urgent tasks for biogeographers during the current century.

14 Looking at the past – human impact through time

It is generally understood that there are probably very few areas left in the world where there has been no impact from human action. Modification is usually, but not always, greatest where human occupation has been longest. In this chapter we will examine the consequences of human activities on the environment within a temporal dimension, bearing in mind the lack of baseline evidence for most of the time that humans have inhabited the earth. It is possible to demonstrate that even with this restriction, the extent to which humans have modified their immediate, and more recently, their global environment can be seen.

15 Environmental change and conservation

In this chapter, we look at conservation as one of the more significant elements in the dynamic interactions between species and their natural environment. This has particular importance as it seeks to maintain a desired range of species and habitats often against prevailing ecological and environmental forces, particularly global climate change. It will inevitably mean that if conservation is to remain true to its origins, it must rethink some of its current strategies.

16 Looking forward – prospects for biogeography

The role of biogeography is probably more important at the beginning of the twenty-first century than ever before – and in this chapter we seek to show how, in an increasingly competitive world, it justifies itself as a relevant and timely area of study. In various guises, it has been a part of study and scholarship for many hundreds of years but only recently has it become a discrete academic topic. Biogeography is a dynamic subject area with many topics of relevance to current conditions and a growth in significance in public, business and political dimensions.

1.7 Web resources

There is also our website, www.pearsoned.co.uk/ganderton, which has a lot of resources for you to use. It includes a glossary, links to many other useful and relevant sites, book reviews dealing with anything which we can get hold of that has a reputable and preferably peer-reviewed biogeographical or environmental aspect, free or shareware computer software for downloading and 'trying before you buy', together with suggestions for projects and coursework. Even more importantly, it will be regularly updated with reviews, topical issues and links to current issues of relevant journals and magazines. We will be including guest articles from internationally known ecologists and biogeographers to provide a cutting edge to the material which is on offer.

There are many other good websites and among others, we suggest:

http://ecospace.newport.ac.uk is maintained by Malcolm McElhone and is a treasury of interesting and often valuable resources. Its primary function is as a web-based learning environment for biogeographers, ecologists and environmentalists.

http://www.landscape-ecology.org is the website of IALE (International Association of Landscape Ecology). While not specifically directed at biogeography, there is a lot of useful material concerning the science underlying landscape ecology.

http://www.nearctica.com/ecology/ecology.htm is a wide-ranging academic site which concentrates on global biodiversity, island biogeography and ecological issues.

There is a very useful way of finding out more about topics mentioned in this book – by using the immense power of Internet search engines such as Alta Vista, Google and Lycos, to name but a few. Of these, Google seems to provide the most relevant responses when you type ecological or biogeographical terms or keywords into the search box. It is important to check that material that you intend to use comes from a reliable source, such as a university or well-known and respected NGO. Some sites are less than honest and you need to satisfy yourself that the information you have is both accurate and reliable (and up to date where possible). If you cannot initially find what you are looking for, click on the 'Advanced search' label and retype your keyword(s) and search again. Try it!

There is no doubt that an environmental view of biogeography does provide fascinating and valuable insights into the behaviour of plants and animals and often, by implication, humans in their local and wider environment. It is not an easy subject to teach well or even to learn if you require definitive answers or 'cut and dried' solutions, but loosen your mindset, be prepared for interactions (mostly good) between disciplines and we are sure that you will undergo one of the best learning experiences ever!

References

Brown JH and Lomolino MV. 1996. *Biogeography*. Sinauer Associates.

Carson S McB. 1978. *Environmental Education, Principles and Practice*. London: Edward Arnold.

Christopherson RW. 1994. *Geosystems: An Introduction to Physical Geography*, 5th edn. Prentice Hall.

Coker PD. 1978. Environmental education in higher education. In Carson S McB. (ed.) *Environmental Education, Principles and Practice*, pp. 238–43. London: Edward Arnold.

Dahl E. 1951. On the relation between summer temperature and the distribution of alpine vascular plants in the lowlands of Fennoscandia. *Oikos*, **3**: 22–52.

Thomas HSC. 2004. In Holden J. (ed.) *An Introduction to Physical Geography and the Environment*. Pearson.

CHAPTER 2

Issues in biogeography

Key points

- Biogeography is a constantly changing subject;
- The issues arising from this change in knowledge can impact at individual, species and community levels;
- Issues at the individual level focus on changes in organism survival;
- Issues at the species level focus on changes in ecosystem structure and distribution;
- Issues at community level focus on changes in global biodiversity;
- Human interest in biogeographical matters is increasing although there is also no other time when we have had a greater impact upon the natural environment.

2.1 Introduction

The 'biogeography' represented in this book is not the whole story. It is a selected portion captured at one moment in time. One key aim of this book is to provide a framework with and through which a greater understanding of the subject can be gained rather than to attempt a definitive work. Biogeography as a subject has changed radically over the past 30 or so years. Early studies like those found in texts such as Eyre (1968) where it was seen as synonymous with vegetation mapping have long gone. Today, the development of the subject, mirroring developments in ecological theory, has moved it out of the purely spatial to consider elements such as the impact of behavioural ecology of species on their distribution (Berger 1994). If our research perspective has changed then so has our attitude to the subject. It is no longer a pure information-gathering exercise but a way of considering the future of the world. Studies in distribution can now be used in debates ranging from ethics to conservation (IGBP 2004). The aim of this chapter is to highlight some of the current debates surrounding all aspects of the subject (Figure 2.1). The sub-headings form the framework within which debate has taken place for many years. The actual cases mentioned are those easily found in a literature review of current books and journals. These cases can be subsumed under four headings: individual, species and community levels and human response.

Figure 2.1 Smallholding near Camembert, France. This rural scene illustrates many of the key issues concerning biogeographers today, including selective breeding and species reduction (both affecting the cattle) and conservation (of the unimproved grassland)

2.2 Issues in biogeography at the individual level

Although much of biogeography deals with the distribution of large-scale features such as ecosystems there is an increasing realisation that small-scale changes are important. Put simply, anything which affects an individual organism will have repercussions for the whole species. At this scale most work has been focused on genes and genetic changes (although more can be done – Holderegger 1997). If we accept that a species distribution is determined by its responses to physical and biological conditions and that these responses grade from optimum to suboptimum, etc. then it follows that all the individuals in a population are also distributed along a response curve. Intuitively, one can see a survival benefit in this. If conditions change slightly then there will still be members of that species for whom the conditions are suitable. These ideas formed the original basis for revegetation of mineral waste tips – areas so acid that only a few grass species would be able to tolerate conditions. Conversely, as we see here, the reduction of genetic ranges (phenotypic variation) in a population can cause serious consequences for its viability. Three examples serve to illustrate this point.

2.2.1 Genetic modification

The first example involves the use of genetically modified organisms. This controversial technique is being used as an extension of selective breeding. The main reason for using such techniques is to produce a 'better' crop, usually one that has specific pest resistance. However, as research carried out by the Ecological Society of America discovered (Highfield 1998), care must be taken. It appears that a modified oilseed rape crop (usually referred to as a *transgenic* crop) can hybridise with weed relatives. This would give a hybrid with the characteristics of a weed and the resistance of the transgenic crop. It would spread more easily and have fewer natural enemies. Whereas this might not be too much of a problem away from agricultural settings (as other natural predators might well control the plant's distribution) there are other concerns. For example, if one trait (i.e. pesticide resistance) were modified then others might also be unintentionally altered, e.g. protein types to which some people might be allergic.

2.2.2 Reduction in genetic variation

The genetically modified crop is an example of an increase in genetic information, compared with an unmodified crop. There is also concern at the other end of the spectrum – reduction of genetic variability through loss of older varieties. As noted above, every population has a range of genetic responses which confer a range (albeit a limited one) of responses to given environmental conditions. If conditions were to change then it should be possible for some members of the species still to be represented. From that, a new distribution would arise, using the new, better-adapted members as the norm. However, if the population is reduced then those genetic resources are also reduced. This has been of great concern to conservation biologists in calculating the viable population size of a threatened species (e.g. 200 animals might appear to be more than sufficient but what if they all come from the same population or gene pool?). One area that has seen this aspect take on a greater importance is agriculture, particularly in the European Union. The drive for conformity of crop type has led to the dramatic reduction of varieties. A number of major food types (e.g. potatoes, sheep and apples – Tibbetts 2004) have seen a dramatic decrease in the varieties allowed to be marketed under EU rules. Opponents of this logically argue that since most of the crops have yet to be fully screened then we might be removing vital genetic traits (e.g. for pest resistance) without being aware of our actions. Lest this should be seen as a purely European problem, Holden *et al.* (1993) demonstrate that it is a global one.

2.2.3 Pesticide management

We might be more aware of our use of pesticides today but there are still problems to be discovered. Currently, integrated pest management (IPM), the use of natural enemies against pest species, is gaining in importance in both developed and developing countries. This has been heralded as the natural, 'safe' way to stop pests. However, it does have important implications as Waage (1997) has demonstrated. The natural predator (or biological control) might be useful against the pest but it is also new genetic material. In addition, the pest might itself be an introduction for which no native predator exists. In either situation the potential exists for this new genetic material to be introduced into the natural environment and create changes at the individual level.

2.3 Issues in biogeography at the species level

Many of the issues seen at species level involve the alteration of the existing community with the addition or reduction of species, irrespective of the genetic implications. At this level we are more concerned about potential changes in the structure of the community. Our current interest in mosaic theory – that ecosystems are made up of a myriad of sub-ecosystems – would be influenced were it known that the sub-ecosystem's structure had been changed. A survey of the current research identifies four main areas of concern: the addition and reduction of species, remediation (conservation of species through retention of seed banks, etc.) and alteration (changes in species through breeding).

2.3.1 Species addition

The addition of species is a common feature in many ecosystems especially urban ones (Berkowitz *et al.* 2003). The spread of European exploration has often meant the spread of European species. Often, these new species have no natural predators and can cause significant damage to the existing ecosystem. For example, around Sydney, Australia, the introduction of *alien species*, i.e. animals such as the red fox or domestic cat, have decimated the marsupial populations of the surrounding wooded areas. This in turn has led to a build-up of leaf litter which has created hazardous fire regimes. At the same time *exotic species*, plants such as *Lantana* have taken over wetlands and created serious damage along urban waterways. The problems of these additions have not been confined to large cities. Swenson *et al.* (1997) researched the difficulties experienced on the Juan Fernandez Islands. These Chilean islands are of international importance, designated as Biosphere Reserves by the IUCN. Despite their relative isolation no fewer than 227 species had been either introduced or were naturalised. Two garden ornamental plants *Lantana camara* and *Lonicera japonica*, were reported as creating a serious threat to indigenous species. To restore the ecosystems their immediate removal was recommended.

2.3.2 Species reduction

At the other end of the scale there are those who try to stop species removal. Along with species introduction one has seen a concomitant reduction in native species. CREO – the Committee on Recently Extinct Organisms – argues that it is crucial to keep data on losses so that we can try to find ways of reducing it. Numerous texts (e.g. Groombridge and Jenkins 2002) catalogue the changes to biogeography due to loss of species. Although loss is indisputable it is by no means clear-cut. Extinction in one area might still leave other related populations intact. The large blue butterfly (*Maculinea arion*) became extinct in southern England despite considerable conservation effort. Areas have now been successfully restocked with the closely related Swedish race (Marren 2002).

2.3.3 Remediation

To what extent should we try to save species that are in danger of extinction? Taking a global view, one could argue that throughout geological time species have changed. Even taking the longest-term view about our current global situation we can hardly be talking about ecosystems more than a few hundred thousand years old, possibly less. Irrespective of this, wildlife conservation is an increasingly important part of our lives, possibly because we recognize that although changes occur they are occurring at a far faster rate (Magurran and May 1999). One way of stopping this rate of loss is to create ways of keeping species. There are three ways commonly seen at present. Seed banks, as the name suggests, are collections of seeds held under carefully controlled conditions (Holden *et al.* 1993), including moisture reduction and temperature control. Given this control, long-term viability of seed stores could be expected. Alternatively, one could aim for gene banks. These would be similar to seed banks but would have regard for genetic diversity. Some crop plants have global networks of stores to allow for variations. Finally, there is *ex situ* conservation. This new term covers the conservation of species away from their natural habitats, e.g. zoos. Commonly used for rare animal breeding programmes, their use is controversial. However, there are notable successes for this method. Marwell Zoo near Winchester in southern England has an international reputation for several *ex situ* species projects including the golden lion tamarin.

2.3.4 Species alteration

Finally there is the example of species alteration through breeding programmes. Although this could be seen as genetic alteration of individuals, Laikre and Ryman (1996) comment on a little-discussed aspect of this work. They argue that harvesting and breeding of species can affect biodiversity and that this is more marked for economically important species where changes are probably going to be larger scale. Taking their argument, it would not be difficult to see how fish breeding and release programmes for sport fishing could soon diminish the intraspecific diversity of species. This paper shows, incidentally, another issue in biogeography – that of gaps in research and literature (see below).

2.4 Issues in biogeography at the community level

Changing to the community level, one finds issues that have an impact at the ecosystem or biome level. Usually these are matters of global significance. They attract a great deal of interest both from biogeographers and the general public. Although these issues are important, there is a tendency for them to overshadow issues at other levels which is a problem for two reasons: that smaller-scale issues such as changes in genetic composition can impact on the entire ecosystem and that these large-scale issues tend to hide the complexity of the case possibly leading to a poorer understanding (Ehrlich 1996). The literature relevant to this is considerable but can be grouped under four headings for ease of discussion: conservation, habitat loss, biological and physical alteration of habitats.

2.4.1 Conservation

One way of keeping biodiversity and reducing the loss of species is by carrying out some programme of remediation, i.e. to stop further loss or reverse that loss. Two of the more common ideas here are preservation and conservation. Strictly speaking, preservation is the maintenance of species without any outside interference and should be restricted to activities such as seed banks. However, confusion between it and conservation (simply, the active management of a resource) is such that the terms are seen as almost synonymous. Nature conservation is one of the great forces in human usage of the biosphere with considerable impact upon species (11.3 per cent of the earth is under some sort of conservation agreement – WRI 2003 – see Figure 2.2). It seeks to produce a (human-desired) landscape usually acting counter to a force or forces which would promote change. To this extent, by managing woodlands – for example, one is likely to spend most time felling trees and clearing areas to promote diversity which is actually anti-nature in attempting to stop succession! Whatever one's perspective, it should be noted that this *in situ* conservation (to distinguish it from *ex situ* conservation) is crucial in the protection of key habitats.

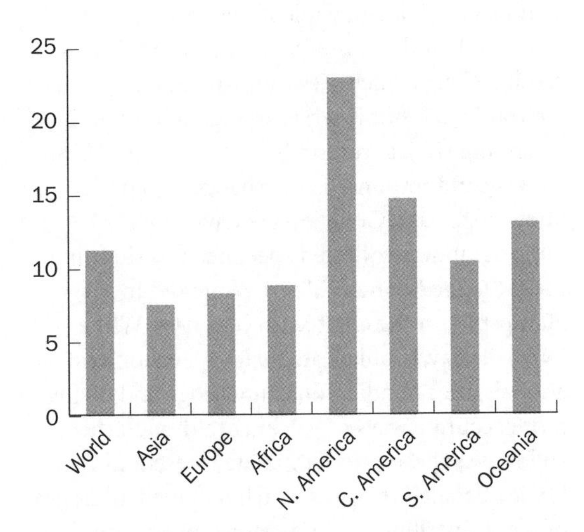

Figure 2.2 Percentage of area in each region designated as a biosphere reserve. Data adapted from WRI Earth Trends Biodiversity and Protected Areas Data. The key point is not the actual area involved in hectares but the overall amount of land given (totally or partially) to conservation measures making this a key land-use tactic

2.4.2 Habitat loss

Habitat loss is one of the most important issues facing human usage of the biosphere. Numerous studies have highlighted the rate of change in habitats as diverse as tropical rainforest and coral reef. Usually this loss is described in terms of the loss of area and the potential (or actual) extinction of species. While this is true in many cases it does hide a greater malaise. For example, if a rainforest is clear-felled then the resulting ecosystem (either grassland or agriculture) will have a completely different set of ecological characteristics. The rate of productivity will alter as will the amount of above-ground biomass. It is possible that there would be knock-on effects in other, nearby, ecosystems. Habitat loss also suggests loss of area as well as species. This would result in fewer, smaller areas (usually referred to as fragmentation). Studies have shown that fragmented areas have a far poorer chance of providing a sufficient genetic variety for a species. The gaps between fragments do not have to be large; 100 m would be far too much for woodland mice, for example. Finally, habitat loss could also affect interspecific relationships. Holmes' (1996) study of parasites illustrates this aspect. He argues that fragments decrease the perimeter : area ratio which allows more 'foreign' species to invade and colonise the area. Some of these might be parasites which could either develop new strains or be more effective because the stress of reduced area reduces the hosts' immune system response. Whatever the mechanism, there is the potential for an area to have more epidemics, etc. reducing still further the community.

2.4.3 Biological alterations

Changes to ecosystems do not mean that areas have to be lost. It is likely that changes will incur an adjustment in community structure. Today we are aware increasingly of the way in which people seek to alter habitats. Sometimes this alteration is almost total as in the case of farming, sometimes it is intermittent and dependent on other factors being present. For example, the phenomenon of 'acid rain' occurs throughout much of Europe and Scandinavia. It is only really a problem when it combines with a set of physical factors to create loss of species. So acid rain in the south of England's chalklands has a different impact from the same rain in more acid ground conditions such as Scotland (or the Black Forest in Germany to note a crucial European area). Where the effect is greater so there will be a concomitant effect on community structure. Pollution can also affect the distribution of species or varieties. Land degradation and desertification can also alter the biology of the area. Areas of semi-desert subject to intense pressure from grazing can change to desert with a subsequent change in the flora and fauna that can survive. Given the increasing extent of the Sahara Desert at the moment then there is the potential for considerable community disruption. In Plate 1, the effects of landscape modification, initially for farming and later for tourism, have had a significant effect on the ecology of this habitat.

2.4.4 Physical alterations

A final example of issues at the community level can be seen in the physical alteration of areas and the impact this has on wildlife. Wilson (1997) highlights the problems of species loss and community structures due to urban development in Hong Kong. Of a number of streams sampled in the area only two showed significant dragonfly populations and these were under threat. Alteration of space can also bring opportunities. World War II bombed sites in London contained significant numbers of rare species which would have been threatened by renewed development!

2.5 Human response to issues in biogeography

As we continue to gather information so our attitude changes. This means that our reactions to biogeographical issues are not constant. For example, we might change paradigms, i.e. our notion of biogeography changes as our knowledge of it does. The paradigm is a key scientific concept. It sums up what we believe about a subject and how we believe it. Thus at any one time a subject may have only one (or at the most, two) paradigms accepted within it. Two ideas are beginning to take hold in biogeography and might well become major forces: chaos theory and mosaic theory. Chaos theory describes the way in which small changes can have large implications. Mosaic theory (Forman 1995) suggests that ecosystems are made up not of uniform areas but a variety of specialised sub-ecosystems. Putting these notions together could mean that we view ecosystem change as being far more susceptible to small alterations than we supposed (suggesting a need to be more cautious of change), while if the area is full of mosaics then the loss of one patch is unlikely to result in harm. Such views are, of course, contradictory, but it serves to highlight the way in which the subject responds.

A second response is the change in our knowledge base. Gone are the days when biogeographers collected specimens by shooting them (as the famous Gilbert White did in Hampshire in the eighteenth century – White 1978). Today, sophisticated data collection and analysis are extending our knowledge and giving us more accurate ways of working (although there are still large gaps, e.g. in temperate forests). Ehrlich (1996) detailed the gaps which included awareness of plant distributions, indicator groups, sampling techniques, edge effects, etc. Despite the large list, we do have a considerable range of data, although such work does highlight the fact that we always operate on incomplete information. Addressing directly this need for management information, Smythe *et al.* (1996) devised the BURN project

(Biodiversity Uncertainties and Research Needs). The aim was to turn around the usual research pattern and ask decision-makers what they needed rather than produce research to an ecological agenda and then see where it fitted in. Although this might lose some scientific background, the power of such a method to actually achieve practical results is impressive. Equally useful approaches can be found in allied subjects where, for example, geographical information systems can produce large databases for biogeographers.

Increasing interest in biodiversity and wildlife conservation has led to changes in sociopolitical/economic systems. The Green vote in elections can be seen in nations from Germany to Australia. The rise in interest in organisations such as the National Trust and the Royal Society for the Protection of Birds in the UK (where the largest organisations can boast memberships in excess of 2 million people) has put pressure on governments to change. However, as in all issues the picture is not so straightforward. The emphasis on 'indigenous peoples' in forests has led Richards (1996) to question the whole notion. In West Africa forests belong to a far wider group than just the forest inhabitants. Although public understanding of, and support for, conservation measures is crucial in successful implementation, Holl *et al.*'s (1995) work in Costa Rica has shown that many local groups lack a detailed understanding of biodiversity processes. To show how even this idea can be challenged, Decher (1997) proposes that since local people in Ghana have been managing sacred groves for centuries they should be made part of any conservation initiative.

Finally, there is need for the appreciation of time and space. Although succession (ecosystem changes through time) is a common field exercise for students, ultimately one can only see zonation (community distribution in space) and *infer* succession. To take just one example, Catsadorakis and Malakou (1997) have argued that the past development of the Prespa National Park in Greece is vital to understanding how it should be conserved now. Ignorance of this could actually reduce rather than enhance biodiversity.

2.6 Summary

Biogeography is a particularly dynamic and important subject at the present time which gives this chapter two objectives. The first is to highlight current interest in the fate of species and the distribution of biological resources. In this regard, biogeography is ideally placed to provide the academic rigour needed. The second objective is to highlight the responses of biogeography to current interests from both public and research arenas. To illustrate this, examples chosen here have come from recent reviews of the literature. In common with other areas of environmental and ecological research, we see cases covering the three scales: individual, species and community.

APPLICATIONS – USING BIOGEOGRAPHY

Using ideas from this chapter in real-life situations

On 9 March 2004, the BBC published a story about the debate surrounding genetic modification of crops (http://news.bbc.co.uk/1/hi/sci/tech/3545659.stm). The idea behind the story was that many people did not believe the scientists' claim that genetic modification is basically 'safe' in environmental terms (in fact 80 per cent of people polled were against it). In tests genetically modified (GM) maize performed better and was judged to be friendlier because it had more wild plants in the crop area. However, other crops were sprayed with a soon-to-be-banned herbicide.

Where does biogeography fit into this? Firstly, the study of ecological science should enable us to determine the validity of the claims, i.e. we use knowledge to critically evaluate the claims of others. More specifically, biogeography studies the distributions of species. If, as opponents claim, GM crops can spread into wild populations we should be able to monitor this. Also, based on our knowledge of how populations behave we should be able to determine if GM crops are a problem in the first place. Working with genetic and molecular ecologists would allow biogeographers to see if there was any interbreeding (a key issue) and, if so, the rate of change. In terms of jobs, external sites noted in the story range from pressure groups (Friends of the Earth) to government departments (DEFRA) and professional associations (The Soil Association), all with a high degree of input from biogeographical studies.

Review questions

1. To what extent does scale affect the nature of the issue in biogeography?
2. Research a detailed case study illustrating one of the issues at each of the scales.
3. Differentiate between conservation and preservation.
4. Account for changes in our knowledge of biogeography in the last 50 years.
5. Given recent changes in public interest in the natural environment, describe how this might affect biogeographical research in the next 20 years.

Selected readings

There are very few texts which focus just on key issues in biogeography and those that do often have a particular bias. To gain a global overview of key topics there are two series which report regularly on biogeographical (and other) issues. The World Resources Institute 'World Resources' biennial (latest 2002–4) and related website are excellent, as is the Worldwatch Institute's annual *State of the World* text (latest 2003) published by Earthscan. A more academic perspective is the recent *Ecology: Achievement and Challenge*, published in 2001 by Blackwell and edited by Press *et al.*

References

Berger J. 1994. Science, conservation and black rhinos. *J. Mammalogy*, **75**(2): 298–308.

Berkowitz AR, Nilon CH and Hollweg KS. 2003. *Understanding Urban Ecosystems*. Springer-Verlag.

Catsadourakis G and Malakou M. 1997. Conservation and management issues of Prespa National Park. *Hydrobiologia*, **351**: 175–96.

Decher J. 1997. Conservation, small mammals and the future of sacred groves in West Africa. *Biodiversity and Conservation*, **6**(7): 1007–26

Ehrlich PR. 1996. Conservation in temperate forests: what do we need to know and do? *Forest Ecology and Management*, **85**(1–3): 9–19.

Eyre SJ. 1968. *Vegetation and Soils: A World Picture*, 2nd edn. Arnold

Forman RTT. 1995. *Land Mosaics*. Cambridge University Press.

Groombridge B and Jenkins MD. 2002. *World Atlas of Biodiversity*. University of California Press.

Highfield R. 1998. Researchers reveal risk of creating super weeds. *Daily Telegraph* online (7.8.98).

Holden J, Peacock J and Williams T. 1993. *Genes, Crops and the Environment*. Cambridge University Press.

Holderegger R. 1997. Recent perspectives in conservation biology of rare plants. *Bulletin of the Geobotanical Institute Eth*, **63**: 109–16.

Holl KD, Daily GC and Ehrlich PR. 1995. Knowledge and perceptions in Costa Rica regarding environment, population and biodiversity issues. *Conservation Biology*, **9**(6): 1548–58.

Holmes JC. 1996. Parasites as threats to biodiversity in shrinking ecosystems. *Biodiversity and Conservation*, **5**(8): 975–83.

International Geosphere–Biosphere Programme. 2004. http://www.igbp.kva.se/cgi-bin/php/frameset.php. Page accessed 23/1/04.

Laikre L and Ryman N. 1996. Effects on intraspecific biodiversity from harvesting and enhancing natural populations. *Ambio*, **25**(8): 504–9.

Magurran AE and May RM. (eds) 1999. *Evolution of Biological Diversity*. Oxford University Press.

Marren P. 2002. *Nature Conservation*. (New Naturalist Series). HarperCollins.

Richards P. 1996. Forest indigenous people: concept, critique and cases. *Proceedings of the Royal Society of Edinburgh Section B (Biological Sciences)*, **104**: 349–65.

Smythe KD, Bernabo JC, Carter TB and Jutro PR. 1996. Focussing biodiversity research on the needs of decision makers. *Environmental Management*, **20**(6): 865–72.

Swenson U, Stuessy TF, Baeza M and Crawford DJ. 1997. New and historical plant introductions, and potential pests in the Juan Fernandez Islands, Chile. *Pacific Science*, **51**(3): 233–53.

Tibbetts G. 2004. Bringing the Bloody Bastard back to historic orchards. http://portal.telegraph.co.uk/news/main.jhtml?xml=%2Fnews%2F2004%2F01%2F22%2Fnfruit22.xml (accessed 28/1/04).

Waage J. 1997. Global development in biological control and the implications for Europe. *Bull Oepp.*, **27**(1): 13.

Wilson KDP. 1997. The Odonate faunas from 2 Hong Kong streams, with details of site characteristics and developmental threats. *Odonatologica*, **26**(2): 193–204.

White G. 1977. *The Gilbert White Museum Edition of the Natural History of Selborne*. Shepheard-Walwyn.

World Resources Institute (WRI). 2003. *World Resources 2002–2004*. WRI.

Websites

Homepage of the IGBP, an international organisation working on topical global biosphere research.
http://www.igbp.kva.se/cgi-bin/php/frameset.php

One of the key global conservation organisations whose website has a wealth of data.
http://www.wri.org

CHAPTER 3

The physical environment

Key points

- The environment can be thought of as a series of interconnected cycles;
- Any change in one cycle will affect all the others and alter the total environment;
- The cycles which make up the physical environment exert considerable influence over plant and animal distributions;
- The physical environment operates at three different scales or levels. At each level the impact and its implications differ;
- At the individual level, the physical environment controls the existence of the organism;
- At the species level, the physical environment controls the distribution of the population;
- At the community level, the physical environment controls the composition of the ecosystem;
- These physical factors can be translated into a series of key concepts which help explain plant and animal distributions;
- It is usually assumed that the physical environment is in equilibrium with plants and animals. When changes occur through time there are two responses that can be made: organisms change their responses (evolution) or their locations (succession).

3.1 Introduction

The physical environment dominates plant and animal distributions. Early work on global vegetation patterns (e.g. Köppen) produced maps which were almost identical to climatic ones. Despite its importance, the physical environment has often not been integrated into biogeographical studies in the same way that ecological theory has. Part of the problem is the sheer amount of physical environmental material; studies will cover topics from global geology to microclimates. Rather than attempt a complete review of every key factor, this chapter aims to give an overview of those physical factors which affect biological distributions directly at a number of scales. This has the advantage of focusing the study and allowing it to be more closely integrated with ecological concepts.

Figure 3.1 shows a river in flood. Looking at it we can argue that the physical environment there operates at three scales (refer to Table 3.1). At the individual level the physical environment determines the success or failure of that organism. The main physical factors are those which affect the individual directly – light, temperature, chemical environment (toxicity, salinity, etc.) and microhabitats (small variations in the physical conditions, e.g. pools and riffles). These are the survival concepts; the basics needed to allow an organism to survive. Ecologically, this can be

Figure 3.1 Impact of the physical environment at a local scale. A river in flood conditions, Minnesota, USA. The river carries considerable sediment downstream although it is already discoloured by the presence of acid runoff from moorland. Some bankside trees are tilting due to loss of soil around the roots. Such environments are high energy, often with specific plant and animal communities living there

expressed in the notions of tolerance and optimum range. One result of this would be a range of individual characteristics of that species seen in terms of phenotypic variation.

Moving to the species level, the same elements of the physical environment are present (atmosphere, hydrosphere and lithosphere – see below) but the organisation and concepts differ. Light and temperature become the climate for the area. Similarly, the chemical environment scales-up to become the biogeochemical cycles. Here, the ecological concepts move upscale to encompass environmental gradients, ecotones and patch dynamics.

Similarly, the global-scale community level deals with these three spheres but in global terms. Thus weather becomes climate, etc. Onto this pattern we can overlay time – temporal changes which bring about a change in distribution of a

Table 3.1 The impact the physical environment has differs with the scale we are looking at. One of the problems in biogeography is that although we are considering the same physical elements, the results can differ, depending on the scale. What keeps an individual alive might not be the same when we look at species or community level

	← Anthropogenic changes →		
Scale **Level**	**Local (micro-)** **Individual**	**Regional (meso-)** **Species**	**Global (macro-)** **Community**
Physical environment	Light Temperature	Weather	Climate Glaciation Latitude/altitude
	Chemical environment	Biogeochemical cycles Aquatic environment	Marine environment
	Microhabitat	Soils Aspect	Geology Plate tectonics Barriers Landforms
Ecological concept	Optimum range Tolerance Adaptation Phenotypic variation	Environmental gradient Ecotones Patch dynamics Genotypic variation	Zones Biomes Disjunction/endemism Vicariance/dispersal
	Succession ←	Temporal changes →	Evolution

species (succession) or a change in its responses to the environment (evolution). This natural system has continued up to the present when anthropogenic (human) influences have altered (radically in some areas) the ecosystems at local, regional and global scales.

3.2 What is the physical environment?

What is the physical environment and how does it vary between places and times? A simple question maybe, but the answer to which has changed throughout the course of human history. At one level, it is very simple: the physical environment consists of the interactions of the three 'spheres': atmosphere (air), hydrosphere (water), lithosphere (rocks and soil) which together with the fourth 'sphere' – biosphere (plants and animals) – make up the environment. At another level we find that the way we perceive the physical environment impacts considerably on our ability to understand it. Continental drift is a useful example here (see Box 3.1).

As noted above, at the broadest scale we recognise four key aspects (usually referred to as spheres) whose operation and interrelationships define the physical conditions of the earth. These are related to each other with a series of two-way connections. Anything affecting one direction will have a countervailing reaction (see Figure 3.2 and Box 3.2). In the physical environment, it is the relative proportions of these three 'spheres' in each place which create the unique physical environments (the abiotic conditions of ecologists) that provide the living conditions for the fourth sphere, the biosphere.

BOX 3.1

Alfred Wegener and the Continental Drift Theory

The continental drift idea started with Alfred Wegener in the early part of the twentieth century. Wegener was a German scientist who noticed that some continents (especially South America and Africa) seemed to fit together like a jigsaw puzzle. Up to this time, biogeographers had not been able to satisfactorily explain the distribution of similar fossils across the southern hemisphere continents. Wegener's idea was simple and matched the need at the time. Continents 'floated' upon the oceanic crust and as they moved so populations of species were separated (or met!). From shaky beginnings this became the leading theory until the development of a rival 'plate tectonic theory' which started in the early 1970s. This theory was based on scientific analysis of rock compositions and ages. Slowly, the evidence against Wegener mounted until this new theory became accepted. Some would say it is the ultimate theory, but we ought to note that it had a shaky beginning with few adherents and was not too well received initially!

Why do we bother with such ideas? Because biogeography is about the distribution of plants and animals and we need to know about these changes and how they happened. This new understanding (which may, of course, be replaced in time) has led our drive to quantify and map species which has, in turn, led to far greater understanding of areas previously seen as remote (e.g. New Guinea – Flannery 1995a). Is this relevant today? Yes, if we do not understand how something like global warming works, how can we predict accurately its effects? More data = better theories!

BOX 3.2

Spheres, interrelationships and feedback loops

The simple diagram in Figure 3.2 is a popular way of introducing the subject but it does have some significant aspects. The first is that all parts are interrelated. It means that changes in one affect changes in all the others. It is also usual to accept that as one aspect changes so its ability to change further might also change! As a forest grows so its floral composition changes in response to light. For example, an English oak–ash forest would have ash as the first tree (fast growing and open leaf structure) which would eventually be completely shadowed out by the slower-growing oak.

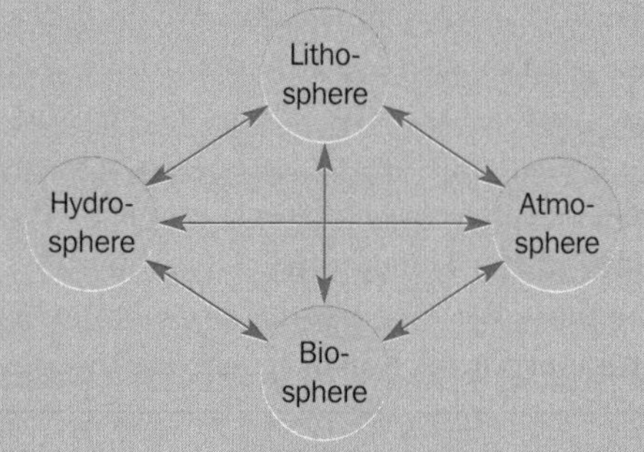

Figure 3.2 The four 'spheres'. The complexity of the environment is such that it is usual to study each of these in turn but, since they interact with each other, there will always be some overlap

Just a UK woodland phenomenon? Far from it! Studies of temperature changes in the Eocene (about 55 million years ago) suggest that there was considerable loss of methane from the earth's crust. This set up an increase in ocean carbon storage which reduced the impact (Bains *et al.* 2000). This is one example of a feedback loop where one factor in turn affects others. Where this impact increases the other factor we talk about a positive feedback (success breeds success); where it decreases the factor, it is a negative feedback. Both types of loop are extremely common in the environment.

Empirically, it is important to measure all aspects of the physical environment. There is the temptation just to map distributions without regard to the climate, etc. While this was seen as acceptable when the ecosystem was the basic unit of study, today our micro-scale approach demands a more refined analysis. Micro-scale work involves studies at the sub-ecosystem level. It argues that the ecosystem is a mosaic of smaller systems linked by common plants (vegetation being easier to analyse and appreciate than animal distributions). Work by Forman (1995) best exemplifies this approach. He argues that measurement (including measurement of the physical environment) is vital to appreciate the complexities of the landscape.

On a more practical level, Jones and Reynolds (1996) discuss the range of physical measures available, noting:

> The distribution and abundance of all plants and animals are determined to some extent by the abiotic features of the environment. Thus, measurement of environmental variables forms an integral and planned part of most ecological field work. The choice of measurement technique is rarely straightforward, since even the simplest variables such as temperature can be measured with an array of methods differing in precision, practicality and expense. (p. 281)

As they rightly argue, the range of analyses and techniques is very broad. They mention a few of the simpler ones which address key elements of the physical environment. Table 3.2 lists some of the key factors we need to look at.

An appreciation of the interrelated nature of the environment is important. Likewise, thorough study of any ecosystem requires a range of physical

Atmosphere	Lithosphere	Hydrosphere
Wind Light Temperature Solar energy	Slope angles Altitude Soil characteristics Soil oxygen Latitude	Rainfall pH Turbidity Water flow Conductivity Salinity Water chemistry

Table 3.2 Key environmental factors. Most studies would start with these factors although as our studies become more sophisticated so does the range of measurements we take. The aim is to get the key data with the minimum of cost in time and effort

measurements. The next stage is to see how the physical environment relates to organisms at a variety of scales (individual, species and community) and to examine the implications for biogeography and ecology.

3.3 The physical environment at the individual level

The presence or absence of an organism at a site is due to a complex interplay of factors both biotic and abiotic. Biotic factors can even produce abiotic effects. For example, 'nurse plants' can provide the shelter needed for another plant to exist (Crawley 1997). As Peterken (1993, 1996) has demonstrated, loss of ancient woodlands (i.e. loss of shade) has often led to the decline of shade-loving plants such as dog's mercury (*Mercurialis perennis*). Further, one must consider whether the presence of a plant or animal at any given place is due to its *tolerance* of conditions there, its *avoidance* of unfavourable conditions elsewhere or as part of the *trade-offs* between its various physiological responses (Figure 3.3).

This might suggest that there are an almost infinite number of factors involved in the location of an individual. Whereas it might be possible to measure every factor, in reality the majority of the explanation can be reduced to a few key physical and chemical factors. Physically, an organism has need for amounts of light, temperature, water and shelter (or microhabitat). Light (or more correctly, solar radiation) is a vital component of

Figure 3.3 The number of physical factors influencing the location of an individual organism are considerable and can include both biotic and abiotic components. In addition, some factors may be more important than others. Here, on the shores of the Great Lakes, USA, a bellflower survives in a small crack, showing that factors need only be present in minute amounts for survival

photosynthesis in plants. In order to exploit this resource, plants have evolved separate photosynthetic pathways (usually referred to as C_3 and C_4 – tropical and salt-marsh plants tend to have C_4 mechanisms and temperate plants, C_3). This response means that separate species are better adapted to given light levels. Shade-tolerant plants function best at low light levels, while beech (*Fagus* sp.) and wheat (*Triticum* sp.) respond rapidly to increases in light level.

Temperature is the most studied of the factors. This is probably due to the importance that temperature has on a range of physiological responses. Studies have demonstrated that the survival of individuals and species is related to a specific isotherm. For example, Begon *et al.* (1996) link isotherms not only to distribution of organisms but also to storage, dormancy and germination of seeds. Shelter is another important part of the physical environment. Animals use shelter to provide respite from conditions outside their preferred ranges. Some plants, especially those in marginal habitats such as exposed mountain areas, will only survive if they can avoid prolonged exposure to high wind. Shelter does not need to be physical. Crawley (1997) described the interaction between a rush (*Juncus* sp.) and a shrub. Removal of the rush allowed chemical changes in the soil: the increase in salinity reduced the shrub's biomass. Water is a key factor in the survival of species. Whereas some plants are drought-tolerant (xerophytic), e.g. cactus, others require almost constant waterlogging to survive. In biogeography, there is less concern for actual water supply and requirements than for general climatic conditions (for which temperature is often considered the key factor). However, in agriculture water demand is crucial in getting crops to optimum condition. Irrigation is supplied according to the requirements of the plant and the supply of water at any given location.

The chemical requirements of plants and animals are similarly split between those elements they required for survival and those avoided for similar reasons. Carbon dioxide and oxygen levels are important in photosynthesis. These gases, linked to radiation levels and water supply, help explain the distribution of plants in North America (Begon *et al.* 1996). In this case, one photosynthetic pathway (C_4 – associated with tropical conditions) is strongly correlated with evaporation. A greater percentage of C_4 plants is found in the southern United States as one might expect given the requirements of C_4 photosynthesis. A range of nutrients of which nitrogen, phosphorus and potassium form the major requirements for plants are supplemented by trace elements or minerals such as copper and zinc (and which are dealt with more fully below). Against this must be set a range of other chemicals to which the plants may well be less tolerant such as the heavy metals.

What are the ecological and biogeographical implications of these factors? For an organism to exist in a given place it must be able to survive within the physical conditions of that place. The ability to do so is referred to as *tolerance*. Typically, we assume that for any given factor, e.g. light, temperature, oxygen, there are limits below and above which the organism cannot survive (see Figure 3.4). In theory it should be possible to take every factor and find the limits which the organism can tolerate. Such a diagram would outline the *range* of the organism. This can be mapped to give a distribution map of the species (a fundamental concept in biogeography). Further, one can distinguish a number of different ranges. The organism will have a set of conditions within which it is perfectly adapted (producing an *optimum* range). It should be able to outcompete all other species to be dominant. Outside that there is a *suboptimum* range where the organism can survive but may be outcompeted by better adapted organisms. By the time the limits of the range are reached, the organism cannot survive. The implication of this simple idea is that ecosystems are collections of species whose physical ranges are very similar. This also implies that any given ecosystem is a mosaic of individual responses to microhabitat conditions.

The range of the organism is not produced by chance. Previous generations have passed on

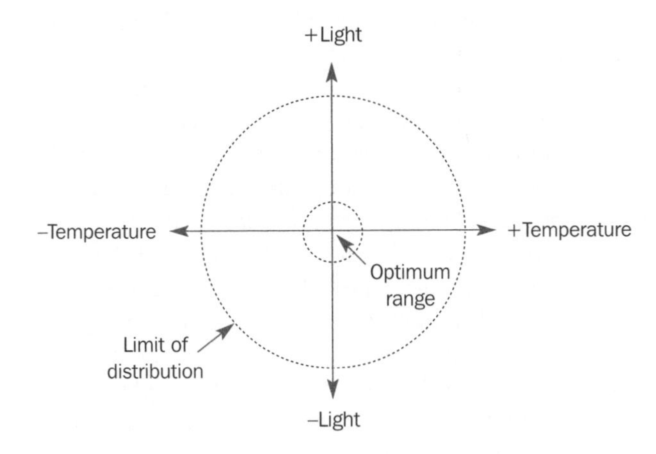

Figure 3.4 A highly simplified model of the impact of physical factors on distribution. Assume just light and temperature are involved. There are levels above and below which the species is not adapted. The 'best' conditions would be the optimum range. Outside that, conditions would worsen until the species dies out. In reality these lines would be highly complex interweavings of species where an individual might survive outside its range but be unable to breed

genetic characteristics which result in the organism being *adapted* to current conditions. Of course, if the current conditions change rapidly then the organism will have to change further (i.e. adapt) or die out. Within any individual there are slight variations on the basic genetic information. Such variations, referred to as *phenotypic variations*, might allow the individual to survive (for the species response, see below).

3.4 The physical environment at the species level

Whereas the existence of an individual is a matter of its specific genetic make-up and the microhabitat, the factors controlling an entire species are more broadly based. The individual is only of 'value' if it furthers the existence of the species in that location. As we change the scale to the regional level then the physical factors are broader. The light and temperature that control the individual become the weather pattern for the species. Similarly, the chemical environment focuses on nutrient (biogeochemical) cycles and the aquatic environment and the microhabitat becomes soil and aspect (topography).

The division between weather and climate is a question of scale. Climate is said to be the long-term mean for the weather although the precise timescale is unclear. This might lead one to assume that the weather is of no significance but this ignores the life strategies of some organisms. Studies have shown that for certain groups of organisms the annual weather pattern becomes crucial. Typically these are organisms with a very short lifespan which produce large numbers of young, few of which survive to maturity (referred to as 'r' strategists). Studies by Andrewartha and Birch (reported in Begon *et al.* 1996) on a species of thrips found that a large variation in the mortality and survival was due to weather patterns. Similar results were found for the winter moth. The severe droughts in the UK in 1976 created very favourable conditions for the population explosion of the ladybird. It is not only insects that are affected by seasonal weather patterns. Severe winters such as that in the UK in 1986/87 left numerous bird populations depleted. Watkinson (1997) gives further examples when he quotes research into annual plants in California and the Sonoran Desert. Both places had examples adversely affected by weather fluctuations but this was not universal. The implication is that weather affects some species more than others which would lead to a change in species composition. In other words, although these could be seen as localised and short-term events it is not beyond possibility that they could prove a final difficulty for a species in a marginal habitat.

At the species level, the chemical requirements of the organism while still being required could be subsumed under the broader heading of nutrient (or biogeochemical) cycles. Both plants and animals require a set range and amount of nutrients. The supply (or otherwise) of these nutrients will control the distribution of a species. Traditionally, plant species and nutrient demands are used because they more easily demonstrate the requirements of organisms. Plants require three essential nutrients – nitrogen, phosphorus and potassium. In addition, a range of minor nutrients and trace elements is also required. Sources of these nutrients vary but would include all the major spheres – lithosphere, hydrosphere and atmosphere. The complexity of the situation can be seen by studying just one example, the nitrogen cycle at the Hubbard Brook Experimental Forest (see Box 3.3).

This means that plants must be able to take up nutrients to survive. However, the supply of most nutrients is dictated by the overall chemical environment. Some nutrients (especially phosphorus) are very sensitive to changes in soil acidity. It limits the available habitat range for plants and requires them to adapt to the prevailing conditions. Further, it is possible that nutrient availability varies through time giving 'ancient' areas different nutrient status from 'younger' places: Australia is particularly deficient in this regard (Crawley 1997 p. 530, Specht and Specht 1999). In some circumstances the lack of a vital nutrient will determine the nature and composition of the ecosystem. Northern Canadian streams are very low in phosphates. This limits the plant and animal range. Experiments adding phosphate fertiliser to the streams showed considerable changes in species composition.

BOX 3.3

The Hubbard Brook Experimental Forest

Image courtesy of John Campbell, USDA Forest Service.

Hubbard Brook (see image, above) is 3,160 ha of New Hampshire, USA, forest used as one site in the Long-Term Ecological Research programme (LTER). The project started in 1955 and was aimed at finding quantitative details of forest ecology. Of the many studies carried out, one of the most important was the deforestation of one catchment to see what happened to the physical environment. The changes were quite marked. Nutrient loss was over 12 times normal although there were considerable variations in this. Nitrogen (as soluble nitrate) showed up to a 40-fold loss while there was little change in sulphate. Nutrient flow and plant species are in equilibrium. One of the objects of the study was to alter initial conditions. By deforesting part of the forest it was possible to study the effects. Massive soil loss and translocation of minerals radically altered the habitat. Some plant species would be badly affected and might even die out in a location. It is not just a matter of having the right chemicals present, it is having them available.

For further details visit www.hubbardbrook.org.

Aquatic ecosystems (usually regarded as comprising wetlands, lakes, ponds, oceans, seas, estuaries and rivers) provide an interesting comparison with their terrestrial counterparts. There are areas of great biodiversity (such as coral reefs and coastal areas) and virtual deserts in terms of biomass (e.g. open ocean at depth). As a physical environment it provides a range of habitats that link individual, species and community – Figure 3.5. Recent discoveries in the cycling of nutrients and energy (the conveyor belt theory) link deep ocean and surface water in a way not seen in terrestrial systems. Despite this complexity all aquatic systems are part of the hydrologic cycle and as such share common features (though often with different emphases). The biogeography of aquatic systems depends on the interplay of these factors in any given area (see Dobson and Frid 1998):

Figure 3.5 This river system in Indonesia illustrates both physical and human impacts. Near the sea the river has maximum volume and chemical content but with relatively low energy. In contrast, this river's headwaters would be high energy but with low nutrient level. Each aspect of the river's development will, in turn, affect (and be affected by) the biosphere. In addition, the impact of settlement in this area creates a risk of pollution, further altering the river's biota

1. *Landform*: the shape of the land surrounding the aquatic system determines a range of physical and biological parameters. For example, rivers can be zoned into three areas corresponding to erosion, transport and deposition zones (similar to the Davisian idea of youth, maturity and old age). Each of these zones has its own physical characteristics in terms of energy input and water output. Estuaries also show marked responses to their landforms. Estuaries such as the Fowey in Cornwall, UK, are the result of drowned valleys caused by relative shifts in sea level after the Ice Age. Norwegian fjords characteristically have shallow entrances (lips) which constrain the movement of water which dominates the sea floor. In the deeper ocean seamounts create areas of shallower water which can be exploited. Recent research along the mid-ocean ridges have shown a unique assemblage of plants and animals that seemingly thrive in this hostile environment (Wharton 2002);
2. *Energy*: for any system to exist it must have energy flow. In rivers, this is seen in the longitudinal section from headwater to sea. The high-energy headwater environment contains a different range of species from the estuary. There is some connection through continual loss of animals downstream (termed drift). For oceans and lakes there is the use and recycling of energy through the process of stratification – see below. Energy also enters the system through tides and waves in seas and oceans;
3. *Water chemistry*: just as in terrestrial systems where the composition of the air is important (especially if the organism is susceptible to specific chemical or biological agents) the chemistry of the medium dominates all aspects of the aquatic environment. Leaving aside the actual chemicals and amounts involved, there is broad division between marine and fresh water. Fresh water is dominated by acidity and

impurity, both of which can have profound effects on the biotic environment. An increase in acidity (often linked to acid rain episodes) allows the mobilisation of aluminium ions which are toxic to a wide range of organisms. An increase in impurity through rainwater, storms or human action such as pollution reduces light levels and increases oxygen consumption (which in extreme cases can deoxygenate the water in a process known as eutrophication). Marine water is far more chemically constant. The key aspect is the regulation of water and salt levels by plants and animals. Even small changes in salinity can have a profound effect on organism distribution (see Chapter 9);

4. *Water physics*: the specific properties of water create a unique habitat. Unlike similar small molecules it is a liquid at room temperature. Its molecular bonding gives it an unusual set of physical characteristics such as specific gravity (i.e. density) and a high specific heat capacity. The former may change the distribution of some species but the latter has the greater impact on species. The reaction of water to a heat source is to set up a series of layers (referred to as stratification). The sun's rays heat the upper layer of the water body. As the temperature rises it forms a layer. The different physical properties of this layer means that it is unlikely to mix with other layers. By the summer, there are three layers. As these cool in the autumn the top layer becomes colder and denser and sinks. This brings nutrients to the surface (the autumn overturn) and takes oxygen to the lower layers. Although the mechanism differs, the same energy-driven concept of stratification is seen in anything from ponds and lakes to oceans. Further, in oceans the connections between sea bed and surface allow a global-scale energy and nutrient shift (upwelling);
5. *Substrate/sediment*: the sediment or rock making up the floor of ocean, lake or river (called the substrate) has a great influence. Chemical properties such as acidity and composition can create a range of habitats. Physical properties such as particle size determine which organisms can survive. For example, shellfish often require a rocky substrate. Wetlands provide a variation on this theme in that the concern is not only with availability of oxygen and nutrients but also the natural presence such as toxins from undecomposed matter;
6. *Nutrient input*: all systems require nutrients. Whereas much of this can come from upstream there is also decomposition in the system itself. A study by Minshull (quoted in Dobson and Frid 1998) demonstrates the changes in species composition following the input of organic matter. Studies show that nutrient uptake can be highly complex. Work in the Arctic (McKane *et al.* 2002) suggested that species differentiated nutrient uptake in terms of timing, depth from which chemicals were used, and chemical form. Were this to be repeated elsewhere it would mean that nutrient relationships are far more complex than we think.

Soil is one of the thinnest of all ecosystems and yet the most vital (Charman and Murphy 1991, Miller 1998). In most areas the active depth of soil is mostly less than 1 m (measured against the earth's radius of 6,000 km) and some would argue that it is the top 25 cm where gas, nutrient and water exchange take place that is the most crucial (see Figure 3.6). For terrestrial ecosystems it links the biosphere with the abiotic components providing many 'services' from nutrient supply for plants to anchorage for trees. It is worth considering that although it is often regarded as part of another ecosystem, soil is an ecosystem in its own right. It has a plant and animal assemblage and can exist without other ecosystems (unlike, say, forests which need soil for survival). The key difference is that soil does not use sunlight for photosynthesis so the role of primary producer is taken over by decomposers such as fungi.

Soil formation is controlled through two agents: soil-forming factors (which control the types of material available) and soil-forming processes (which determine the way in which the soil is

Figure 3.6 Soil is one of the most overused and least regarded ecosystems. Its surface permits the growth of crops. It also regulates the transfer of gases and nutrients between the lithosphere and the atmosphere. Although this scene in the Atherton Tablelands, Australia, seems peaceful, such farming activities impact directly upon the Great Barrier Reef

made). There are generally agreed to be five soil-forming factors, usually expressed as the Jenny equation – **CLORPT** (or **CL**imate, **O**rganisms, **R**elief/topography, **P**arent material and **T**ime). Climate is a key factor because of its ability to limit soil formation. It controls the amount of precipitation and evaporation which determines both the rate of formation and the movement of minerals through the soil. Dry environments support little life so there is less organic matter to add to the soil. Cold environments limit growth rate but increase the chance of organic matter accumulating. The physical movement of soil components brought about by cycles of freezing and thawing in extreme environments is shown in Plate 2. Dead organisms and plant/animal waste provide the material (humus) needed for the active component of the system to be developed. Further, live animals can help create soils (or retard their growth). Plants clinging to rocks reduce the rate of breakdown into soil. Alternatively, animals digging at soil and loose rock aid the breakdown process, e.g. rabbits and badgers. The relief of the area (topography) will determine the development of the soil and its components. One example here is the catenary succession based on the chalk downland slopes in southern England. At the top of the slope, soil is thin (because it is exposed) and nutrient supply is fair. Downslope the soils are thinner due to gravity erosion. At the slope bottom, soils are most likely to be deeper and more fertile as they collect all the upslope nutrients. Where nutrients are likely to be in short supply and under poor climatic conditions, woodlands, even in temperate parts of the world, tend to be restricted in growth. This is particularly well shown by the example of dwarfed climatic climax oak woodland shown in Plate 3.

In addition, the topography controls the weather patterns and thus rainfall in the area. Parent material provides the minerals and chemical nutrients for soil development. This can be rock or a range of transported materials (sediments). The key factor is the mineral composition of the parent material. Sands and sandstones tend to produce acid soils with very few nutrients, whereas granite can be weathered to produce a wide range of minerals, most notably clay. Mineral composition is also linked to ease of weathering. Harder minerals such as quartz (sand) weather more slowly; granites decompose more slowly than volcanic rocks. Time is the final factor. The greater the amount of time, the more developed and deeper the soil (as a generality). Thus tropical soils can be millions of years old (and 20–30 m deep), while the highly variable development of European soils through the last Ice Ages means that, in the UK at least, 1 m is a good soil depth. Such a link does not always hold; Australian soils are old but arid conditions and poor parent material mean that development is reduced.

Soil-forming processes (pedogenesis – see Figure 3.7) describe the ways in which solid rock and sediment can be reduced to mineral fractions which, with the addition of organic material, make up soils. Perhaps it is best to view this as a form of

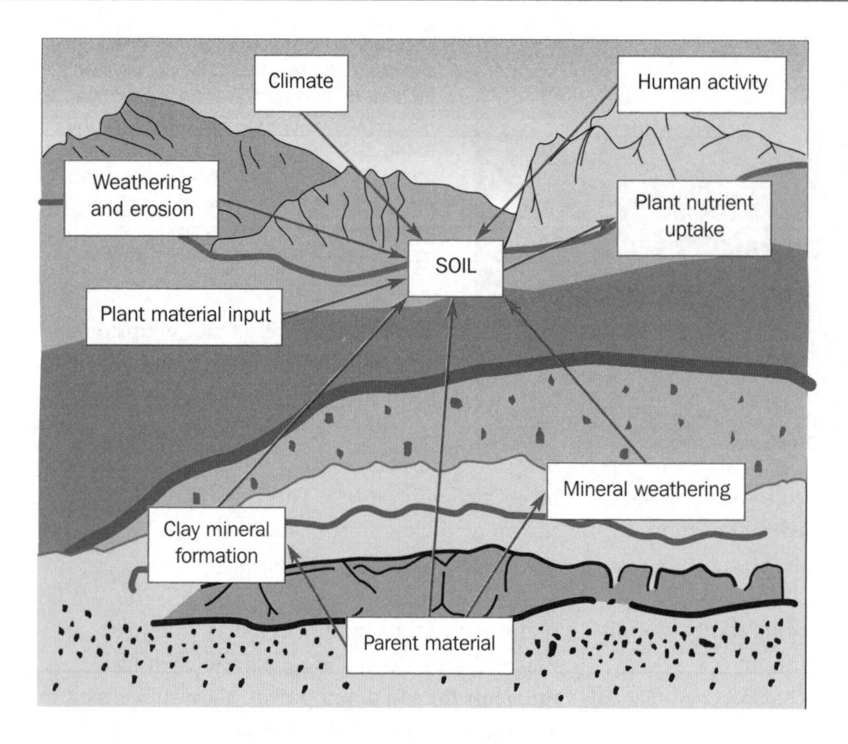

Figure 3.7 Soil formation is a complex interaction between the four spheres. The lithosphere provides the rock and sediment (parent material) which may be further transformed before becoming the soil's mineral fraction. The atmosphere and hydrosphere both provide and remove minerals through weathering, erosion and transport. The biosphere is not an idle recipient of this, but actively takes and returns chemicals and can also provide feedback mechanisms (both positive and negative) in soil formation. The one element missing here is time, which generally allows the soils to develop further (although this may not provide a more nutrient-enriched soil)

succession. The initially bare rock is broken down by weathering and various water-based processes. Defined as the breakdown of rock *in situ*, weathering provides both the mineral fraction of the soil and the nutrients which can be taken from the rocks. Weathering can occur through physical (e.g. alternate freezing and thawing), chemical (e.g. solution) or biological (e.g. burrowing animals) agencies. Water-based processes offer a secondary way of removing and relocating materials. Water moving through the profile can redistribute clays, leach anions and cations and aid the movement or accumulation of organic material. Soil drainage can also influence the movement of key minerals in the soil. For example, a very free-draining soil is more likely to have leaching of iron down the profile. All these processes are subject to different rates of change due to the topography and climate of the area and the amount of time for pedogenesis. From the early formation of the mineral fraction, some coloniser species such as lichens and mosses will grow on the surface and continue the breaking down of material. Slowly the mineral components will increase and some organic materials build up. This early soil will be able to support some grasses. The process of rock breakdown and plant development continues until the soil is as fully developed as conditions permit.

Soil factors and processes combine to produce a range of soil types. For biogeographers these factors and processes are important in helping to understand distributions of plants and animals. Of more immediate use are the soil components and soil types generated by these processes. Soil components are the features that make up the soil – essentially the mineral fractions, structure and texture. The mineral fractions – sand, silt and clay – are used in a tri-plot diagram in an attempt to broadly classify soil types. The structure of the soil refers to the way in which it aggregates, while the texture is the relationship between particles.

The variability of soil types is such that even a relatively small field can have three or four distinctive soil types. However, classification at this scale would render any system unwieldy. This situation is not helped by the range of classifications available, each with their own system and rationale. Whichever system is used, the main objective is to describe the individual layers (horizons) of the soil, their components and structure and the comparison with the other horizons in the soil profile and with other soil profiles. This has led, in the UK, to the description of the four basic soil types – each one associated with specific climatic/vegetational types (see Table 3.3): brown earth, podsol, rendzina and gley (sometimes spelt glei).

The final feature at species level is the aspect or direction of slope. This can have a significant effect on species distribution at a regional level. One example of this is the small side valleys of the Rhine in southern Germany. Here, the slope aspects of the steep-sided valleys are north/south. The north-facing slopes are usually covered with coniferous forest with the lower slope by the village devoted to pasture. The south-facing slopes are almost completely given up to viticulture.

What are the ecological and biogeographical implications of these factors? One of the key ideas presented at the species level is the gradual change in abiotic conditions. Referred to as an environmental gradient, this is a key feature in the study of the change of species (see Chapter 9). One obvious expression of this is the zonation from open lake to woodland where water level would be the environmental gradient. More subtle gradients can be seen, e.g. moisture content of soil and the distribution of brambles. As conditions change so it is possible to see one ecosystem give way to another. The changeover is gradual; there is a time where ecosystems coexist. Such boundary areas are called ecotones and their ecology is becoming of increasing interest. Not only do they illustrate competition principles they also represent more dynamic systems than that found either side. Estuaries could be seen as ecotones between fresh and marine water. Species existing in such places face a doubly hazardous environment as they change from completely fresh water to salt water to aerial exposure on a regular basis. It is possible that the gradient operates in more than one direction (or that there is more than one gradient). Under such circumstances then, it is likely that patches or mosaics of similar sub-ecosystem species would develop. The (patch) dynamics of these areas can be used to explain the distribution of species.

Table 3.3 Overview of key UK soils and their main characteristics. The key point for biogeographers is to be able to identify key soil types and their contribution to species distribution. Alternatively, it is possible to have identifiable soil changes within a small area, e.g. field, and have little obvious change in species

Soil type/characteristic	Main characteristics	Typical rock type	Main ecosystem
Brown earth	Deeper forest soil common in S. England	Clay	Temperate woodland
Podsol	Shallow stony soil with few nutrients. Commonly a 'pan' layer is formed due to mineralisation	Sands and gravels	Heathland
Rendzina	Thinner soil, alkaline	Chalk	Chalk downland
Glei	Waterlogged soil, found near rivers and marshes	Clay	Wetland

Human impact in tropical rainforests where small areas are cleared for subsistence agriculture and subsequently abandoned demonstrate the successional processes associated with patches. That an approximation to the original rainforest would not return for centuries was one of the more remarkable discoveries.

Finally at the species level there is the concept of genotypic variation. It is possible to link some of genotypic variation to environmental conditions. Here the range seen at the individual level becomes the parameters for the species. One practical use of this is the revegetation of heavy metal mine workings. Experiments showed that some of the grass species were more tolerant of pollution than others. Thus the normal curve of characteristics would have a few members at the extreme end capable of living in such conditions. By repeated selection through the generations the median characteristic would be skewed towards heavy metal tolerance.

3.5 The physical environment at the community level

At the community level the physical factors which shape plant and animal distribution (and development) are global in scale. Here, the basic rules which govern the range of an individual or species do not hold because the timescale coupled with these factors means that concepts are not constant. For example, a major element of biogeography was that the climatic belts were stable. However, recent research on plate tectonics suggests that climatic belts have altered significantly during the course of geological history (see Box 3.4). Thus, study of the global physical environment involves analysing the development or evolution of geology, atmosphere, oceans and landforms and the concomitant ecological concepts that result.

BOX 3.4

Oceans, climatic belts and biogeographical regions

Continents did not wander across fixed belts and the ocean is not stagnant water. Although we have known this for some time evidence of the size and extent of changes has been less forthcoming. Recent studies have shown that what we have taken as being relatively static may be deeply affected by global movements of continents and ocean currents:

- Oceans are connected by a series of currents usually referred to as the conveyor belt hypothesis. Ocean currents do not, it appears, just move slowly as the name suggests. Studies in the North Atlantic and Pacific oceans show that these currents are variable, with potentially considerable influence on terrestrial climates (Dickson *et al.* 2002, Clark *et al.* 2002, McFaden and Zhang 2002);
- Continental plate movement can alter seaways (or even cut them off). Such a change in the seas around Indonesia may be linked to increasing aridification in Africa (Cane and Molnar 2001) and plate activity in the Himalayas may have affected monsoonal distributions (Zhisheng *et al.* 2001);
- The idea that islands are natural laboratories with isolated populations may have to be reassessed if work by Calsbeek and Smith (2003) is repeated elsewhere. Their study of lizard populations off the Floridan coast suggests that storms may aid distribution of populations thus altering the genetic composition of disparate islands.

Geology is the key to understanding the historical development of biogeography. Since the creation of a relatively solid planet some 4.5 billion years ago (at current estimates) changes in the composition, distribution and disposition of land masses, atmosphere and hydrosphere have shaped the development of the biosphere. We recognise the length of time development has taken place through the geological timescale (see Table 3.4). Time is divided into eras corresponding to key biological/physical events. Further, these eras can be subdivided into periods again with a key biological theme. Boundaries between periods are constant for only a small area – the geology of Europe and that of America is sufficiently different for different timescales to be produced. Often, boundaries may be set because of the absence of deposition, i.e. at the time the area under consideration was a land mass. From the time of the Cambrian onwards there have been sufficient remains of organisms to allow us to see the evolution of major classes or even orders. Tying in fossil remains with time periods is the 'fossil record' (see Table 3.4); a set of remains (or impressions or casts of remains) that is used to trace the evolution of species.

Table 3.4 The geological timescale and an outline of physical and biological developments

Time (millions of year BP)	Era	Period	Main physical characteristics	Main biological characteristics
0.1		Holocene		
1.5	Quaternary	Pleistocene	Ice Ages	Early hominids
5	Tertiary	Pliocene	Cooling and drying. Ice caps formed	Spread of land species
24		Miocene	Changing global climate patterns	
34		Oligocene	Alpine orogeny	
55		Eocene		
65		Palaeocene		Development of mammals
144	Mesozoic	Cretaceous	Continued break-up of Pangaea	Flowering plants, end of dinosaurs. Massive extinctions
206		Jurassic	Development of N. Sea oilfields	
248		Triassic		Start of dinosaurs
290	Palaeozoic	Permian	Near-single land mass – Pangaea	Massive extinction event
354		Carboniferous	Hercynian orogeny	Development of land animals
417		Devonian		Development of land plants and animals
443		Silurian	Climate stabilised. Caledonian orogeny	Corals and fish evolve. Evidence of land plants and animals
490		Ordovician		Marine life. Major extinction at the end of the period
543		Cambrian		Numerous life forms developed
2,500	Precambrian	Proterozoic	Continents are formed	Development of algae
3,800		Archaean	Reducing atmosphere	Some bacteria developed

Despite its importance the fossil record is an imperfect tool. To become a fossil or impression some sort of physical evidence of sufficient size must be left behind. This precludes numerous soft-bodied organisms, especially microscopic forms (although recent evidence in Australia questions this with chemical remains of life forms stretching back into the Precambrian). More importantly, a fossil can only be produced in areas where conditions are suitable. Normally this would mean being trapped in sediment. Since sediment is not the only material produced it follows that the fossil record is dependent upon the rock cycle – a model illustrating how and why the lithosphere changes. Basically, any rock type can be turned into any other rock type. For biogeographers this means that organisms can only be preserved as fossils in conditions favourable to fossil production which implies that during geological time, many areas have been incapable of either supporting life or of preserving it. Whereas the fossil record and the rock cycle are crucial parts of historical biogeography, one of the most fundamental changes has been brought about by plate tectonics (see Box 3.1).

The earth's crust is composed of a series of solid plates. Each plate changes shape during geological time through the addition of material at some boundaries and the removal of excess at others (a third boundary type allows slippage of plates past each other so that the constant volume of the earth is preserved). The continents are found on top of part of some plates. They are fixed in position – they only appear to 'move' because of the movement of the plate beneath them. The relative disposition of these continents changes through time. Likewise, plate construction and destruction (at mid-ocean ridges and ocean trenches) contributes to the changing size, shape and location of the oceans. It is possible that when two plates each with land mass meet, the resulting collision creates mountain chains. It is on this movable platform that organisms have developed and spread. As will be shown below, the relative positions of land masses and oceans influence the spread and evolution of species.

Given the distribution of global climates it is tempting to consider that they have been permanent. Indeed, it was one of the 'fixed' factors in describing palaeobiology – the present is the key to the past. Of course, with plate tectonics and greater data we can no longer hold this view. There is overwhelming evidence that the ancient land masses did influence ocean circulation and climate. As continents split up so ocean circulation patterns were changed. Land masses shifted northwards, altering the distribution of land and climate patterns. This, in turn (with plate tectonics), radically altered sea levels which are thought to be one of the prime causes of species extinction.

Leaving aside variations due to plate tectonics, the driving force of both weather and climate is solar radiation. The amount of solar radiation falling in any given place is determined by the time of day, season and latitude. This radiation heats up the earth and since the amount between any two places will differ it follows that there will be thermal gradients. Movements to equalise (and thus remove) these gradients will mean the movement of energy which we experience as wind. Further, localised gradients can produce clouds, precipitation, etc. On a daily basis we refer to this as weather but over a sufficient period of time it would be seen as climate (i.e. the generalised weather pattern). We can divide the world into climatic belts whose presence is very similar to that of biomes. On to this general picture must be put some important variations:

- Firstly, the amount of radiation received appears to be more crucial polewards. For example, the tundra is a marginal habitat with several months of short day length months each year. Ability to obtain enough radiation during the summer months can often determine the boundary of tundra;
- Secondly, altitude can substitute for latitude in some cases. For example, as one climbs higher in mountain chains such as the Andes the vegetation pattern will change in a similar fashion to that which one would expect if one travelled polewards;

- Thirdly, variation can be seen in the shorter-term climatic cycles and their influence on vegetation distribution – notably the Ice Age. Although the most recent Ice Age is one of the better known, more recent climatic phenomena, it is certainly not the most dramatic or of longest duration. Climates have changed throughout geological history and not just because of plate tectonics (Bradley 1981). It is one of the great challenges for palaeobiogeographers to understand both the distribution of plants and animals and their ecologies. However, in terms of recent events, the Ice Age (actually a series of warmer periods interspersed with colder ones) has had the greatest effect. At their greatest extent the ice sheets covered most, if not all, of the British Isles, destroying vegetation and soils. That the most recently deglaciated parts of the British landscape has been able to revegetate within the space of about 12,000 years is a demonstration of the speed with which recolonisation can take place.

The causes of major climatic changes throughout geological history can be linked to plate movements. Such an explanation is not possible with the relatively recent changes of the Ice Age (*c.*2million years BP to 10,000 BP). There is an increasing amount of evidence that links these climatic changes to variations in the earth's orbit coupled with a feedback mechanism linked to the amount of solar radiation reflected from the earth. The effect of these was to reduce the amount of radiation striking the earth (effect of orbit) and to reflect much of that which did reach the surface (effect of feedback). The result was to change the climatic zones around the world. Most noticeably, the more even climate of earlier times was replaced by a steeper temperature gradient between pole and equator. The geographical effects can be summarised as:

1. Shift in climatic zones. For example, during the maximum European glaciation the polar areas had shifted south between 10° and 20°S. Generally the changes brought about the reduction in size of more tropical ecosystems and the expansion of grasslands;
2. Changes in patterns of air/water circulation as a result of (1). This might be broad scale;
3. Changes in sea level. Perhaps the most important aspect, as it led to the creation of land 'bridges' which allowed the migration of animals and plants. Major examples include the linking of Britain to Europe and the exposure of much of the sea bed around the northern coast of Australia linking it to Papua New Guinea. During warmer periods such land bridges were inundated.

Given these changes there were five biogeographical responses:

1. Some species could keep to their habitats by migrating with them although they would be unlikely to move very fast. Some species might lag behind and so new habitats would be opened up lacking 'slower' species;
2. Vegetation zones shifted towards the equator. This would also affect subsequent successions. There is evidence to suggest that even at the height of the glacial periods there were still areas (refugia) where plants and animals could exist. This would refer to glacial areas, but other places such as rainforests whose ranges were diminished would also 'retreat' to refugia;
3. Some species adapted to the cold, e.g. mammoths. During this time animal extinctions were relatively few. A more common response was adaptive radiation, i.e. creation of new subspecies/species;
4. Some species became extinct. There is some debate about this because losses were not even across the board but linked to certain groups such as the megafauna of Australia. Arguments continue about whether such extinctions could be linked to climatic changes or human influence, the so-called overkill hypothesis, e.g. hunting (see Flannery 1995b for arguments on human-based extinctions);

5. Massive post-glacial lake systems were created (only to drain again a few thousand years later) which influenced the distribution of freshwater species but which also acted as barriers.

Oceans are paradoxical areas for biogeography. They provide barriers to the spread of terrestrial species. At the same time they permit the distribution of marine flora and fauna (although even here small changes can create effective barriers). There are three questions here: how and why have oceans developed; how do they operate and what are the implications for ecosystems? Much of the answer is bound up in plate tectonics. We tend to think of plates as terrestrial features (remember it was first thought of as *continental* drift) but it is just as much oceanic. We can frame plate tectonics to look at it from a hydrologic perspective. From this, we can see the development of oceans as the changing distribution of large areas of habitat (see e.g. Thurman and Trujillo 1999). The how and why of ocean development is centred on constructive plate margins (often called mid-ocean ridges). Here, an area of magma weakens the oceanic crust forming a rift valley. This sinks and the water invades the area. From here, the rift keeps spreading and deepening eventually forming the ocean (see Segar 1998).

Water acts much like air with the exception of its higher density and specific heat making it an ideal energy transport. To create such a mechanism one needs to remember that if energy reaches two areas in different amounts an energy gradient will be set up and a flow will start. Recently, advances in oceanography have linked all energy flows in the oceans into one – the conveyor belt theory (Segar 1998 p. 234). This flow of energy around the globe has an ameliorating effect in terms of surface currents (for example, the Gulf Stream keeps Britain warmer than it would otherwise be). Over time it leads to the development of ocean 'climates' (Segar 1998 p. 195).

The implications for ecosystems are considerable. Shifts in ocean climate belts, currents and distributions could create both barriers and opportunities. The break-up of Gondwanaland about 300 million years ago left relict animal assemblages in South America, South Africa and Australia. Modern-day current patterns in the mid-Pacific are responsible for the climatic shifts referred to as *El Niño*.

Finally there are the physical impacts of landforms. In terms of biogeography there are two main areas of study. The first relates to the size of the landform. As a rule the largest continents have changes in climate as you go inland. The idea (referred to as continentality) is that oceanic air masses bring rain to coastal areas. Go inland and the amount of moisture decreases. Thus you get rainforests by the coast and deserts inland (e.g. Africa, eastern Australia). The second relates to the distribution of mountain ranges. These large features create changes in global atmospheric conditions whereby circulation patterns and thus precipitation (and to a lesser extent, temperature) are influenced.

What are the ecological and biogeographical implications of these factors? There are four major concepts which use the global scale as part of the explanation: zones, biomes, endemism and vicariance/evolution.

The concept of the zone seems simple enough at first glance – an area of specific environmental conditions within which a homogeneous area of vegetation can be identified – but it does beg the question of how specific either conditions or range of species need to be. There does not appear to be universal acceptance of what constitutes a zone or how it relates to other spatial units. The major problem lies with the location of the boundary. What might seem like a sharp line at a distance becomes a ragged mosaic close to (called an ecotone). A study of the zonation of mountain flora provides a good example of this (Tivy 1993). Mount Kosciusko in south-east Australia is one of the only truly alpine areas on the continent (Williams and Costin 1994). As one rises from the grasslands surrounding the mountain range there are a number of distinct zones. These, with the corresponding physical environment are:

Vegetation	Physical environment
Montane forest	Relatively deep soils, sheltered
Subalpine woodland	Relatively deep alpine soils
Tall alpine herb communities	Valleys and basins with cold air drainage
Short alpine herb communities	Snow-patch meadow soils. Snow cover >8 months per year
Feldmark	Very wind-exposed, above permanent snowline

Although this is a simplified version it does illustrate the divisions that can be made. The advantage is that it helps study of environmental gradients.

Often the differences between zones are small. For ease of description, study, etc. it is possible to group numerous zones together to make our largest global ecosystem, the biome. Currently, there are about 14 recognised biomes (depending on how one classifies them). Each one may consist of several subdivisions, and as a result their use in ecology is limited. However, they do provide a useful overview of key vegetation types which helps greatly in conservation work. Originally, these global divisions were used in distinguishing broad climatic belts. One of the first biogeographers to do this was Köppen whose maps are still used today.

Early natural scientists such as Charles Darwin and his contemporary Alfred Russell Wallace noticed that there were some species which seemed to exist nowhere else and others where they existed in two or more otherwise separated areas. The former are referred to as endemic species while the latter form a disjunction pattern – occurring in more than two places at once. Some areas such as Australia have more endemics than one would expect and the causes of this are of interest to biogeographers. One obvious explanation would be the splitting up of the large southern supercontinent Gondwanaland leaving Australia isolated. Unfortunately there is another reason for endemics to be of interest – their very rarity makes them vulnerable to outside (i.e. human) interference. Whereas endemics occur in only one place, provincial species take this one stage further and live in only one habitat-range of the area. The distinct nature of certain regions' vegetation led people to draw specific maps based on provinces (see Brown and Lomolino 1998, p. 304). A final example of the distribution of communities comes from those species we find in different places – the concept of disjunction. Ratites (an order of birds which includes the emu and ostrich) are found today in three separate locations – South America, South Africa and Australia. The argument here is that they migrated to these areas before the break-up of Gondwanaland and that the subsequent development of similar features could be predicted by simple Darwinian theory as parallel evolution.

All of these ideas are trying to answer the question of where and why certain species developed. Initially, biogeographers try to find the centre of origin of the species. Once this has been established it is then possible to address the main point – how did the species spread and develop? There are two competing theories – vicariance and dispersal. Vicariance biogeography assumes that the development of an organism occurs before the division/creation of a barrier and that the barrier just splits up a range into two or more disjunctions. On the other hand, dispersal theory suggests that organisms migrated over pre-existing barriers.

3.6 Summary

The distribution of plants and animals is dominated by physical conditions. It is vital therefore to know what physical factors are important and what their measures are. Each species has a range of conditions within which it will thrive. An ecosystem is thus a collection of species with similar conditions for survival. In studying physical conditions it is useful to see what is important on an individual, species and community level: factors will change according to the scale. The most significant contribution to species distribution has been plate tectonics and the physical implications of ocean and continent size and location.

APPLICATIONS – USING BIOGEOGRAPHY

Using ideas from this chapter in real-life situations

Mining is a key aspect of our modern economies and yet it is possible for it to drastically alter plant and animal distributions. One classic case was the impact of metalliferous mining waste on grass distribution in Anglesey, Wales, UK. Here, some areas were grassed and others were not. During the study, it became apparent that certain grass phenotypes were more tolerant of mine waste than others. This led to the expansion of study in remedial ecology and the revegetation of areas with special heavy metal tolerant cultivars.

A more recent example comes from the UNEP/UNCTAD organisation called Natural Resources and Sustainable Development (www.natural-resources.org). A study by the Australian Centre for Mining Environmental Research discussed the impact of mining on biodiversity. It acknowledged the potential for harm but it also suggested ways forward. Examples of mining companies involved in helping rare plant conservation were given. The practical side of this is that mining specialists (especially geologists) are well aware of the chemical and physical nature of the material they are dealing with and how it can affect biogeography. Many companies are now proactively dealing with environmental issues which are seen as part of their 'bottom line' of responsibilities. Without work in biogeography (both theoretical and applied) it is unlikely that such a situation would have occurred.

Review questions

1. What is the significance of tolerance to species distribution?

2. Why does scale matter when considering key physical factors?

3. What is the value of studies such as Hubbard Brook?

4. Describe the development of plate tectonics in relation to climatic belts.

5. Account for the change of species composition with change in both altitude and latitude.

Selected readings

Texts tend to be either introductory or in considerable depth, with many studies of individual regions, ecosystems or species left to academic papers. One of the better physical texts is Ernst (1999), while the best for more advanced work has to be Rosenzweig's (1995) text.

References

Bains S, Norris RD, Corfield RM and Faul KL. 2000. Termination of global warmth at the Palaeocene/Eocene boundary through productivity feedback. *Nature*, 407: 171–4.

Begon M, Harper JL and Townsend CR. 1996. *Ecology*, 3rd edn. Blackwell Science.

Bradley RS. 1981. *Quaternary Palaeoclimatology: Methods of Palaeoclimatic Reconstruction*. Allen & Unwin.

Brown JH and Lomolino MV. 1998. *Biogeography*, 2nd edn. Sinauer.

Calsbeek R and Smith TB. 2003. Ocean currents mediate evolution in island lizards. *Nature*, 426: 552–5.

Cane MA and Molnar P. 2001. Closing the Indonesian seaway as a precursor to east African aridification around 3–4 million years ago. *Nature*, 411: 157–62.

Charman PEV and Murphy BW. (eds) 1991. *Soils and their Management*. Oxford University Press.

Clark PU *et al.* 2002. The role of thermohaline circulation in abrupt climate change. *Nature*, 415: 863–9.

Crawley MJ. 1997. Life history and environment. In Crawley MJ. (ed.) *Plant Ecology*, 2nd edn. Blackwell Science.

Dickson B *et al.* 2002. Rapid freshening of the deep North Atlantic over the past four decades. *Nature*, 416: 832–7.

Dobson M and Frid C. 1998. *Ecology of Aquatic Systems*. Longman.

Ernst WG. (ed.) 1999. *Earth Systems*. Cambridge University Press.

Flannery T. 1995a. *Mammals of New Guinea*. Reed Books.

Flannery T. 1995b. *The Future Eaters*. Reed Books.

Forman RTT. 1995. *Land Mosaics – the Ecology of Landscapes and Regions*. Cambridge University Press.

Jones JC and Reynolds JD. 1996. Environmental variables. In Sutherland WJ. (ed.) *Ecological Census Techniques*. Cambridge University Press.

McFaden MJ and Zhang D. 2002. Slowdown of the meridional overturning circulation in the upper Pacific Ocean. *Nature*, 415: 603–8.

McKane RB *et al.* 2002. Resource-based niches provide a basis for plant species diversity and dominance in Arctic tundra. *Nature*, 415: 68–71.

Miller GT Jnr. 1998. *Living in the Environment*, 10th edn. Wadsworth.

Peterken G. 1993. *Woodland Conservation and Management*, 2nd edn. Chapman & Hall.

Peterken G. 1996. *Natural Woodland*, Cambridge University Press.

Rosenzweig ML. 1995. *Species Diversity in Space and Time*. Cambridge University Press.

Segar DA. 1998. *Introduction to Ocean Sciences*. Wadsworth.

Specht R and Specht A. 1999. *Australian Plant Communities*. Cambridge University Press.

Thurman HV and Trujillo AP. 1999. *Essential Oceanography*. Prentice Hall.

Tivy J. 1993. *Biogeography: A study of plants in the ecosphere*, 3rd edn. Longman.

Watkinson AR. 1997. Plant population dynamics. In Crawley MJ. (ed.) *Plant Ecology*, 2nd edn. Blackwell Science.

Wharton DA. 2002. *Life at the Limits*. Cambridge University Press.

Williams RJ and Costin AB. 1994. Alpine and sub-Alpine vegetation. In Groves RH. (ed.) *Australian Vegetation*, 2nd edn. Cambridge University Press.

Zhisheng A *et al.* 2001. Evolution of Asian monsoons and phased uplift of the Himalaya-Tibetan plateau since late Miocene times. *Nature*, 411: 6.

Websites

A key site for the long-term ecological research programme which gives comprehensive datasets for physical and biological aspects.
www.hubbardbrook.org

Part of a large collection of material dealing with all aspects of geological history including climates, rocks and species.
www.ucmp.berkeley.edu/help/timeform.html

CHAPTER 4

The biological environment

Key points

- Although often considered a less significant component, biology plays a key role in species distribution;
- This can be seen at three scales: individual, population and community;
- At the individual level, genetics plays a key role in determining the characteristics (and thus survival) of the species;
- Populations consist of members of the same species in a specific area. At this scale, key issues are those which affect the range as a whole, such as human impact;
- Community-scale responses have to do with species interactions which are often considered to be governed by a set of 'rules';
- Changes in distribution can also occur as sets of species interact together; succession is one of the most prominent examples of this;
- The impact of these biological reactions in both individuals and ecosystems has theoretical and practical implications for us.

4.1 Introduction

The physical environment can be a powerful force in determining species distribution. For this reason, many biogeographical maps have a physical origin, e.g. climate. Recent research suggests that the addition of iron to sea water could trigger a considerable growth in biomass which would help combat global carbon build-up (Schiermeier 2003). This gives the impression that the biological component is a passive recipient of whatever physical process is dominant. In most ecosystems this is far from the case: species, especially plants, have a complex feedback mechanism with the physical environment where each influences the other. A simple illustration is the oak–ash woodland where open populations of ash allow oak seedlings to flourish which eventually shade out the ash.

In this chapter we aim to bring the debate back towards the biological aspects of biogeography and show how species, from individuals to communities, can influence population distributions. Species exist at three scales: individual, population and community. At each scale there is a set of biological controls on organism distribution. This is the focus on the first part of this chapter. The second focus is on how change can occur through biological mechanisms. Finally, there is consideration of the implications of this for biogeography.

4.2 Biological controls on organism distribution

4.2.1 Individuals

In terms of individuals we are concerned about three aspects of biology: reproduction, adaptation and dispersal. The first is important because it determines the survival of the species. Although there are a range of mechanisms available, the common goal is to continue the species. A second way of keeping the species alive is for it to adapt to the environment. This also involves genetics but the result is usually more dramatic than reproduction. If conditions are sufficiently altered it might result in a new species (for example Darwin's Galapagos finches – see below). Finally, dispersal mechanisms should be considered as this is the key to species range change.

Sex is important. With a few exceptions, no individual is capable of self-replication, i.e. breeding without a partner. Some female insects, notably aphids, produce daughters by live birth without fertilisation throughout the summer. Each daughter is an exact genetic copy of her mother and there is no chance for any variation to occur. A similar process occurs in some less advanced animal groups such as the protozoa, and some members of other groups such as the coelenterates (hydroids and jellyfish) will use 'budding' as a means of producing more individuals when environmental conditions are satisfactory, reverting to sexual reproduction when conditions become less favourable (it might be worth considering the biogeographical significance of this). The 'budded' individuals will be identical to the parent while the offspring of sexual reproduction will have some differences in their genetic constitution. It is not just sex *per se* that is significant but the sex ratio of offspring. A number of species (e.g. tuatara – Nelson *et al.* 2002) produce more males/females according to a range of environmental conditions. In the case of the tuatara (an endemic New Zealand lizard) higher temperatures produce more males (with interesting implications for survival in terms of global warming!).

It follows that sex is undoubtedly important in preserving a diverse genetic base for all species (although the degree of importance is a matter of debate – see, for example, Morrow *et al.* 2003). Any group of organisms that has little genetic variation between individuals will be more at risk of succumbing to adverse circumstances than one in which there is more variation. This is one of the reasons why animal breeders are concerned to keep records of the parentage of their stock since inbreeding reduces genetic diversity and may permit the development of unwanted characteristics (especially in terms of breeding endangered species in zoos for example – see Pullin 2002).

Plants employ a range of techniques for ensuring their survival. Some have developed a technique of producing genetically exact copies of themselves – such as the plantlets produced along the leaves of *Bryophyllum*. Others, like cultivated strawberries, produce horizontal stems (runners) on which develop small plantlets, which can root very easily. Other plants such as onions or tulips produce small underground bulbs as offshoots from the main bulb. This is referred to as vegetative reproduction and in the case of strawberries and similar species provides a rapid method of establishing new individuals, and frequently used by gardeners.

The more conventional production of seeds occurs as a result of pollination, usually from another individual or occasionally from another flower on the same plant. A few plant groups, such as dandelions (*Taraxacum*), may self-pollinate, while others will not set seed unless the pollen comes from a completely different individual. This combination of random wind or insect pollination might be expected to produce unusual hybrids between grasses and bluebells, but in practice this does not happen since grass pollen is unable to fertilise a bluebell or vice versa. The species are thus reproductively isolated from each other. Such isolation can occur as a result of differing flowering times or more usually it is that the plants are not genetically similar enough (see Box 4.1).

BOX 4.1

Breeding and primroses

Three closely related species from the primrose family (Figures 4.1 (a)–(c)) are commonly found in fields and woodlands in Europe and show an ability to interbreed and produce fertile hybrids because they are genetically close and compatible and flower at about the same time:

Figure 4.1(a) *Primula vulgaris* is typically a plant of woodland edges and clearings which flowers in late spring throughout much of the British Isles. The pale sulphur-yellow flowers are produced in large numbers, each on a single stalk

Figure 4.1(b) *Primula veris* is found in chalk and limestone areas where it occurs in grasslands and hedgerows and banks. The butter-yellow flowers are smaller than those of the primrose, and are borne in a cluster on a tall flowering stalk; the flowering period overlaps that of the primrose

Figure 4.1(c) *Primula elatior* is quite widespread in woodlands throughout Europe, but has a very restricted area of occurrence in England where it occurs in a small number of woodlands on base-rich clay soils in East Anglia. The drooping, pale yellow flowers are larger than those of the cowslip, but are borne in a moderately tall stalk

Cowslip	*P. veris* – fields, usually on base-rich soils, quite common in the UK
Oxlip	*P. elatior* – ancient woodlands in East Anglia, very rare in the UK
Primrose	*P. vulgaris* – woodland edges and hedge banks, widespread in the UK

The commonest hybrid is *P. veris* × *vulgaris*, known as the false oxlip, and usually found in open fields, rather than in ancient woodlands which are the normal habitat of *P. elatior.* Primroses do occasionally hybridise with oxlip (*P. elatior* × *vulgaris*), but this form has no common English name. The hybrids resemble each other quite closely and bear a strong resemblance to *P. elatior*. The false oxlip is prone to back crossing with either parent and is, as a result, quite variable in flower size and colour.

Two other naturally occurring species of *Primula* are found in the UK, known as the bird's-eye primrose (*P. farinosa*) and the Scottish primrose (*P. scotica*). The two species are very similar in appearance although *P. farinosa* has rose-lilac flowers and *P. scotica*, purple flowers. In the UK, *P. farinosa* occurs in northern England and the Scottish borders while the much rarer *P. scotica* is endemic to coastal grasslands in the far north of Scotland and the Orkney Islands. A study of the chromosome numbers shows that *P. scotica* is a hexaploid, with three times the chromosome complement of *P. farinosa*. The fact that the Scottish primrose is an endemic (which occurs nowhere else in the world) could possibly be attributed to the isolation and genetic drift of its ancestors as a result of the last glaciation, which would have constituted a major barrier to communication with populations of the bird's-eye primrose. A related species, *P. scandinavica* (Figure 4.1 (d)), is known only from Norway and Sweden, where it was first differentiated from *P. farinosa* in 1938 as a result of chromosome studies. It is an octoploid, closely resembling *P. farinosa*, but with 72, rather than 18, chromosomes; it occurs in similar open calcareous montane grassland, often as quite isolated populations. The genetics of *Primula* species is a complex and interesting area for research that has strong biogeographical links and much experimental work has been undertaken. Hambler and Dixon (2003) have summarised the relevant work on the three species mentioned in this paragraph. In common with other groups within the genus, the development of such genotypes can be accounted for by conventional models of allopatric speciation in combination with hybridisation and polyploidy (Kelso 1988).

Figure 4.1(d) *Primula scandinavica* from the Hardangervidda in Norway. It appears as though isolated populations of *Primula* species such as *P. farinosa* in montane or similar environments have a tendency to develop polyploid forms through self-fertilization and the ready availability of new habitats as a consequence of retreating ice sheets

A species of plant or animal has a particular genetic constitution, and it is this factor which differentiates it from organisms with different genetic constitutions (see e.g. Ridley 2004). Individuals of the same species may interbreed freely and fertile offspring will normally result. Closely related species may interbreed but the chances of a successful outcome depend on the nearness of the relationship. Domestic dogs (*Canis familiaris*) can interbreed with wolves (*C. lupus*) but not with foxes (*Vulpes vulpes*) and domestic cats (*Felis catus*) successfully interbreed with European wild cats (*F. silvestris*).

Closely related animal species can be interbred although the hybrids are almost always sterile. The commonest example is probably the horse–donkey cross, known as a mule, a strong and hardy animal which was often used to carry goods and equipment in rough terrain unsuitable for vehicles or even pack ponies. Failure to interbreed among animals may be due to a number of factors of which the more important are:

- Behavioural differences in courtship;
- The more extreme examples of structural incompatibility of genitalia that occur among insects;
- Physical or behavioural barriers that tend to enhance the differences between species, which may once have been closely related.

Over long periods of time, the genetic constitution of populations of a particular species may change through chance mutation and where the population is isolated from others of the same species by virtue of a mountain range, or because it is on an isolated island, the differences may, over time, lead to the development of different species (i.e. adaptation). This is certainly the case in the finches (*Geospiza*) of the Galapagos Islands that were studied by Charles Darwin. Interfertility between isolated populations declines with increasing duration of isolation and long isolation tends to produce a situation where gene flow diminishes and reproductive isolation exists (see Chapter 6). Allied to this is the idea of *gene flow*. As species move so their genetic characteristics also move with them. However, as with other aspects of genetic change this factor is not constant. Not all genetic material will move (which might eventually lead to speciation) and some mechanisms may prove less useful. A study of dispersal via sheep (Willerding and Poschlod 2002) indicated that this was a less significant factor than others.

Finally, there is the issue of dispersal. In terms of biogeography, dispersal refers to the movement of organisms to extend/change the size and/or shape of the species' range. Thus dispersal *within* the range does not count. Dispersal, along with speciation and extinction, are the key elements in changing species distribution patterns. Such range changes can have considerable impacts on the new areas. There are six aspects we need to consider when looking at dispersal:

- **Types of dispersal:** refers to the ways by which a species changes its range. We recognise three main types: long-distance, e.g. by birds where the dispersal jumps over some areas to utilise other ones; slow, diffusion dispersal where the species gradually moves to a new range, and secular migration where species do not so much change range as evolve along the way (see Brown and Lomolino 1998);
- **Mechanisms:** the actual process used to move the organisms. Generally divided into active or passive, it determines which routes are available for the species. Active dispersers, i.e. those that move themselves such as animals, have one set of options whereas passive (e.g. wind-dispersed seeds) organisms have another. Recent studies show an additional complication in that studies of mechanisms may be scale-dependent (e.g. Burns 2004);
- **Barriers:** the physical environment can provide a range of options to help or hinder the movement of species (see Chapters 3 and 8). In addition barriers have considerable significance in the debate about speciation although their actual role may be more problematic (e.g. McDowall 2002);
- **Aids:** the opposite of barriers – those aspects which assist the movement of species. One of the most common ideas here is the *corridor* – a routeway along which a range of organisms can pass. Alternatively, a *filter* is a more selective route, while there are random opportunities that can move species (sometimes referred to as *sweepstake* routes). Direction is also a factor whereby some movements occur almost exclusively from one place rather than another (e.g. Mátics 2003);
- **Establishment:** the final element is successful colonisation and reproduction (some colonists might survive at the edge of their ranges but be unable to breed). Alternately, some areas might be too small to colonise – a point noted in recent island biogeography research (Zalewski 2004) where population age was also a factor;
- **Social ecology:** studies of an Australian bird species (Cale 2003) demonstrated that population structures and movements were dictated, at least in part, by the social organisation of the species. As behavioural ecology becomes more established this might become a more significant factor.

4.2.2 Populations

The individual scale is concerned with the survival of the organism and the movement of its offspring both inside and beyond the current range.

Populations are concerned with the aggregate effect of groups of organisms. A population is usually defined as a group of individuals belonging to a particular species within an area, which enables them to breed with each other. Because a population can only consist of one species – the commonly accepted breeding unit – two or more species cannot belong to the same population.

Populations can be spread over considerable areas but the genetic constitution of the species involved remains constant throughout the area of distribution. A meadow buttercup from England is genetically more closely related to a meadow buttercup from Germany than it is to a creeping buttercup. Geographical separation of populations is widespread and enables maps to be constructed which indicate the extent of a particular species distribution on a local or even global scale. This is particularly useful in planning for the conservation of less common species where precise information on the spatial distribution of what may be isolated populations informs the conservation and management process.

Loss of habitat through agriculture, transport or housing development may lead to the fragmentation of populations of many species in heavily settled parts of the world. Isolated populations may decline and die out because there is insufficient appropriate habitat for them and because physical barriers such as the clearance of woodland for agriculture or road building prevent some species from moving out of suboptimal habitats to larger fragments. The isolation of populations in this way gives rise to the idea of *metapopulations* where the isolated groups are considered as subpopulations linked by migration between them. Physical and other barriers will limit the rate of migration and in some cases populations will disappear altogether as a result of natural causes – such as when birth rates fail to match or exceed the death rate of that population. Where links between subpopulations are good, numbers may increase as a result of migration adding to the natural increase of the population.

Woodland management in the UK for endangered mammals such as the hazel dormouse (*Muscardinus avellanarius*) provides natural, high-level bridges of living branches across wider paths which enable this arboreal animal to move easily through its range. A similar approach has been used to provide links between metapopulations of endangered red squirrels in northern England where their preferred habitat, coniferous woodland, has been bisected by road construction.

4.2.3 Communities

Communities of plants or animals are composed of a number of populations of different species in a particular habitat. The habitat (literally, the place where something lives) can be very small, as, for example, an anthill in grassland, or quite extensive, as in tropical savannahs and forests. Within a community, population numbers will fluctuate on a seasonal as well as a daily basis, as will the numbers of individuals. This may be due to the influence of climatic or other environmental factors (see Chapter 9), to interactions with other species or to harvesting by humans.

Within a community, species have roles which depend to some extent upon their lifestyle. There must be plants and usually animals, which use the plants for food or shelter – or both, and there may well be other animals, which utilise herbivorous animals as food, and even carnivores, which regard other carnivores as prey items. A range of organisms is usually present (except in the very earliest stages) which utilise dead and decaying plant and animal material and help recycle the essential nutrients. This concept of a community is well known since it includes primary producers, herbivores, carnivores and detritivores operating as part of a complex food web.

Plant and animal species differ in their tolerance of environmental factors and different species have different roles and requirements in relation to the surroundings in which they occur. Sometimes the roles may be separated in time or space – for example, the temporal separation of carnivorous birds such as owls, which tend not to hunt during the hours of daylight, which avoids conflict for food resources (small mammals) with day-flying birds such as kestrels or buzzards. An example of spatial separation is shown by the tendency for small insectivorous birds such as *Certhia* (tree creeper) to work over the trunks of trees whereas members of the *Paridae* (tits) spend most of their foraging time on the smaller and higher branches. Such separation, whether in space or time, avoids competition for resources and enables species to coexist within the community, provided they could cope with the local environment.

The advantage to an immigrant species of being different in some respect is that its chances of survival may be improved as it finds an appropriate niche which is not already occupied by a species with similar requirements. This is particularly well demonstrated by the varying beak lengths and shapes of wading birds, which feed on tidal mudflats. In this case, the design of beak allows the species to make use of a particular prey resource; birds with long beaks (such as the curlew (*Numenius*)) can extract their prey from lower down in the substrate than is the case with birds such as the dunlin (*Calidris*) whose beaks are short. The avocet (*Avocetta recurvirostris*) has a thin, upwardly curved beak, which it uses to agitate and filter the top few millimetres of the mud. The three methods of feeding ensure that there is no competition for food, although space may become a consideration if population sizes are large relative to the available habitat.

Some species are undoubtedly more important than others for the continuation of a community. The species may be responsible for maintaining the spatial structure of the community or in preventing it from developing at some point into a climax vegetation type. In the first example, an animal such as an elephant is crucial to the maintenance of the savannah biome in Africa, by its browsing and the resultant suppression of scrub on the grasslands. In this respect it is even more important than the large number of grazing animals such as wildebeest or zebra, and probably more important than the other major browser, the giraffe. The reason for this is that giraffe tend to feed on higher vegetation whereas elephant tend incidentally to trample and crush scrub and small trees that they may not actually eat. In its controlling capacity, the elephant must be regarded as the *keystone* species in whose absence the scrub will inevitably reduce the area of grassland, which in turn will have major consequences for the herbivores and carnivores in this ecosystem. Keystone species occur in other habitats – grazing animals such as sheep or rabbits have an equally important role in preventing scrub colonisation of grassland and could be regarded as being keystone species. The notion of keystone species is becoming increasingly important especially in conservation.

4.3 Change due to biological mechanisms

The ideas described above are those key factors which affect the distribution of organisms by virtue of the genetic/biological construction of that organism. This section moves on to consider how biological interactions can influence species distribution. Here we are more concerned with community and population scale changes.

4.3.1 Changing communities 1: altering species composition

The composition of communities varies over time as species arrive and either flourish or disappear after a brief period of residence because of local environmental and biotic conditions (one of the foundations of island biogeography theory – see Chapters 12 and 13). Surprisingly, the dominance of a species within a community may not necessarily be due to the fact that it is extremely successful as a colonist but may also reflect its ability to avoid the attention of grazing or browsing animals by virtue of poisonous or bad-tasting leaves, exudations of gum or a surfeit of spines. The success of the introduced rhododendron (*Rhododendron ponticum*) from Turkey in colonising much of the deciduous woodland of the British Isles over the past 250 years is a case in point, as is the unwise introduction of the bramble (*Rubus fruticosus*) to New Zealand in the nineteenth century. The problem is that in the British Isles at least, rhododendron has few, if any, animal herbivores, and there are no suitable herbivores which would keep the bramble at bay in New Zealand. Some tropical species, such as the *Acacia* of the African savannah, survive because of the existence of an apparently symbiotic relationship between the plant and extremely aggressive and territorial ants whose defence strategies tend to put off all but the most determined herbivore. In cases where a plant is eliminated from a community because of overgrazing, the vacant niche is eventually colonised by another species. Species change within habitats through this mechanism can eventually lead to the development of a climax community, which bears little relation to the original pioneer assemblage. Similarly species change can result from a change in human land use. This is well exemplified by the way in which areas of calcareous grassland in western Europe, originally derived by the removal of woodland to increase the area available for sheep grazing, have changed over time (see Box 4.2 and Figure 4.2).

Figure 4.2 In the valley of the River Seine, near the city of Rouen, France, are some spectacular inland chalk cliffs and grasslands which are noted for the high biodiversity of their flora. Areas which are no longer grazed or managed are beginning to show signs of scrub development

BOX 4.2

Medieval farming and calcareous grasslands

In the medieval period, most of these grassland areas were grazed by large flocks of sheep and the resulting wool was a major source of wealth and export trade for England. Although the wool industry gradually diminished in importance towards the beginning of the seventeenth century, the escape from captivity of another grazing animal, the rabbit (originally imported from the Mediterranean area), and its spread throughout the country over the next few centuries ensured that grassland areas remained well grazed.

Unfortunately, rabbits did not confine their attentions to grasslands alone and by the mid twentieth century, it is estimated that up to 30 per cent of cereal production was lost. Conventional control methods (hunting, trapping or shooting) had little effect on the population, and it was only the introduction of myxomatosis in 1953 – a highly infectious viral disease spread by rabbit fleas – that control was established (Armour and Thompson 1955). It is estimated that mortality rates exceeded 99 per cent and rabbit populations have not recovered nearly 50 years later due to recurrences of the disease. The particular virus only affected rabbits, both wild and tame, and the native hare was not affected. The recent decline in hare numbers is probably related more to changes in agricultural practice. Well-grazed calcareous grasslands such as the sward shown on Plate 4, have among the highest levels of biodiversity in any anthropogenic ecosystem in western Europe.

Since the decline of rabbits, many areas of what were once described as high-quality calcareous grassland have begun to revert to scrub and eventually to woodland. Rabbits and sheep are selective grazers whose jaw action and feeding behaviour restrict the food plants available for consumption. The removal of these herbivores meant that tree and shrub seedlings as well as the coarser grasses and herbs that would not have had a chance to become established, could now do so (Figure 4.3). Fine-leaved grasses and low-growing or rosette plants whose growing points were not easily exposed to damage by nibbling were able to spread vegetatively since most of the flowers would have been grazed off. Once this grazing pressure was removed, the immediate response was a great increase in the number of flowers, an effect that persisted for several years. When grazing did not resume, scrub and weed species began to assume dominance and within 10–15 years, many areas that had been open grassland had developed into hawthorn scrub and are now quite dense woodland. It is worth noting that some rabbit populations began to show resistance to the virus in the 1970s and the virus is now less virulent (see Trout *et al.* 1992, Manchester and Bullock 2000).

Figure 4.3 This field boundary on the chalk slopes of the southern Boulonnais in France is very rich in species such as orchids (eight species were recorded within 10 m of the site of this photograph). It shows the gradual colonisation of chalk grassland with scrub and coarser grasses once the area is not regularly grazed

4.3.2 Changing communities 2: responding to species development

The development of a community depends upon a number of factors, environmental and biotic. For example, the sequence of communities that arise as a result of the colonisation of sand dune habitats is well understood for appropriate coastal sites in western Europe. Pioneer species are eventually replaced by other species and as the community becomes more stable over time, further species may replace those that arrived earlier. The 'rules of assembly' (see also Chapter 5) for such a community are well understood and as a result, vegetation types within the community sequence can be readily distinguished, as can the associated animal species. Even minor changes in the environment may produce quite substantial changes in species composition, depending upon the ecological requirements of the keystone species. You would be most unlikely to find plant species normally occurring in the stressful and dynamic environment of the fore-dunes (nearest the sea), as major components of the more sheltered and stable dune slack communities further inland. From this point of view, communities are changing distribution due to biological constraints (even if these are responses to abiotic changes).

There are situations where the rate of community change is much more gradual – such as the gradation from grassland through scrub to woodland and eventually to high forest. The precise constitution of communities varies according to the geographical distribution of the species involved. This 'functional type' approach is an interesting one since it seeks to identify functional groups of species, which as a consequence of shared adaptations carry out key roles in the community, without needing to identify and produce extensive species lists. It also means that by identifying the presence of groups with particular functions, it is possible to compare, for example, the role of fire in maintaining the lowland heathlands of western Europe and its effects on the *fynbos* heaths of South Africa. Functional types are based upon the life form and appearance of the plants, which the Danish botanist, Raunkiaer, described about 100 years ago. The system is based upon the position of the 'perennating buds', which survive from one year to the next, and the nomenclature is logical if complex. Some species' 'buds' are below ground level as in the case of bulbs or corms; others at or slightly above ground level and trees have their perennating buds well above ground level. European and South African heathers may belong to different species but are morphologically similar and characterised by an ability to resist and recover from drought and fire.

4.3.3 Changing communities 3: succession

In any habitat, over a period of time, it is possible to see a sequence of species colonising, growing and being replaced in turn. Some will persist for a few years while others remain for many years, surviving the competition for space and nutrients that characterises all living communities. This process, known as succession, is usually a fairly dynamic process in most parts of the world. The rate at which succession occurs in a particular zone depends upon such features as climate – the cooler a habitat is, the slower is the likely rate of change. Long-established communities are usually stable and are termed 'climax communities'. The rate of species turnover, originally quite high, gradually declines but never reaches zero, even in the most stable communities. Species are always dying out as a result of small changes in the environment, which will quite probably permit the successful establishment of another species. Typically, succession from bare ground to mature climax woodland takes several hundred years in temperate climates, and much longer in cooler or subarctic areas. This is termed *autogenic succession* because it is driven by processes that originate within, rather than external to, the community, such as the accumulation of leaf litter in a forest or an increase in shading due to tree canopy development. Succession that takes place on developing sand dunes or on glacier forelands where bare ground is left by retreating ice is referred to as *primary succession*. It can also occur, for example, when ejected volcanic material is colonised by plants as on the Icelandic island of Surtsey (or Mount St Helens – see Chapter 12). Changes can also occur to established ecosystems (and are usually referred to as *secondary succession*).

The essential difference between primary and secondary succession is that secondary succession involves the recolonisation of an area after some sort of disturbance. The critical point is that secondary succession relies upon the availability of some sort of soil with humus and its associated microbial and invertebrate life. The rapidity with which colonisation occurs and which species colonise is dependent upon the type of soil available, the presence of any propagules or a seed bank, and the proximity of sites from which colonisation can occur. Sites where secondary succession occurs are normally much quicker to develop than those which rely upon primary succession. The earliest stages of both types of succession are characterised by the presence of short-lived (often annual) plants, which tend to include grasses and weeds, which can tolerate a wide range of habitat and resource issues.

Over a long period of time, the differences in the development of primary and secondary succession under similar circumstances tend to blur and it becomes very difficult to distinguish between, for example, ancient woodland and secondary woodland which might have existed for 300–400 years or so. The only reliable way is to look for specific indicators or keystone species which will probably be present in ancient woodland but not in the secondary.

Sometimes external intervention may halt a succession for a period of time. In this context, many areas of calcareous grassland in the British Isles represent a form of interrupted secondary succession on the site of destroyed woodland, maintained as grassland through grazing over many centuries. This interrupted succession is sometimes referred to as a *plagioclimax* or arrested succession, since removal of the restraining factor, in this case grazing, will allow the original succession to proceed. The present-day state of heathland plagioclimax in lowland Europe is due partly to grazing but also to the effects of fire as a controlling agent, which prevents the community from reaching a stable state. Frequent disturbance by fire, grazing or other management allows many communities to become more diverse in the short term since immigrant species may be able to colonise gaps left by fire or uprooting, but will not always be able to gain dominance.

In all cases, the successional process implies that plants must become established in a habitat before any animal colonists are able to survive. At its most basic, the lack of food and shelter in the colonisation phase may be unfavourable for most animals but once plant establishment is under way, animal life can colonise and further modify the habitat and fungi and bacteria can assist with the recycling of organic materials. This modification and recycling process is termed *facilitation* (see Box 4.3) and implies that the pioneer species are of great importance in the process of succession (see e.g. Bowman and Seastedt 2001). In successional stages on glacier forelands, bryophytes are often the pioneer species and in a relatively short time they will have facilitated the establishment of grasses and herbs. Within a few years, however, the bryophytes will have disappeared because of competition for light and space with the successor species. Facilitation in later stages of succession is often responsible for greatly increased biodiversity; this is particularly so when, during the later stages of woodland succession, mature trees begin to dominate the habitat, with the development of many more niches that can be exploited by birds and insects, for example. Earlier species will, of course, be lost because of environmental change but the overall effect of facilitators appears to be one of steadily improving the biodiversity of their habitats.

BOX 4.3

The impact of facilitation

Facilitation is a particularly potent factor in areas where the environment is severe. In the Jotunheimen, Norway, studies of the vegetation on glacier moraine chronosequences (datable deposits of mineral material left behind as glaciers melt) have shown that within a few years of deposition bryophyte and lichen species begin to colonise, together with free-living colonial cyanobacteria such as *Nostoc*. Many of the bryophytes (such as the thalloid liverwort *Blasia pusilla*) have symbiotic cyanobacteria in their tissues and so, typically, do the lichen species such as *Stereocaulon* and *Solorina* which are characteristic pioneers. Cyanobacteria are known to be able to fix nitrogen and in this environment, the plants act as facilitators from the earliest stages of the succession. If the mineral material is even slightly calcareous, mountain avens (*Dryas octopetala* – Plate 5) is an early colonist and this species has a fungal symbiont, *Frankia*, which also fixes atmospheric nitrogen. Thus the initial severe soil environment is greatly improved by a build-up of nitrogen in the soil and the boost in plant growth that this provides produces a build-up of humus in the soil. Later stages in the succession include the development of willow or birch scrub, and open birch woodland in less exposed sites. More exposed sites tend to develop low-growing shrub cover, known as lichen heaths, wet areas develop into *Sphagnum* bog or other bryophyte communities and undulating ground develops a mosaic of dwarf willow or *Vaccinium* species, depending upon the duration of snow cover (Coker 2000). The Scandinavian glaciers reached their maximum post-glacial extent in about 1750 and since then, have begun to retreat as a result of climatic warming. There have been a number of episodes when the period of retreat has been interrupted by short periods of advance, and Matthews (1977) has dated the moraines which have been pushed up by ice advance by the use of lichenometry which correlates the diameter of individual colonies of the yellow map lichen *Rhizocarpon geographicum* (see Figure 4.4), which grows on siliceous rocks, with datable material such as parish records, historical photographs or maps. As most of the sites involved lie above the tree line (about 900m, Coker 2000), tree-ring dating (dendrochronology) is not an option. It is also important to note that local topography has a significant effect on the type of vegetation which develops. Vegetation that develops in sheltered areas will be less badly affected by adverse environmental conditions and succession will tend to proceed at a faster rate than on exposed sites.

This is not the only work in such areas. A classic study by Crocker and Major (1955) at Glacier Bay in south-east Alaska showed similar trends for species (see also Figure 4.5).

Figure 4.5(a) This photograph shows part of the path of the lava from the 1944 eruption of Vesuvius, near Naples in Italy. Some colonisation is visible, with grasses, bushes and shrubs. The successional process is probably facilitated by the lichen *Stereocaulon* which is an early coloniser of harsh environments such as recently deglaciated areas or lava fields

Figure 4.4 The successional process is probably facilitated by the lichen *Rhizocarpon* which is an early coloniser of harsh environments such as recently deglaciated areas

Figure 4.5(b) The group of people examining vegetation on the lava flow are biogeographers who were attending the International Association for Vegetation Science meeting in Naples in 2003

Succession should also be considered in relation to scale. An autogenic succession over a glacier foreland, several hectares in extent, will appear to show gradual transitions from pioneer to more established species as the distance from the glacier snout increases, but within this slowly changing landscape, there will be numerous places (microhabitats) where the succession has been interrupted by some outside influence such as a rock fall or landslide, or where an animal has defecated (see Figure 4.6). Such events can interrupt or even destroy an existing succession and bring about the start of another.

It is important to note that the final stage of succession in almost any environment will depend on local conditions, and that the time taken to reach the final stage can range from a few tens of years to many hundreds. The reproductive strategies of trees which are keystone species in succession from bare soil to mature forest change dramatically from the fast growth, abundant seeding and wide dispersal of pioneer species such as birch to a much more restrained production (with larger and less easily dispersed seeds) such as oak. The pioneer tree species such as birch are not likely to live more than a few decades after establishment, whereas the later species may live for many centuries and come to dominate by their individual extent rather than numbers. Many seeds are produced during the lifespan of these trees and some will remain viable for long periods in the soil seed bank, awaiting a chance to take advantage of a break in the canopy caused by the death of an ancestor. The lighter seeds of the pioneer species can be spread over large areas by wind and the species will colonise suitable habitats by chance rather than design.

Figure 4.6 In periglacial areas and similar mountain environments, frost heaving of soil (caused by the expansion of soil moisture as it freezes) churns up the soil and initially restricts the species which can effectively grow under such conditions. Typical frost heave vegetation consists of lichens and mosses, possibly with grasses. Later stages in succession are often characterised by low-growing shrubs such as dwarf willow

During the successional process, there may be evidence of inhibition of the growth of one species by another. Inhibition does not normally show in the earliest stages of colonisation of a newly available habitat, but later on when the true pioneer species are replaced by later pioneer species. The precise form of inhibition is often difficult to perceive although competition for space, both below and above ground, is one form. Another more subtle form of inhibition is produced when one species is able to interfere with the metabolic processes of another by chemical means. Such an effect, known as *allelopathy*, occurs when birch seedlings colonise areas of moorland or heathland on which heather (*Erica* and *Calluna*) species grow (see Figure 4.7). The inhibition occurs through the diffusion of a

Figure 4.7 Dwarf shrub heath. The dominant species in this vegetation type are the dwarf birch (*Betula nana*), crowberry (*Empetrum*) and lichens – in particular several species of *Cetraria*. The vegetation rarely exceeds 20–30 cm in height

chemical from the roots of the birch into the soil which affects the mycorrhizal fungi which occur in the roots of the heathers. The heather declines in vigour and may well die while the birch is able to make use of the space so created.

It is worth noting that the mechanisms of succession are likely to be complex, depending upon the type of community involved. Connell and Slatyer (1977) have produced an overview in which they propose three possible models – facilitation, inhibition and tolerance. An alternative perspective has been given by Tilman (1985). He looked at the resource ratio model where emphasis is placed upon the changing competitive abilities of species as environmental conditions alter over time. His model suggests that at any time within a terrestrial succession, the dominance of species is subject to the relative availability of light and a limiting nutrient, usually nitrogen. It is often the case that early stages of succession take place in habitats where there is ample light but little in the way of nutrients (e.g. sand dune – see Figure 4.8).

Over a period of time, nutrient availability increases as a result of the accumulation of leaf litter and the activities of soil animals and bacteria. This in turn allows plants to grow and biomass to increase, thus beginning to restrict the amount of light which can reach the soil surface. Species respond differently to particular combinations of light and nutrient levels and in this way it is usually possible to see how species partition themselves along a successional gradient. As a general rule, vigorously growing species such as grasses do well in early stages of succession where nutrients are unlikely to be limiting, while shade-loving species are generally found as ground cover in later stages of succession where competition for space is less severe. Few grass species can tolerate the shaded conditions of a typical woodland and not many woodland herb species will be found growing outside this habitat. Tilman's model suggests that the sequential change over time of dominant species in an ecosystem is due to changes in relative competitive abilities as first one species then another comes to dominate as a result of particular environmental conditions. Rates of succession may also be influenced greatly by the way in which herbivorous species operate. Experimental work on early stages of plant succession on newly ploughed fields, in which the earliest colonisers – annual species of herbs – are eventually replaced by grasses, has shown that leaf-eating insects will tend to delay the successional process. This is achieved by reducing the vitality and hence, the rate of colonisation by grasses, whereas insect larvae which tend to eat plant roots will tend to speed up the rate of succession because the dominance and persistence of the non-grass species are reduced.

Figure 4.8 Sand dune systems are extremely dynamic environments and species must be able to tolerate the dehydrating effects of salt spray, the erosive properties of windblown sand grains as well the chance of being buried. Foredune species are fast-growing species which can tolerate some drought through morphological adaptations such as the ability to reduce transpiration by rolling up the leaf blade, or through having reduced numbers of stomata or waxy leaves. The dune blowout occurs where an area of weakness develops in a previously stabilised dune (marram grass (*Ammophila arenaria*) is frequently planted or may naturally occur to achieve this). Such weakness can be caused by human and animal trampling or grazing which damages the stabilising vegetation. When this dies, the sand is less well fixed by the grass roots and can either be blown or washed inland. The photo shows the deep and extensive root systems of the marram grass and the wind bedding of the sand

In addition to the autogenic model, there are two other types of succession which should be considered. *Autotrophic succession* occurs when an area is opened up for colonisation and the colonists are green plants – hence the 'autotroph' label. An example of this type of succession is where soil is exposed through a landslip or as a result of road works. The exposed soil surface is more or less rapidly colonised by plants and the habitat is not degraded or lost, merely occupied.

Allogenic succession occurs where the process is driven by external processes such as changes in climate, the physical environment or grazing regime. The outcome of an allogenic succession is not as easily predictable as with the more conventional autogenic succession, and the external process may operate on a small scale, such as the shading effects of a mole or ant hill, or even the nutrient enrichment of the soil by animal faeces. Examples of large-scale allogenic succession might include the silting up of an estuary and the seawards extension of salt marsh, along with colonisation of what was once salt marsh by deciduous woodland as the salt content of the soil decreased over time (Ranwell 1974). Ranwell's study of the Fal estuary in Cornwall showed that over the last 100 years, the seaward edge of the salt marsh had extended by 800 m and the landward edge of the marsh was colonised by woodland. The vertical range of the succession was just over 2 m.

4.4 Implications

The work described above highlights the significance of biology in biogeography. As we research more so we uncover more about the importance of the biotic component in species distribution. This has implications for our study and analysis of biogeography (theory and practice) as these examples show:

- **The impact of dispersal and barriers.** The work noted above could give the impression that dispersal is an accepted fact, especially in the modern world. The rise of 'alien' species (i.e. those brought in from other ecosystems) has reached such a pitch in many areas that it threatens the native ecosystems and also challenges our ideas of the nature of barriers to dispersal;
- **Assembly rules.** Although the notion of sequential development might seem to have a certain logical appeal it is still a controversial subject with arguments both ways on the topic (see also guild structure). The more we research the more complex the situation is seen to be, which has implications for wildlife conservation among other aspects (Watkins and Wilson, 2003);
- **Conservation practice and ecosystem theory.** The example of English medieval downland is a particularly appropriate one given that one of the first modern nature reserves was a downland. Sheep had been removed because they were trampling the downland flowers. Upon removal of the sheep, scrub species flourished because they were not being selectively grazed. Such fine detail can make or mar a project. As conservationists we know how close to the limits of knowledge we often are! In a more recent case (Caspersen and Pacala 2001) woodland composition and function were seen to vary according to successional stage, with implications for conservation of specific elements (seres) in each successional stage;
- **The nature of the climax composition.** In theory, there comes a time when the end point of a succession might be reached – where the rate of change of species over time is very low and where replacement of individuals when they die occurs on a pro rata basis by younger individuals. This end point is conventionally known as the *climax* – a term which has been in use for more than 80 years since it was first proposed by Clements in 1916. He proposed a *monoclimax* – a single final composition regardless of starting point. Never overwhelmingly supported, the idea had critics such as Tansley (1939) while Whitaker (1953) proposed a *polyclimax* model. Present-day opinion is divided among the polyclimax and

climax pattern supporters. Recent studies on UK vegetation as part of the National Vegetation Classification (Rodwell 1992) show the importance of the continuum approach as a means of explaining differences in vegetation types along climatic gradients and seem to support the climax pattern hypothesis.

- **Climax stability.** Does a stable climax community exist anywhere in the world? It seems unlikely unless it is possible to demonstrate that the rate of change is very small. Successions can take anything from a few years to several centuries to complete according to local climatic conditions. The process may be interrupted by fire or flood, or by animal activities. Many communities are never able to complete a succession to a stable climax because the likely chances of interference are high, as for example in grasslands or woodlands.

The impact of humanity on the biosphere is such that there are few areas in the world where human (anthropogenic) influences are not felt, and where such activity has not seriously affected many natural ecosystems and the plant and animal communities which exist within them. Human activities are not good for natural succession.

4.5 Summary

One of the most important aspects here is the role biology plays in biogeography. It is more than just the addition of a prefix: there is a need to appreciate fully the complex interactions between (and within) individuals, populations and communities. There is something to the argument that biogeography is about dispersal, speciation and extinction (all three biological ideas), but this then misses the interactions just noted and the impact of the physical environment which, although not part of this chapter, is very much a key component.

Biology plays a vital role. If the object of the species is to survive then reproduction is crucial not just to the next generation but to successive ones where small genetically induced alterations may make the subsequent adaptations fit the organism better for the changes in environmental conditions. One of the first major concepts was dispersal. Organisms do not just roam around at random but often have restricted avenues open to them. This suggests limits to species distribution above those set by environmental conditions (introduced species soon show that new environmental conditions can be tolerated – and thus demonstrate the impact of the barrier). Beyond the individual scale we find that interactions between and within groups play an important role in species distribution in both time and space.

Outside the interactions between individuals we have those processes operating on a larger scale, e.g. succession – changes through time (although zonation – changes in space – can also be considered). We must also add human action into this because we can see the way in which, through nature reserves and agriculture for example, we are able to shape population distributions.

Finally, there are the implications of these concepts for our knowledge of biogeography. Perhaps the most important thing to remember is that our knowledge is increasing all the time and that this could (and often does) cause us to make significant reassessments of our understanding of the role of biological processes. This is nowhere clearer than in our next chapter where we examine the way in which seemingly random sets of distributions are turned into ecosystems.

APPLICATIONS – USING BIOGEOGRAPHY

Using ideas from this chapter in real-life situations

Using biogeography to help monitor global change! The International Geosphere-Biosphere Programme (IGBP) is a major research body set up in 1986 by the International Council of Scientific Unions. The aim is to monitor global change. One group within this is the Global Change and Terrestrial Ecosystem project (http://www.gcte.org/index.htm). This group works on five areas with sets of sub-themes within those. For example, the section on change in ecosystem structure had projects dealing with patch dynamics, landscape processes and global vegetation dynamics. The aim is to co-ordinate international work on these topics and to provide expertise to fill in the gaps where these are shown to exist.

One study (http://www.gcte.org/inchausti.pdf), examined links between biodiversity and ecosystem functioning. The study showed how small changes in ecosystems can have considerable effect on biodiversity. Other aspects of the work tried to assess the importance of losing key organisms in the ecosystem. It was accepted that some species are more valuable than others in ecosystem functioning and that we need to find ways to identify the traits of these key organisms so that we can better monitor any changes. Other work has tried to scale up small patch dynamics principles to the broader landscape picture. This is another case of biogeography research being at the forefront of knowledge.

Review questions

1. Which is the greatest and why: reproduction, adaptation, dispersal? Does this vary with scale?

2. What are the relative merits of sexual and asexual reproduction in terms of biogeography?

3. Compare and contrast a range of dispersal strategies. To what extent is this life form/genus-specific and what is the biogeographical implication of this?

4. What is the value of precise spatial distributions to nature conservation?

5. What are the genetic limitations of metapopulations?

6. Explain the role of keystone species in terms of both species distribution and nature conservation. Can we link these two perspectives and if so, how (or why not)?

7. Why could 'rules of assembly' be seen as being controversial?

8. Using one of the implications noted above (or considering a new one), describe, for a range of named species, how such an implication might affect our understanding of its biogeography.

Selected readings

There is any number of good texts dealing with the biological aspects of biogeography. For the ecological side, Begon *et al.* (1996) is a good start especially at the individual level and with succession/zonation. Ridley (2004) covers much of the ground in an accessible way in terms of population change, while Morin (1999) looks at the community-scale concepts.

References

Armour CJ and Thompson HV. 1955. Spread of myxomatosis in the first outbreak in Great Britain. *Annals of Applied Biology*, **43**: 511–18.

Begon M, Harper JL and Townsend CR. 1996. *Ecology*, 3rd edn. Blackwell Science.

Bowman WD and Seastedt TR. (eds) 2001. *Structure and Function of an Alpine Ecosystem*. Oxford University Press.

Brown JH and Lomolino MV. 1998. *Biogeography*, 2nd edn. Sinauer.

Burns KC. 2004. Scale and macroecological patterns in seed dispersal mutualisms. *Global Ecology and Biogeography*, **13**(4): 289.

Cale PG. 2003. The influence of social behaviour, dispersal and landscape fragmentation on population structure in a sedentary bird. *Biological Conservation*, **109**(2): 237–48.

Caspersen JP and Pacala SW. 2001. Successional diversity and forest ecosystem function. *Ecological Research*, **16**(5): 895.

Clements FE. 1916. *Plant Succession. An Analysis of the Development of Vegetation*. Washington, DC: Carnegie Institute.

Coker PD. 2000. Vegetation analysis, mapping and environmental relationships at a local scale, Jotunheimen, Southern Norway. In Alexander R and Millington AC. (eds) *Vegetation Mapping from Patch to Planet*, pp. 135–58. Chichester: John Wiley & Sons, Ltd.

Connell JH and Slatyer RO. 1977. Mechanisms of succession in natural communities and their role in community stability and organization. *American Naturalist*, **111**: 1119–44.

Crocker RL and Major J. 1955. Soil development in relation to vegetation and surface age at Glacier Bay, Alaska. *Journal of Ecology*, **43**: 427–48.

Hambler DJ and Dixon JM. 2003. *Primula farinosa* L. *Journal of Ecology*, **91**(4): 694–705.

Kelso S. 1988. Evolution in the genus *Primula* sect. *Aleuritia*: isolation, secondary contact, homostyly, polyploidy. Abstract A602, 1988 meeting of the American Institution of Biological Sciences at Dacis, California.

McDowall RM. 2002. Accumulating evidence for a dispersal biogeography of southern cool temperate freshwater fishes. *Journal of Biogeography*, **29**(2): 207.

Mátics R. 2003. Direction of movements in Hungarian Barn Owls (*Tyto alba*): gene flow and barriers. *Diversity and Distributions*, **9**(4): 261.

Manchester SJ and Bullock JM. 2000. The impacts of non native species on UK biodiversity and the effectiveness of control. *Journal of Applied Ecology*, **37**: 845–64.

Matthews JA. 1977. A lichenometric test of the 1750 end-moraine hypothesis, Storbreen gletschervorfeld, Southern Norway. *Norsk Geografisk Tidsskrift*, **31**: 129–36.

Morin P. 1999. *Community Ecology*. Blackwell Publishing.

Morrow EH, Pitcher TE and Arnqvist G. 2003. No evidence that sexual selection is an 'engine of speciation' in birds. *Ecology Letters*, **6**(3): 228.

Nelson NJ, Keall SN, Pledger S and Daugherty CH. 2002. Male-biased sex ratio in a small tuatara population. *Journal of Biogeography*, **29**(5–6): 633.

Pullin A. 2002. *Conservation Biology*. Cambridge University Press.

Ranwell D. 1974. *Ecology of Salt Marshes and Sand Dunes*. London, Chapman & Hall.

Ridley M. 2004. *Evolution*, 3rd edn. Blackwell Publishing.

Rodwell J. (ed.) 1992. *British Plant Communities*, vol. 1: *Woodlands*. Cambridge University Press.

Schiermeier Q. 2003. Climate change: the oresmen. *Nature*, **421**:109–10.

Tansley AG. 1939. *The British Islands and their Vegetation*. 2 vols Cambridge University Press.

Tilman D. 1985. The resource ratio hypothesis of succession. *American Naturalist*, **125**: 827–52.

Trout, RC, Ross J, Tittensor AM and Fox AP. 1992. The effects on a British wild rabbit population (*Oryctolagus cuniculus*) of manipulating myxomatosis. *Journal of Applied Ecology*, **29**: 679–86.

Watkins AJ and Wilson JB. 2003. Local texture convergence: a new approach to seeking assembly rules. *Oikos*, **102**(3): 525.

Whitaker RH. 1953. A consideration of the climax theory; the climax as a population and pattern. *Ecological Monographs*, **23**: 41–78.

Willerding C and Poschlod P. 2002. Does seed dispersal by sheep affect the population genetic structure of the calcareous grassland species *Bromus erectus*? *Biological Conservation*, **104**(3): 329–37.

Zalewski M. 2004. Do smaller islands host younger populations? A case study on metapopulations of three carabid species. *Journal of Biogeography*, **31**(7): 1139.

Websites

Such a diverse area as this does not produce obvious sites dedicated to it other than university department notes. However, these are useful starting places – portals (places with lists of other sites rather than organisations):

The WWW guide to ecology.
http://pbil.univ-lyon1.fr/Ecology/Ecology-WWW.html

The genetics virtual library.
http://www.ornl.gov/sci/techresources/Human_Genome/genetics.shtml

General biology resource.
http://www.biology-online.org/

CHAPTER 5

Putting it together – classification, biodiversity and ecosystems

Key points

- Classification is the basis of all biogeographic study because it enables us to make a complex environment more understandable;
- Classification has numerous uses beyond the simple grouping of species;
- Classification is not without problems. These can range from the separation of two similar life forms into species to the theory of evolution;
- Classification is controlled by three factors: the features used for classification, the methods used and the scale upon which the classification is to operate;
- Biodiversity is a commonly used (and misused) word implying the range of species present in an area but in fact dealing with diversity from genetics to communities;
- Biodiversity is held to be important partly because of the value people place on having a diverse range of species in the biosphere and partly through its connections with ecosystem functioning;
- Although a simple term to use, the actual concept of biodiversity is beset with practical and theoretical differences;
- Ecosystem theory can be used in the preparation of larger-scale classificatory methods;
- In practice, classification can be seen as one of the most fundamental tasks of the biogeographer and is crucial for successful analysis.

5.1 Introduction

Organisation is at the heart of biogeography, for only when we can group, compare and contrast physical and biological characteristics can we move forward (at any scale) to understand and explain the distributions we see. Classification is fundamental to biogeographic research (see Figure 5.1). From the huge numbers of plant and animal populations it seeks to simplify the situation so that by naming any given life form all other similar ones have similar properties. It allows others to understand the precise nature of the organisms under discussion, leading to the production of a set of characteristics for each species. These species can then be grouped to produce an ecosystem (in classification terms a set of definable species characteristics in an area) which also has a set of characteristics. Into this comes biodiversity – a relatively recent concept which acts as a shorthand for the species richness of the planet. Although seen as something that classification can focus on

Figure 5.1 Classification is fundamental to biogeography. It is only possible to gauge species loss if you know how many species are present. Classification tends to be more complete for larger species such as mammals but lacking in the more numerous life forms such as insects. Such gaps create problems in understanding distributions and testing ecological concepts

(i.e. in identifying and recording species) the whole nature of biodiversity issues a great challenge to biogeography. It is a challenge to theory because it questions the fundamental nature (and use) of the whole notion of species – particularly at the genetic level. It is a challenge to the practice of biogeography because by focusing on the nature of the ecosystem it questions the idea of a homogeneous area. Rather, it puts forward the notion of patch dynamics – that each area is made up of a mosaic of small 'sub-ecosystems'. Thus the ecosystem map (one of the fundamentals of biogeography) becomes both theoretically and practically a more difficult task.

This chapter starts by considering the basic question – why classify? Although it might seem self-evident, there are many compelling reasons why we should classify and some equally compelling reasons to be cautious in our approach! From this it is possible to look at some common techniques in classification and the scales at which they can be used. The next section explores the concept of biodiversity. It has come to play an increasingly important role in classification both by challenging our notions of the concept of species and our value in so doing. Finally, there is a brief overview of major classificatory models and ideas which provides both a guide to common practice and a historical look at the development of the subject.

5.2 The basis for classification – why bother?

Classification is so much a part of modern biogeography that it almost seems to go unchallenged. However, at a time when the costs and benefits of any course of action are being questioned it seems appropriate to examine this concept from the most fundamental principles. So, what is the use of classification? Of the few writers who have explicitly argued this point, the work of Hawksworth (1988) and Nelson and Platnick (1981) provides a useful summary. In essence, they support classification because:

- **It is a fundamental part of a wide range of scientific disciplines.** The ability to name a species so that *that* name is readily understood by others to indicate a unique type of organism is crucial in such areas as conservation, environmental monitoring, agriculture and biotechnology, to name just a few common examples, as well as in the fields of education and training. This is not a facile point – it acts as precise biological shorthand and simplifies discussions between practitioners; (Balaenoptera is a 'reserved' term used for a particular group of marine animals – including blue and finback whales, while 'whale' can be anything marine and large to the untutored eye – which is what makes historical data difficult to interpret accurately);
- **It is possible to make a set of common characteristics**. The Linnean binomial system which is at the heart of species naming is a classification system (e.g. *Quercus robur* or pedunculate oak). The full name is given exclusively to one tree species. It is possible to study this tree and work out its physiological responses to environmental conditions (or, more

accurately, *range* of responses since individuals will differ within a population). The results would hold true for all members of the species populations. At the genus level (i.e. *Quercus*) it is possible to group similar characteristics together for a range of similar tree types. From here, it is easy to compare oak woodlands wherever they occur in the world because they have a shared set of similar characteristics;

- **The classification system can be used for the storage and retrieval of information.** While this idea is easy to see at the species level where just the name can convey a whole range of ideas and meanings, it is probably more useful at the ecosystem level. One common system in use today is to divide ecosystems according to their dominant species, e.g. oak–ash woodland. Not only does this instantly identify key species it also provides a wealth of other biotic and abiotic information. This system, currently being used in major projects like the UK NVC programme (Rodwell 1992–98), shows the value of such an approach;
- **It is possible to predict new information.** With many ecosystems insufficiently described and thousands of species not even classified it is possible to extrapolate from existing species and systems to new situations. A common example of this comes from the fossil record. Palaeontologists often use the characteristics of existing species to predict the habits of new fossil groups just discovered;
- **It provides order and allows for accurate hypothesising.** Classification is ultimately a simplifying tool. This allows for the ordered study of species rather than the random gathering of information on individuals. Once we know that this is correct then we can move on to create hypotheses which can be tested against the species or ecosystem.

Classification can be justified but this does not mean that the system is perfect. In fact increased research in areas such as biodiversity and landscape mosaics means that the classificatory system is under increasing strain. To be used effectively one must be able to see both the advantages and the limitations of the methodology. In addition to the authors noted above, others have also investigated this crucial area (see e.g. Scott-Ram 1990, Quicke 1993 and Salthe 1985). Their ideas which range from the practical to complex theoretical constructs appear to be focusing on a few key problem areas:

- **What do we need for an individual to exist?** An oak tree or a lion obviously exists as a single entity but what of a symbiosis? Are there two species or one? How do we classify coral? At the ecosystem level where is the boundary? To the palaeontologist, where is the time limit for the species? These are all different areas of work but they illustrate the same key idea – we need to know the limitations of what we are studying. In biogeography, the time element is perhaps the most vital. If we are studying species across the recent Ice Ages how are we to know if the ecological tolerances have changed when the organism looks the same? Is there any guarantee that similar structures serve similar functions in dissimilar organisms? For the geneticist how much information must change before a new species can be noted (i.e. is it a new species or just a variant of an existing species). Current work on biodiversity (see below) has made this more than just a question of semantics – Gaston and Spicer (1998) list no fewer than seven different types of approaches to the concept of species;
- **How much can theory be involved in classification?** Above, it was argued that classification can be used as a shorthand way of conveying information about a species or ecosystem. But where does the boundary lie (see Box 5.1)? How much information can be put to any one unit before it becomes so specific that it fails in the task of simplifying the situation? At the other end of the argument, how much information can be left out of classification, i.e. what is the key fundamental information that separates one species from any other?

BOX 5.1

The nature of the individual

Where is the boundary between one species and another and does this include sub-species? In medieval times in East Anglia, cockles (*Cerastoderma* sp., bivalve marine molluscs) were a key part of the diet. A study at Castle Acre Priory in Norfolk found that rather than one species (*C. edule*) there also appeared to be another (*C. lamarki*). The key difference between the two was the number of ribs in the shell. Some specimens had intermediate numbers of ribs! Since the study was a dietary one it is probable that there was no difference between the two species. However, with an increasing interest in the speed of evolution and the physiological response to environmental change such questions become crucial. Great accuracy in classification is crucial if changes are as small as noted here.

Source: Ganderton (2001).

- **What is the structural stability of the species or area?** This argument has been seen elsewhere. In species, the old concept of uniformitarianism ('the present is the key to the past') is under fire now that we know species can adapt rapidly to new situations but still look outwardly similar. The rate of change of evolution is imperfectly known for most genera. Further, some writers (e.g. Rasnitsyn 1996) take this further and argue the case for a more structured approach to classification because taxa are on a continuum rather than being discrete species.
- **What are we classifying for?** Here the answer must be as much a question of resources as it is of science. Does the system need to provide an easily accessible guide (e.g. for education) or is it required for some more detailed work? Do the same criteria need to be applied for species classification as for environmental impact statements?
- **What system can we use?** It might seem that there is only one system for each classificatory exercise but in reality there can be several valid models. Work by Salthe (1985) is interesting in this respect. In dealing with the classification of hierarchies in nature, he argues for two perspectives – ecological and genealogical. Both look at the same organisms but with a different view in mind (organisms and lineages respectively).
- **What features do we use to identify and classify?** Some (e.g. Nixon and Wheeler 1990 and Lipscomb 1984) suggest the careful choice and use of theory implying that different systems require different identifying features, whilst others (e.g. Bachmann 1995) suggest that whatever system is used, physical characteristics (the most common classificatory method) might fail in accurate identification.

As with any simplification, it is vital to know the parameters of the methodologies applied so that one can both follow the technique and criticise it when it fails to address adequately the strains put upon it. Given these constraints it is now possible to appreciate the difficulties that we have in trying to find a good classificatory system. The next stage in the discussion is to describe the construction of a classification. There are three key areas that need to be examined: features, methodology and scale.

The features are those elements of the individuals which can be used to identify it uniquely and separate it easily from closely related ones. One of the most common features is physical appearance. It is easy to see (unless dealing with micro-organisms) and relatively simple to distinguish between individuals. An obvious use of this would be to classify (and identify) plant and animal species. While such a feature would be ideal for individuals the problem becomes more complex when dealing with groups of individuals. Although early work on ecosystems classification used plant species as the key factor (e.g. woodland, grassland), a greater refinement in knowledge necessitated a change in approach. One common method was to look for relationships between sets of species which could be used to find patterns in otherwise seemingly homogeneous areas (e.g. oak–ash woodland). Work in woodland areas in the UK has been particularly useful in this regard. Current research has increased this to include other ecosystem types (Rodwell 1992). Once a range of features has been established it is possible to devise a methodology. According to Scott-Ram (1990) this can be derived from either a theoretical or practical perspective (Figure 5.2). This is not the only way of looking at methodologies. In a practical sense in the field most biogeographers are working in one of three areas:

- **Exclusive** – the choice of features which classifies by finding the unique in each species. Commonly, plant keys are set up to exclude species until the 'correct' one is found;
- **Inclusive** – more common with ecosystems and communities, these methods (e.g. phytosociology – see below) seek explanation by bringing together species which help explain distribution patterns and environmental gradients;

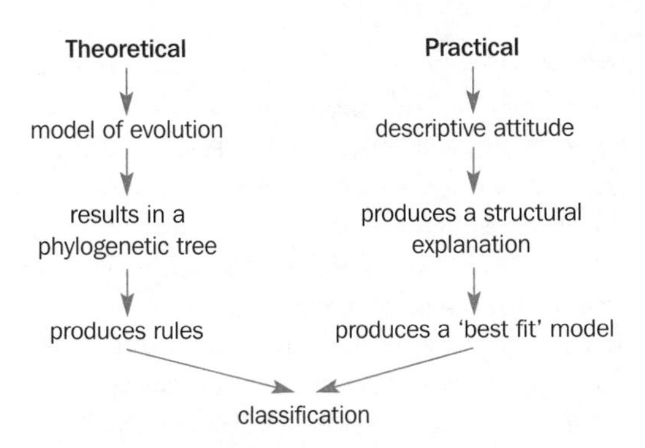

Figure 5.2 Classification by different methods

- **Exploratory** – using techniques, often mathematical, to find underlying structures to group species, e.g. numerical taxonomy, canonical variation.

Finally, there is the question of scale. Classifications have scale limits. For example, plants and animals are more commonly classified using keys to find unique characteristics whereas ecosystems are usually community phenomena.

5.3 What is biodiversity and how can it help our understanding of biogeography?

Biological diversity was a term first used by scientists such as E.O. Wilson to describe the range of organisms on earth. Although it was originally coined to look at more than just species (i.e. to include genetic and ecosystem variations), its shortened version 'biodiversity' has often been used to imply just species. This has led to an interesting line of debate whereby the idea is not seen as science but as an extension of public and political argument on the future of our resources:

> Although biodiversity has concrete biophysical referents, it must be seen as a discursive invention of recent origin. This discourse fosters a complex network of actors, from international organisations and Northern NGOs to scientists, prospectors and local communities and social movements. This network is composed of sites with diverging biocultural perspectives and political stakes. (Escobar 1998, p. 53)

The notion of introducing this idea is twofold: it continues to focus the mind on the philosophical underpinnings of the subject (as was done, above, with classification) and it reinforces the current utility value (see Chapter 16) on the practical uses of biogeography in terms of future prospects.

If the idea of science-as-politics seems out of place then it should be realised that another great political event (the United Nations Rio conference in 1992) brought the term into being in a format usually accepted today:

> biological diversity means the variability among living organisms from all sources including, *inter alia*, terrestrial, marine and other aquatic ecosystems and the ecological complexes of which they are part; this includes diversity within species, between species and of ecosystems.
> (From the 'Convention on Biological Diversity', quoted in Gaston and Spicer 1998, p. 2)

Interestingly, as Gaston and Spicer subsequently note, this term is hardly sufficient even in terms of restricted ecological usage let alone the value-laden ideas put on it which suggest that loss of biodiversity is 'bad' and keeping it is 'good' even though, from the fossil record alone, such an idea is untenable in the long term. Leaving aside the less scientific ideas associated with biodiversity there are still several reasons to suggest its utility. For example, Guptar and Kocchar (1996) outline the value of biodiversity for crop improvement, noting that it can provide a basic framework for evolution, a collection of biological resources for future use and a supply of germplasm in crop improvement. Ehrlich and Ehrlich (1992) in a wider viewpoint note that biodiversity has four main sets of supporting arguments: ethical, aesthetic, economic and ecosystem servicing.

How is biodiversity measured? Since estimates of the total number of species on our planet vary wildly from 3.5 to 111 million with a reasonable estimate of 13.5 million (Gaston and Spicer), then counting them is hardly an option. One way to proceed would be to use any of the numerous ecological techniques for finding species richness, i.e. the number of species. While this is fine for small areas it would be impossible for larger regions. An alternative is to use an 'indicator' group of species, e.g. a taxon. The argument here is that by using an appropriate taxon to represent the community then the diversity of the taxon reflects the diversity as a whole (see Box 5.2).

Obviously one can counter this by noting the lack of correspondence between the indicator taxon and others (Howard *et al.* 1998). However, as Simberloff noted (1998), there needs to be some simple measure to indicate the richness of an area especially if the biodiversity of the area is linked to other action such as conservation. Even if a measure of biodiversity is found, it is only when it is mapped that it becomes of use to the biogeographer. Since we cannot agree on the number of species or the biodiversity of an area it follows that mapping is similarly fraught with problems.

A map of biodiversity is not the same as one for biogeography (which shows the distribution of species) nor is it the same as productivity (which

BOX 5.2

The taxon cycle

Species evolve and eventually become extinct. This suggests a linearity in the system and it has been suggested that as we move towards the present day so we find that biodiversity is increasing. One counter to this is the argument of the taxon cycle. This idea was originally put forward by Ricklefs and Berminghan (2002) following their work with island species in the Caribbean. It suggested that taxa do not increase in a linear fashion but fluctuate in cycles of expansion and contraction. From this it would follow that changes in biodiversity could be, in part, explained by the natural rhythms of taxa. Although this notion has been criticised and at this time should be seen as a controversial idea, it does illustrate the way in which practical and theoretical studies can combine to challenge existing thoughts in the subject. It also highlights the value of a clear classificatory system.

measures biomass). Ideally, it would show areas of particular species richness which would be contrasted with areas of low diversity. Factors implicated in creating such patterns of diversity include:

- **Species–area** – the species–area relationship was one of the first aspects of island biogeography. The basic idea is that the larger the place the greater the number of species it can contain. This would imply greater diversity in the oceans and continents than on islands, although the validity of the concept is now under review (Rosenzweig 1999);
- **Endemicity** – refers to the number of unique species in an area. It tends to be higher in larger, more isolated regions (Brown and Lomolino 1998);
- **Gradients** – the biophysical environment does not change sharply between ecosystems but grades from one to the other. Species richness will change along gradients, e.g. climate (more species in warmer areas), altitude (more species in lower-lying areas) and latitude (more species near the equator) (Heywood 1995);
- **Extinctions and migrations** – species die out or move which can affect the biodiversity record. Work on the great geological extinctions suggests that this is a more important aspect of change than might at first be thought and that its effects are far from equal for all species (Hallam 1987, Moran 1989);
- **Cultivars and invaders** – not all species are in their natural habitat, human agencies are responsible for considerable changes especially near urban areas. The 'success' of the species is often greater than that of native plants. Work by Lovell (1997) for example suggests that introduced species can make up as much as 40 per cent of all species in an area which has considerable implications for the viability of populations. At the same time, Nevo (1998) details work regarding loss of native species in the area;
- **Time and evolution** – most work on biodiversity keeps to a restricted timescale (generally the present). Despite mass extinctions (see above) biodiversity does appear to increase through time not just due to speciation but also to changes in the landscape. Research by Delcourt and Delcourt (1998) has shown that the Ice Age in northern America allowed the development of a mosaic landscape allowing an increase in diversity matching that in habitats. Evolution might be thought to be too weak to influence natural selection rapidly but work by Smith *et al.* (1997) has thrown doubt on this;
- **Species redundancy** – it is argued that not all species are 'valuable' in ecosystem functioning. Above a certain level, processes run regardless of numbers. Thus it is possible that biodiversity might not be as closely linked to species richness as we thought (Naesem 1998);
- **Methodology** – a major aspect of the scientific method but virtually ignored. Studies suggest that a range of methods will yield better results than single experiments (Field *et al.* 1998). Given this, it is possible that biodiversity results are at least in part influenced by the methodology used.

All these ideas above suggest, in one way or another, that biodiversity exists. However, there are several criticisms of this model (see also Figure 5.3).

Figure 5.3 Although it is crucial to understand the parameters of any classificatory and mapping scheme it should also be realised that it is subject to the considerable *in situ* diversity seen in any ecosystem. Any mapping of this part of the Lake District, UK, would need to take both natural plant distributions and introduced species (forestry plantations, far bank) into account

Firstly, there are those concerns raised by theory. Harvey and Partridge (1998) raise the issue of evolutionary theory which by implication can affect the development of biodiversity. Evolution is not one model but several competing models. We can only know how biodiversity works once we know which model is the most accurate. It is not just in evolution that models are lacking. Loreau (1998) reports on the difficulties of linking biodiversity with ecosystem functioning due to problems with finding adequate models. Another theoretical problem is the role of barriers. In biodiversity, barriers are important because of the influence they have on species development. Work by Dennis and Shreeve (1997) and Myers (1997) questions this assumption and notes other effects such as isolation which could have just as great an effect.

Secondly, there are problems associated with the stability of biodiversity. The fossil record is a notoriously biased source of information due to differential preservation of fossils. There is also no clear idea as to the role played by geology in the growth of species. There is some evidence (e.g. Warheit 1992) to suggest that abiotic factors are as likely as biotic ones in the determination of diversity. Finally, there is a need to find an accurate measure of biodiversity:

> The measurement of the abundance or rarity of species is a central problem facing the study and management of biodiversity. Rarity is complex and multidimensional; species differ in the density of their local populations, in the ubiquity of their populations across a landscape, in the geographical ranges across which they are distributed, and in many other respects. . . .
> (Kunin 1998, p. 1513)

Thirdly, there is the question of the range of species used (despite the breadth of the term it is still species that are at the core of biodiversity). A common technique in conservation is to use one species as an ecological metaphor for the entire area. Umbrella species (one key species automatically includes all others) and endemic/threatened species are two ideas commonly used, but both rely on a stability of system that may not be present (see Roberge and Angelstram 2004 and Bonn *et al.* 2002 respectively).

It should be clear from the above discussion that biodiversity offers both opportunities and problems for the biogeographer. But what are the implications for classification and ecosystems, the other themes of this chapter? In terms of classification it could well continue the difficulties faced by taxonomists in producing a clear, unambiguous catalogue of species. An increase in variety will require even more effort even though we still have no idea about the actual number of species we have now. Biodiversity challenges the models we now use. If species can grow and change then any constant we have has gone. Uniformitarianism has proven to be one stable platform in the past but one can no longer assume it will continue to be so. In terms of ecosystems there are also difficulties. The spread of species means that, in the long term, there could be dramatic change in current ecosystem patterns. Flannery's (1995) comments on Australia with replacement of probable semi-rainforest with eucalypt woodland should be borne in mind.

5.4 The use of ecosystems in classification

When we study animals in biogeography we are concerned with the ethology and individual reactions to the physical environment because it is these reactions that determine the range of the organism. When we look at plants we are concerned about interactions between species because that generates the micro-ecosystem within which the community develops. Anything that affects the distribution or composition of plants at this scale is worthy of study. Ecosystems are used as a way of classifying communities. Plants do not move – it makes them the obvious target for any classificatory system: most ecosystems are classified according to their main plant growth forms, e.g. woodland, grassland.

Biogeographers, in seeking to find patterns in organism distribution, have often used ecosystems as a convenient 'shorthand' to describe soil, climate, etc. This is so marked that many of the earliest global vegetation maps closely resembled climate maps. However, this linkage is not as strong as one might imagine. There are numerous difficulties facing the user of ecosystems:

> Knowledge of regional-scale patterns of ecological community structure, and of the factors that control them, is largely conceptual. Regional- and local-scale factors associated with regional variation in community composition have not been quantified. (Ohmann and Spies 1998, p. 151)

However, later in the paper the authors quote from studies they have carried out to remedy this shortfall:

> Climate contributed most to TVE [total variation explained] (46–60%) at all locations and extents, followed by geology (11–19%), disturbance (6–12%) and topography (4–8%). Seasonal variability and extremes in climate were more important in explaining species gradients than were mean annual climatic variations. (Ohmann and Spies 1998, p. 151)

The aim in this section is to note those aspects which make ecosystems less reliable as classificatory tools. An ideal classification is one where the items are easily distinguishable, obvious to see and where there is no change in the system. Unfortunately, this is not possible at any scale for ecosystems. For example, they develop (succession) and change species in an area (zonation). Changes are seen on a greater level if the time factor is brought in. Research in the Sahara by Le Houérou (1997) has shown several regional-scale changes in ecosystems during the last 500 million years in North Africa. This immediately raises a fundamental question – are we classifying the plant assemblage or its location? In other words, is woodland always a woodland even if the Ice Age destroys old habitats but the species 'migrate' south? This is important – as we continue to examine ecosystems we find that the boundaries between them are often more interesting than the habitats themselves. Landscape studies suggest ecosystems made up of mosaics of sub-ecosystems each with their own community. It is sensible to study those ecosystem factors which most directly affect the integrity of the classification, i.e. those elements which cause the area to alter. Three key aspects stand out – boundaries, stability and abiotic factors:

- Boundaries (or ecotones) are the places where one ecosystem grades into another. There is no sharp boundary – ecotones can be very extensive. Since the ecotone is the (unclassified?) place between two or more ecosystems it follows that the study of the boundary is essential to determine both the ecosystem boundary and the nature of the organisms living within the ecotone. There are some who see the ecotone as a collection of environmental gradients along which plants are distributed independent of each other. However, as Callaway (1997) argues, there are numerous problems with adopting this approach. For example, there are examples of plants acting interdependently with others. Far from being isolated as in conventional theory, there is a chance to interact in new communities. Further, there is evidence to suggest that ecotone boundaries are themselves not symmetrical. For example, work by Stohlgren and Bachand (1997) on lodgepole pine indicates that the factors governing the upper and lower limits of distribution are different. They conclude:

  > Because ecotones often represent the physical or competitive limit-of-distribution of species, they serve to define a species' local distribution. . . .We conclude that: (1) different factors control a species' upper and lower elevation limits; (2) unequal competition for resources occurs between tree species where their species overlap; and (3) generally, soil differences may not substantially restrain the movement of some forest types into neighbouring types. (Stohlgren and Bachand, 1997, p. 632)

- In terms of stability it is the third point that is of particular interest if we are considering using ecosystems in classification. If this is seen in other areas then the whole nature of the stable ecosystem (irrespective of location) is brought into question. If the classification is to hold, the unit must be reasonably stable over the period of time of the study. Work in this area demonstrates that this may not always be the case. A study of an arid ecosystem has shown that in the space of only 20 years there have been dramatic changes in the ecosystem with an increase in woody shrubs and the extinction of some animals (Brown *et al.* 1997). The particular situation in a glacier chronosequence (Plate 6) dating back over 250 years shows that the rate of change over time is slow, but inexorable, from pioneer communities which establish within a few years of the land being exposed, to the local climatic climax vegetation which may not be completely stable even after hundreds of years.
- Finally, there is the influence of abiotic factors. The influence of climate on global patterns is already seen in those early vegetation maps noted above but this can also be seen at the smaller scale. Research on the relationship between the field layer and resources in woodland has brought new patterns to light (Converse 1996). Resources needed were not evenly distributed. Some plants requiring higher nutrient levels were found largely on those microsites which seem to have those nutrients.

5.5 Classification in practice

Classification serves two purposes: the unique identification of species and the naming of groups of plants and/or animals usually set in a specific environment. The value of species identification is paramount; only when we know precisely what we are dealing with can we move forward to other aspects of biogeography. At this level, classification is much like the periodic table of the elements to a chemist: it is not so much the name but the properties that are important. The naming of groups has a wider value in that the information that can be carried by such classifications is greater. For example, the identification of the plant classified as a 'daisy' (*Bellis perennis*) will give us the ecological range of that plant. The identification of the ecosystem that contains the 'daisy', e.g. temperate grassland, gives us not only a complex set of biotic and abiotic data but also some insight into the arrangement of allied plants that are included in the classificatory technique which identifies the area as 'grassland'.

Classification has two elements: *technique*, the way in which the classification is used by biogeographers to understand better the area under research, and *concept*, the underlying theory which states how a classification can be made. There are numerous techniques available depending on whether one is interested in species or groups. Species are best served by keys which are either charts or texts (or a mixture) which seek to identify by guiding the user through a series of questions eliminating physical characteristics until only one species remains (hopefully, the correct one! – e.g. Stace 1991). Classification of groups of plants and animals opens up a huge range of techniques depending on the classification followed (which is beyond the scope of this introductory text – see, for example, Shimwell 1971, Causton 1988 and Dansereau 1957).

Classification as a concept refers to that characteristic (or set of characteristics) used to uniquely identify the item under investigation: species or group.

5.5.1 Species classification

Species classification usually requires the use of some visible part of the plant or animal. From thence, it can use one of three techniques.

(a) Classification by structure

The form used today was devised by Linnaeus in the eighteenth century which was founded on the work of early systematists such as Aristotle. The idea was to divide groups according to their comparative anatomy. Essentially, this system is based on the idea of division. Take a group of organisms. Find one characteristic which can be used to separate them. Take one of the groups formed after this subdivision and subdivide that continuing until only one organism remains. That would be the basic unit, the species. In Linnaean terms, the subdivisions go: kingdom, phylum, class, order, family, genus, species, with kingdom being the loosest grouping and species the most rigid. In theory, each organism would have a unique species identifier, the so-called Linnaean binomial (because it consists of two parts, genus and species – e.g. *Bellis perennis*). Names are assigned according to a mixture of Greek, Latin, people's names, etc. The precise name of a species is not all that important: the fact that each species has a tag, is.

(b) Classification by evolution

The Linnaean system has been in use for over 250 years and is the universally accepted method. That is not to say it is perfect or that there are other ways of approaching the same task. If the object is to provide a unique tag then it is possible to use evolution or phylogenetic link as the basis. A new and controversial technique known as *phylocode* challenges the Linnaean system (Rouse 1999). It is based on the evolutionary relationships between species rather than physical characteristics (which can often be brought about by parallel evolution);

(c) Classification by life form

If relationships between plants and their environment are important then it is possible to consider the life form or growth habit of the plant rather than focus on a name. First proposed by Raunkiaer in the early years of the twentieth century, it classifies plants according to their regenerating parts. The advantage is that it can be used for species and ecosystems. In fact Raunkiaer used the system to demonstrate that each major climatic area had a particular type of predominant life form.

5.5.2 Group classification

Turning to group classification (i.e. ecosystem although, strictly speaking, 'ecosystem' is a classificatory idea!) one faces a greater challenge. Here, technique and classification are more closely bound together. Consider the argument put forward by Dahl (1998, p. 1) in his introduction:

> Phytogeography, as the name indicates, is the scientific study of the geographical distribution of plants. . . . The subject can be approached in two ways:
>
> 1. By studying the geographical distribution of plant assemblages, namely plant communities. This is the *phytosociological* approach.
> 2. By studying the geographical distribution of individual plant taxa. This is a *phytogeographical* approach. . . .
>
> The basic material in phytosociological studies is the relevé. . . . The basic material in the phytogeographical approach ... is data on the occurrences of individual plant taxa. . . . [emphases added]

Taking this as a starting point, it would be impossible to separate technique from classification. This point is important – taxonomic principles must be regarded as apart from method if the philosophical basis of the work is to be made clear (and the basis of any classification – its rules are paramount). Bearing this in mind, there are three main classificatory types seen today.

(a) Classification based on abiotic factors

This was one of the first of the modern systems to be used. Its ease of use coupled with the close correlation between abiotic factors (especially climate) and vegetation patterns has made this a useful system. One of the first classifications using this idea was proposed by Sclater in 1858. Using birds as his example he divided the world into six biogeographic regions. This was taken further by a contemporary of Darwin, Alfred Russell Wallace, who, in 1876, published his map of global plant regions. Increasing knowledge led to the construction of one of the most famous climatic maps proposed by Köppen. This has been used extensively since by those trying to make a more accurate reflection of global plant distribution patterns. Trewartha and Neef were two of the first new schemes but others such as Dansereau have also been seen. Although much emphasis has been placed on plant groupings this approach is still being developed. All of the ideas up to now have used maps as their basis. Holdridge (quoted in Archibold 1995) has produced a classification based on mean annual temperature, precipitation and potential evapotranspiration. It avoids making any continental assumptions because this system is based on a tri-plot graph. A relatively new scheme has been proposed by Bailey (1998). He has divided the world into ecoregions which he defines as:

> [a] major ecosystem, resulting from large scale predicable patterns of solar radiation and moisture, which in turn affect the kinds of local ecosystems and animals and plants found there. (Bailey 1998, p. 145)

This should be compared with his definition of an ecosystem:

> an area of any size with an association of physical and biological components so organized that a change in any one component will bring about a change in the other components and the operation of the whole system. (Bailey 1998, p. 145)

The argument here is that ecoregions are all-encompassing, similar to biomes, where the global-scale features are used to determine the major features. Unlike many workers in this field he has included the oceans in his classification. There are subdivisions to this scheme – divisions – which are used to provide a finer scale for better definition of plant groups.

(b) Classification based on relationships between species

The second great school of classification, developed at the same time as the 'abiotic' schemes, uses plant groupings. At the heart of this system is the idea of phytosociology – the organisation, interdependence and development of groups of plants (usually referred to as associations). The key ideas behind this were developed by Braun-Blanquet (also called the Zurich–Montpellier school after the initial study region) and most current methods derive from this (Shimwell 1971, Causton 1998). The basic unit of the Braun-Blanquet classification is the relevé, a homogeneous stand of vegetation. The species composition of the relevé is noted along with the cover (abundance on the ground) of each species and often a figure for its 'sociability' or clumping of the plants. If this is done often enough then a table can be produced showing relevés and species composition. Examination and re-ordering of the table often reveals groups of species which are found growing together and each group may well be recognisable as a distinct association. Although subjective, the system does produce common groupings of sufficient interest to warrant ecological investigation.

(c) Classification based on relative positions in time/space

If the Braun-Blanquet school is subjective then there are objective methods that can be used. Two examples of this approach are ordination (also called the Wisconsin school) and association analysis. In ordination, the key is not the vegetation but the environmental gradient (i.e. changes in abiotic conditions) between two places and the species changes that result. To find these gradients a series of quadrats is used to gather data on plant distribution. Each quadrat is then compared to every other one. If the resulting data are graphed then it is possible to find groups of species that share environmental conditions in common (Cousens 1974). These become the associations. Although it is a far more complex technique it has the advantage of producing a reasonably objective grouping of vegetation. In association analysis the aim is to take the data and divide the quadrats into successively smaller groups until the final groups are nearly homogeneous. Although rarely used today, it does illustrate the range of techniques available.

There is a tendency to see classification as a less important or tedious area of study and certainly, taxonomy does not have the high profile that ecology or conservation has. Despite this it remains absolutely crucial to our study: unless we know exactly what we are dealing with then it is not possible to make any concrete suggestions after that. An analogy could be drawn between concrete mixing and architecture. Architecture creates useful spaces but unless the concrete is good the whole structure is likely to fall down.

5.6 Summary

Although biogeography is the study of distributions, classification is the foundation upon which all the work is based. This chapter has shown both the need for the study of classification (which goes beyond the naming of species) and some of the theoretical and practical considerations around it. The value of classification comes from a rigorous analysis of its use and an examination of the underlying assumptions. Too often this is taken for granted, but if biogeography is to make the contribution it should then it is of fundamental importance that it is based on sound principles.

The chapter title 'Putting it together' suggests that there is more to biogeography than just naming species. Classification distinguishes between distinct life forms: biodiversity gives some indication of how this might change with time, space and conditions. The nature of ecosystems adds further complication: it suggests that we need to be far wider-ranging in our studies to ensure the full range of parameters is understood. Only once this has occurred do we get to use the final systems that have been, and continue to be, devised.

APPLICATIONS – USING BIOGEOGRAPHY

Using ideas from this chapter in real-life situations

Taxonomy is the classification of entities. It is a precise science because it needs to create a universally accepted name for each species and to place that species in the correct place in relationship to other species. This seemingly simple task has created a number of problems including the characteristics needed to classify a group of organisms as a species. This has obvious links to biodiversity: if we do not know if it is a species then how can we say we are losing species at any given rate? Although taxonomy has remained a relatively quiet branch of science, the Internet is set to revolutionise it. In the last few years a number of organisations have set up websites including the Global Biodiversity Information Facility (www.gbif.org) and the Integrated Taxonomic Information System (www.itis.usda.gov). The aim of these groups is to spread ideas about taxonomy and its uses. For example, two groups might study two groups of an organism and might classify it as two species whereas they might be subspecies. Correct and clear naming of species provides the baseline from which a range of key studies take their data including biodiversity studies and global change analysis. It is clear, studying key journals such as *Nature* and *Science*, that this debate is not going away: just the opposite. The development of molecular and genetic research and new ideas in phylogenetics and cladistics means that species and their relationships to other species are under debate as rarely before. Such work is fundamental to biogeography because we base our studies on such foundation work.

Review questions

1. You have to put forward a research proposal for a biogeographic study. Suggest reasons why part of your initial research should include a detailed analysis of classificatory systems.

2. To what extent is biodiversity more than just species?

3. Genetic biodiversity could be taken as a continuum. What impact does this statement have on the concept of the species?

4. Can ecotones be justified as discrete ecosystems? Either way, how does this reflect upon our notion of ecosystem?

5. Compare and contrast two common classificatory techniques in terms of areas delineated and parameters used to make the system.

Selected readings

Classification has tended to go into specialised directions and so much of the basic work is less accessible. In this regard, some of the older texts are excellent in that they have both a range of diagrams of different systems not commonly found elsewhere and some discussion of new (for them) methods. Dansereau (1957) is particularly good for a range of diagrams and discussions. In terms of more recent ideas, Brown and Lomolino (1998) would be worthwhile and discussions in Bailey (1998) are interesting especially as this is one of the more recent classificatory ideas. To get a greater understanding of one system, Humphries and Parenti's (1999) text and CD-ROM on cladistics are excellent, while Rodwell's (1992) five-volume series on types of UK vegetation is a classic, particularly for vegetation scientists and ecologists involved in environmental assessments.

References

Archibold OW. 1995. *Ecology of World Vegetation*. Chapman & Hall.

Bachmann K. 1995. Progress and pitfalls in systematics: cladistics, DNA and morphology. *Acta Botanica Neerlandica*, **44**(4): 403–19.

Bailey RG. 1998. *Ecoregions*. Springer.

Bonn A, Rodrigues ASL and Gaston K. 2002. Threatened and endemic species: are they good indicators of patterns of biodiversity on a national scale? *Ecology Letters*, **5**(6), 733.

Brown JH and Lomolino MV. 1998. *Biogeography*. Sinauer.

Brown JH, Valone TJ and Curtin CG. 1997. Reorganisation of an arid ecosystem in response to recent climate change. *Proc. of the National Academy of Sciences of the USA*, **94**(18): 9729–33.

Callaway RM. 1997. Positive interactions in plant communities and the individualistic–continuum concept. *Oecologia*, **112**(2): 143–9.

Causton DR. 1988. *An Introduction to Vegetation Analysis*. Unwin Hyman.

Converse G. 1996. Distribution of *Viola blanda* in relation to within-habitat variation in canopy openness, soil phosphorus. *Bulletin of the Torrey Botanical Club*, **123**(4): 281–5.

Cousens J. 1974. *An Introduction to Woodland Ecology*. Oliver and Boyd.

Dahl E. 1998. *The Phytogeography of Northern Europe*. Cambridge University Press.

Dansereau P. 1957. *Biogeography: An ecological perspective*. Ronald Press.

Delcourt PA and Delcourt HR. 1998. Palaeoecological insights on conservation and biodiversity: a focus on species, ecosystems and landscapes. *Ecological Applications*, **8**(4): 921–34.

Dennis RLH and Shreeve TG. 1997. Diversity of butterflies on British islands: ecological influences underlying the roles of area, isolation and the size of the faunal source. *Biological Journal of the Linnean Society*, **60**(2): 257–75.

Ehrlich PR and Ehrlich AH. 1992. The value of biodiversity. *Ambio*, **21**: 219–26.

Escobar A. 1998. Whose knowledge, whose nature? Biodiversity, conservation and the political ecology of social movements. *Journal of Political Ecology*, **5**: 53–82.

Field CB *et al.* 1998. Mangrove biodiversity and ecosystem function. *Global Ecology and Biogeography Letters*, **7**(1): 3–14.

Flannery T. 1995. *Future Eaters*. Reed Books.

Ganderton PS. 2001. Grains and foodstuffs. In Wilcox R. *Excavation of a Monastic Grain-Processing Plant at Castle Acre Priory, Norfolk, 1977–82*. Norfolk Archaeology.

Gaston KJ and Spicer JI. 1998. *Biodiversity: An introduction*. Blackwell Science.

Guptar PN and Kocchar S. 1996. Biodiversity and plant genetic resources. *Crop Improvement*, **23**: 123–30.

Harvey PH and Partridge L. 1998. Evolutionary ecology: different routes to different ends. *Nature*, **392**(6676): 552–3.

Hallam A. 1987. End-Cretaceous mass extinction event: argument for terrestrial causation. *Science*, **238**: 1237–41.

Hawksworth DL. (ed.) 1988. *Prospects in Systematics*. Systematics Association Special Volume 36. Oxford University Press.

Heywood VH. (ed.) 1995. *Global Biodiversity Assessment*. Cambridge University Press.

Howard PC *et al.* 1998. Complementarity and the use of indicator groups for reserve selection in Uganda. *Nature*, **394**(6692): 472–5.

Humphries CJ and Parenti LR. 1999. *Cladistic Biogeography*, 2nd edn. Oxford University Press.

Kunin WE. 1998. Extrapolating species abundance across spatial scales. *Science*, **281**(5382): 1513–15.

Le Houérou HN. 1997. Flora and fauna changes in the Sahara over the past 500 million years. *J. Arid Environ*, **37**(4): 619–47.

Lipscomb DL. 1984. Methods of systematic analysis: the relative superiority of phylogenetic systematics. *Origins of Life*, **13**(3–4): 235–48.

Loreau M. 1998. Biodiversity and ecosystem functioning: a mechanistic model. *Proceedings of the National Academy of Science USA*, **95**(10): 5632–6.

Lovell GL. 1997. Biodiversity: global change through invasion. *Nature*, **388**(6643): 627–8.

Moran NA. 1989. A 48 million-year-old aphid-host plant association and complex life cycle: biogeographic evidence. *Science*, **245**(4914): 173–6.

Myers AA. 1997. Biogeographic barriers and the development of marine biodiversity. *Estuarine, Coastal and Shelf Science*, **44**(2): 241–8.

Naeem S. 1998. Species redundancy and ecosystem reliability. *Conservation Biology*, **12**(1): 39–45.

Nelson G and Platnick N. 1981. *Systematics and Biogeography: Cladistics and vicariance*. Columbia University Press.

Nevo E. 1998. Genetic diversity in wild cereals: regional and local studies and their bearing on conservation ex-situ and in-situ. *Genetic Resources and Crop Evolution*, **45**(4): 355–70.

Nixon KC and Wheeler QD. 1990. An amplification of the phylogenetic species concept. *Cladistics*, **6**(3): 211–24.

Ohmann JL and Spies TA. 1998. Regional gradient analysis and spatial pattern of woody plant communities of Oregon forests. *Ecological Monographs 1998*, **68**(2): 151.

Quicke DL. 1993. *Principles and Techniques of Contemporary Taxonomy*. Chapman & Hall.

Rasnitsyn AP. 1996. Conceptual issues in phylogeny, taxonomy and nomenclature. *Contributions to Zoology*, **66**(1): 3–41.

Ricklefs RE and Bermingham E. 2002. The concept of the taxon cycle in biogeography. *Global Ecology and Biogeography*, **11**(5): 353.

Roberge J-M and Angelstram P. 2004. Usefulness of the Umbrella species concept as a conservation tool. *Conservation Biology*, **18**(1): 76.

Rodwell JS. (ed.) 1992–98. *British Plant Communities* (5 volumes). Cambridge University Press.

Rosenzweig ML. 1999. Heeding the warning in biodiversity's basic law. *Science*, **284**(5412): 276–7.

Rouse G. 1999. Pers. Comm., University of Sydney, Australia. Also Pleijel F and Rouse G. u/d. LITU: a new taxonomic concept for biology. Mimeo.

Salthe SN. 1985. *Evolving Hierarchical Systems*. Columbia University Press.

Scott-Ram NR. 1990. *Transformed Cladistics, Taxonomy and Evolution*. Cambridge University Press.

Shimwell DW. 1971. *The Description and Classification of Vegetation*. Sidgwick and Jackson.

Simberloff D. 1998. Flagships, umbrellas, and keystones: is single-species management passé in the landscape era? *Biological Conservation*, **83**(3): 247–57.

Smith TB *et al.* 1997. A role for ecotones in generating rainforest biodiversity. *Science*, **276**(5320): 1855–7.

Stace C. 1991. *New Flora of the British Isles*. Cambridge University Press.

Stohlgren TJ and Bachand RR. 1997. Lodgepole pine ecotones in Rocky Mountain National Park, Colorado, USA. *Ecology*, **78**: 632–41.

Warheit KI. 1992. A review of the fossil seabirds from the Tertiary in the North Pacific: plate tectonics, palaeooceanography and faunal change. *Palaeobiology*, **18**(4): 401–24.

Websites

The first of two links describing classic papers in biogeography up to 1975: useful for finding ideas in their original form.
http://www.wku.edu/~smithch/biogeog/

World Atlas of Biodiversity – interactive map showing a range of physical and biological features.
http://stort.unep-wcmc.org/imaps/gb2002/book/viewer.htm.

NOAA webpage describing their mapping programme with manual to download.
http://biogeo.nos.noaa.gov/projects/mapping/pacific/nwhi/classification/

Part of a huge site with much for biogeographers. This is a guide to cladistics.
http://www.ucmp.berkeley.edu/clad/clad4.html

BOX 6.1

The paradox of scale

Clematis is only one example. As we change scale so the key factor affecting distribution changes and yet it is the same individual (or set of individuals) that we are looking at. This notion of different factors affecting essentially the same individual is referred to as the paradox of scale. Nor is this notion restricted to plants or even just species. Studies by Li (2002) investigated the patterns of phytoplankton in the North Atlantic Ocean. At the individual level, local disturbances meant that no obvious pattern was found but by moving to the macro-level it was possible to detect distinct patterns of species distribution. It is not just species that are involved. Studies by Chase and Leibold (2002) looking at the relationship between productivity and biodiversity found that the resultant data gave curves that differed between local and regional levels. Their argument was that local scales were influenced by species composition. The key point here is that the scale of study might influence the result, which raises the more contentious point of whether pattern is real or an artefact of scale.

Scale is not just a problem of distributions. Since the key biogeographical tool is the map, it follows that issues of scale will affect the information put on them. One way of looking at this is to consider the 'grain' of the subject. Grain can be equated with size so trees have a greater grain than grasses. This will affect the detail per unit area that can be accommodated on a map. Since many studies are multi-species it means that we are mixing grains (i.e. scales) in the one map with possible problems for interpretation (Zeigler *et al*. 2004).

are far from universally agreed), you would have the earth divided into 9–12 global ecosystems, e.g. tundra, temperate forest;

- **Ecoregions** – a relatively new term, this allows us to subdivide the biome into more reasonable regions. Again, depending on the system used, this takes the number of units in places like North America from 3/4 biomes to over 650 ecoregions;
- **Ecosystem** – the most commonly used term (but also the most misused!). Ideally, this should be a recognisable entity with a more or less contained trophic system, e.g. forest, lake. Of course, these can be further subdivided into both types (e.g. of lake: nutrient-rich or nutrient-poor) or parts (e.g. of a lake: lakeside, shallow water, deep water). Since there are no rules about size and scale it means that we must be very careful when reading about ecosystems to ensure we know the scale of the unit under discussion;
- **Community** – part of an ecosystem with a readily identifiable set of plants and/or animals.

Having examined the ways in which we can group species it is necessary to show we can make maps of these patterns. It is also worth remembering that although some resulting maps are reasonably new some have been around for over 100 years. In fact one of the earliest modern studies in biogeography was the creation of maps – Wallace (who, with Darwin, was responsible for the concept of evolution) was better known in his day for the early biogeographic maps he produced and the boundary named after him (Berry 2002).

The upshot of this dynamic history is that we can find a considerable number of different maps all based on slightly different ideas. They normally fall into one of four categories (see also Chapter 4):

- **Organismic**. Maps are produced by spatially grouping a set of organisms with similar taxonomic or phylogenetic relationships rather than similar adaptations. Wallace and Heilprin's maps are two examples for animals (Spellerberg and Sawyer 1999), while Good's map for flora (in Dansereau 1957) is still being used. Note that Good's map divides the earth into six kingdoms which are, in turn, divided into subkingdoms and provinces where necessary. Some current ideas (e.g. Pielou, noted in Archibold 1995) mix zoogeography and phytogeography to produce hybrid maps. As with other maps and ideas, most of the work is carried out with plants. Plants represent a more measurable, static situation, although this does not hold through time. Thus most maps are concerned with one time period and show variations of plants, whereas in palaeoecology and palaeobiogeography it is common to use animal remains. The resulting differences in interpretation are worth considering if we are to use maps critically (see Box 6.2);

BOX 6.2

Species, remains and maps

What is going to be the basis for your map? The use of species has had a long history but with increased information it is not the only way. With an emphasis on biogeographical histories many attempts have been made to show fossil species distributions. Attempts have been made to link fossil and modern species maps to show a continuous history. This can be done but there are limitations in interpretation that need to be considered. Firstly, not every species can be made into a fossil which biases the fossil record towards skeletal species (although new techniques are overcoming this). Secondly, there is no guarantee that the species have the same ranges past and present. Evolution can be rapid (in the space of a few generations) and easily alter in fossil assemblages.

- **Environmental**. Maps or diagrams are produced using concepts of physical environment. Usually temperature and moisture are used with some reference to altitude or latitude. This has the advantage that we can group plants by physical responses to common abiotic conditions. The linkages found by using taxonomic characteristics are missed but it produces schemes more closely tied in with experience. Two of the most common examples of biogeographic patterns use this concept: Köppen's classification (noted in Archibold 1995) uses climatic and seasonal data to produce a scheme which can be refined from 'main division' – global – sets into second and third division classifications, thus allowing regional detail to be shown against the global background. Holdridge's scheme (noted in Spellerberg and Sawyer 1999) is another common scheme linking moisture regimes with physical regions and altitudinal belts. As with any large scale map or scheme there are problems at the boundary. Further, Köppen's map also includes marine areas where there is little evidence to show that this might be accurate (interestingly, there are virtually no modern maps of marine areas: most tend to look at it from a geological/palaeogeographical perspective);
- **Holistic**. These are maps produced using a range of techniques or concepts. The IUCN has produced a map of global biomes which is used in its conservation work. Of more interest in terms of concepts are the ranges of maps which use the 'ecoregion' as the unit of measurement. This allows both physical and biological characteristics to be used. One of the more common ones has been developed by Bailey (1995) as part of the US ECOMAPS project. Of interest to us here is that it produces a hierarchy of units. The USA is divided into three domains with a greater number of divisions and more provinces. Although work like Holdridge's is more likely to be used by biogeographers it is worth noting that these holistic maps are very common, with increasing numbers of organisations with a public face, e.g. pressure groups, choosing to use

ecoregions and biome information as a way of getting their message across;

- **Conceptual**. These are not maps in the strict sense but an exploration of relationships using a range of concepts (Figure 6.2). Although Takhtajan's ideas (noted in Cox and Moore 2000) can be used to produce a more conventional map it is the data on Liliiflorae linkages and relationships that could provide more interest from both conservation and phylogenetic quarters. Maps are still the most commonly used format to show biogeographical distributions but there is no reason why cladistics and genetic analysis could not be developed to show us different patterns of distribution (Macdonald 2003).

Before moving on to consider the factors that give rise to these patterns it is important that we consider the reality of what we are hoping to describe. Put bluntly, do patterns exist? Is a biome 'real' in the sense that all can recognise it? Rutherford and Westfall (1994) state that: 'The biome concept is clarified according to established definitions to permit objective categorization of the major natural systems.' This categorisation uses a mixture of plant (dominance) and physical characteristics (e.g. rainfall). The argument is put that if we can provide a sufficiently rigorous method then it would be possible to create accurate maps. However, that is far from the case. The same article notes the problems seen in South Africa. Other research has discovered that the changes brought about by environmental gradients do not work for all organisms (Lees *et al.* 1999), suggesting that the biome's delineation might depend upon the species used. Others suggest current global human populations are making such changes that appear 'natural' to any system and might be anything but (e.g. Low 2002). Globalisation might be as much an ecological idea as an economic one (Lonsdale 1999).

Figure 6.2 Areas such as these ranges in SW China illustrate the problems of both scales and mapping. There are obvious differences between ridge and valley species and yet this is all classed as 'montane' except at the local scale. Although divisions aid our understanding they are obviously artificial and may obscure patterns as well as help find them

Ecoregions are relatively new divisions. As such there is still a great deal of work to be done before they can be accepted as viable units. The arguments surrounding them are useful to study because they highlight the difficulties we can face. Firstly, what factors are used to determine the boundaries? Taking the initial definition of a region as a place with quantifiable differences, it is important that we can find suitable measures. Of course, we must always recognise that the broader the scale the more variations that will be allowable. However, even a brief review of the literature shows that, as yet, ecoregions are far from settled. Firstly, which factors to use? A Chinese study uses plant functional features (Ni 2001). Another (Hessburg *et al.* 2000) uses primarily geology and landform features reminiscent of the watershed/catchment studies of the 1970s. Even when similar species are used it is not always going to produce clear results. What works in one part of the world (Feminella 2000, Harding and Winterbourn 1997) does not elsewhere (Marchant *et al.* 2000, Wright *et al.* 1998). The point being made here is not an attempt to reduce the significance of ecoregions (or biome) study but to make the reader more critically aware of some of the real problems facing biogeographers today as they seek to find and then explain the patterns they think they see.

6.3 Analysing patterns: ecological processes

The last section described some of the patterns we can see and highlighted the difficulties we face in trying to accurately find and describe such patterns. What we turn to now are the processes which cause distribution of plants and animals to vary. Maps and similar ideas are a visual way of representing and summarising data with all the advantages and disadvantages that this has. Underlying these items are the fundamental aspects of plant and animal ecology. In this section we turn to look at those factors which cause the distributions that we are so keen to map. The location of any given organism, species, etc. is due to the interplay of three main factors: biological, physical and temporal.

Biological factors are those which govern the development, movement, inertia and organisation of species. Although we are looking for patterns of species we are actually looking at the mass grouping of individuals of that species, each with their own parameters for existence. From the mass of potential variables it is possible to focus on about 12 factors which can be grouped into four categories.

6.3.1 Species development

Like individuals, species have a lifespan. Unlike individuals, species may also change their characteristics during their existence. If we are studying this aspect there are three main areas: speciation, extinction and evolution (see also Chapter 13). Speciation is the creation of a distinct species. One of the better examples to illustrate this are the various finch species on the Galapagos Islands (see Brown and Lomolino 1998). First noted by Darwin, these birds have been the subject of intense study. They represent a number of aspects in speciation. Firstly, they show a range of variation which, in the absence of other explanations, must be considered part of evolution. Secondly, they show sufficient variation within species that it is possible to question where the species boundary actually lies. This part is crucial for biogeography. We must know what a species is in both theory and practice. The former dictates the parameters for acceptance as a distinct species; the latter tells us what we are looking at. Since there is an array of evidence that suggests species are gradational rather than abrupt in their characteristics, then it means we chose to define a species on less than perfect knowledge. Although this does not matter for most cases it is very important if we wish to describe a 'new' species and for our study of change through time in the geological record.

The opposite of speciation is extinction. Mechanisms for both speciation and extinction are not perfectly known but we are coming to appreciate that the species has a lifespan and that loss of a species is not compensated for by another species for up to 10 million years. If we are looking at the geological record (see below) then such changes are important: 10 million years is a very short period of geological time.

The third aspect under this heading is evolution. Evolution is the change of physical characteristics through time which could lead to the creation of a new species. The precise mechanism (and the entire subject) is still controversial, but the range and nature of change are important to us (e.g. Hellberg 1998). Some species can evolve relatively quickly and if this is linked to response to the physical environment then changes in pattern can result. Since one of geology's founding ideas is lack of change within a species, then this potential for change could be important. It is possible that the rate of evolution can also impact upon ecological dynamics (Yoshida *et al.* 2003).

6.3.2 Species movement

This includes: adaptation, diversification and dispersal. An adaptation can be thought of as a characteristic (usually physical) that makes a species more suited to a given set of environmental conditions. Of course, when the conditions change the organism must also change (or perish) but there is the potential to anthropocentricise this – organisms do not 'think' about changing, rather they may or may not survive. One current (genetic) perspective notes that all possibilities for a species are present and that only those suited for the conditions survive. If this continues then some possibilities die out and the surviving organisms are grouped around the new mean for that characteristic. This has been demonstrated to occur fairly rapidly in remediation ecology where grass species can become heavy-metal tolerant in a matter of a few generations (Beeby 1993).

Diversification should occur when a new species is formed. Under some ideas of speciation the new species should closely resemble an existing species. Under ideas of competitive exclusion (the ecological concept that suggests two similar species cannot exist in the same niche) the new species must find a new niche, habitat or food source. Referred to as ecological diversification, studies have shown (Brown and Lomolino 1998) that species can rapidly exploit new areas and that there is virtually no overlap. Other studies have examined the impact and spread of introduced species where in some cases the new species can take over existing areas (contrary to classical Darwinian explanation – Duncan and Williams 2002 – although this can itself be challenged – Kennedy *et al.* 2002).

Another example of diversification is adaptive radiation. Here, a single species rapidly gives rise to a number of new species with subtle alterations allowing exploitation of new environments and conditions. Again we can turn to the Galapagos finches described by Darwin to see the changes in bill shape and size and see how this is related to food sources and locations. Although it could be considered part of speciation it is worth putting adaptive radiation in another category because it both highlights the power of this concept to 'spread' species (even if it means as new species) and it highlights the difficulties we can have in separating the wide range of interrelated concepts that make up the biology and ecology of organisms when it comes to discerning the production of patterns.

Dispersal, the movement of individuals or parts of individuals (e.g. seeds) away from the source area, has been a neglected area until recently (Bullock *et al.* 2002). The problems of dispersal will be noted again, below, in relation to the geological spread of species: it is worth noting here as a way of propagating plant species and spreading the range of animals through migration. Dispersal is not a simple matter of movement. Studies have shown that movement can be rapid over a large distance (jump dispersal) or through slow 'creep' through environments (diffusion). It is also possible that the species might take so long to migrate that it evolves along the way. Such ideas of change are important to us because unless we can be certain of the nature of the species we cannot accurately discuss palaeoenvironmental changes. The dispersal mechanisms deployed depend on the organisms. Plants usually 'migrate' through the production of seeds or spores whereas animals tend to migrate. Providing that the organism can overcome any barriers (physical or physiological) and can survive in the new environment then it will have spread its range.

6.3.3 Inertial factors

These tend to reduce the spread of organisms. As such, these are almost the opposite of the last set. Here, species tend to stay in one restricted range or be too far separated to be seen as the same population. In this group we will look at concepts of endemism, disjunction and relicts.

Endemic organisms are found only in a highly restricted range. For example, the Minnesota trout lily is restricted to the floodplains and adjacent forests of two counties in Minnesota, USA (Tester 1995). Many of the plants in the Karoo of Southern Africa are endemic (Osbourne 2000). The fact that these rarities are often mentioned raises a number of interesting points. Firstly, why are they rare? Often, as in the case of the trout lily, the reproductive system reduces the ease of dissemination (here, propagation is by underground bulb rather than seed). Secondly, it is possible that the actual physiological range requirements are limited – the organisms need very specific conditions in which to live. This implies that wherever the environment is specific, there will be a high degree of endemism. Although there would need to be more work on this area, it is noticeable that tropical coral reefs and alpine areas both have a large number of endemic species. Also, the continents of the southern hemisphere have a greater proportion of endemics than their northern counterparts. What does this mean for biogeographers finding patterns? Almost certainly, it implies that there is a distinct habitat (but whether the scale of study is suitable to record it is another matter). It also suggests that the species might be rare or endangered. This depends on the number of individuals in the population and their connections to other populations (if any). It might be that the plant is endemic and rare. It might also be that the species is endemic but locally abundant. Care needs to be taken when discussing endemism: it is valuable in discussions about conservation and biodiversity but in terms of patterns it seeks only to restrict rather than enlarge. There is also evidence to suggest that it might be the geographical structure of the area that is important (Thompson and Cunningham 2002, Bowman and Seastedt 2001).

Disjunct species occur at two or more highly separated locations. It is usually considered that they were initially part of a general population but have become separated through time. In the case of the ratites (ostriches, rheas, etc.) their disjunction is subject to much speculation of interest to us here. Ostriches, rheas, cassowaries and emus are restricted to the southern hemisphere continents and can be seen as a disjunct group. Each continent has its own set of genera which can be mapped. However, this raises the question of how such a disjunction came about. There have been a range of suggestions from a common southern ancestor which, on the creation of barriers, evolved separately, to a northern origin spreading south and surviving, to the idea that recent birds swam across to isolated areas and then evolved further (Davies 2002). The point here is that there is a good deal of controversy much of which is of direct interest to biogeography. Firstly, we need to be certain about the centre of origin and subsequent spread of the species. Secondly, we should know about the subsequent evolutionary ecology of the species so that we can see what changes have been made and when. We could also add some comment about the fact that these species (in common with endemics) do not seem to have spread more widely.

Closely allied to endemism and disjunction is the relict species. Some texts (e.g. Dahl 1998) treat relicts and disjunct species as virtually synonymous. Relicts have a restricted range and may be seen as endemic. The main difference is that relicts are the remains of a once greater distribution. One example of relict distribution are the 'refugia' of tropical rainforests. The basic idea was that when conditions in other areas got worse (e.g. recent Ice Age), the distribution of plants and animals would change. However, in the refugia there would be a continued existence and evolution. When conditions became favourable, the range would expand. Although this is not universally accepted, there is some evidence to suggest that this might be a better explanation: many of the world's biodiversity 'hotspots' are found where the refugia are supposed to be. In this case, the value of the relict/refuge is twofold: firstly, relicts show previous distribution patterns (useful in palaeoecology) and, secondly, refugia could contain the highest degree of biodiversity in that ecosystem.

6.3.4 Ecological organisation

This refers to the way in which species interact at a community level and so determine the distribution of any given species. Three ideas are useful here – assembly rules, physiological gradient and taxon cycles. Organisms do not colonise areas randomly: there is a sequence. Although we could say that one example of this is succession (change in species through time) it could also be considered to be an aspect of zonation (change in distribution of species in space). Since island biogeography became a viable area of study in the 1970s, ecologists have tried to find out the mechanisms through which islands gained and lost species. One item which emerged from this study was 'assembly rules'. It is argued (and this is still a controversial term – see Morin 1999) that the species found in a given place are not random: species occur in a set pattern. Although we can see this readily enough in terms of succession, the assembly rules go further in trying to establish patterns of community development at order or guild level (see e.g. Stiling 1999, Pakeman *et al.* 2000, Belyea and Lancaster 1999, Hraber and Milne 1997). The existence of assembly rules should be of concern if we are trying to find patterns. We accept that patterns are dynamic at all timescales. However, if we also accept that patterns are subject to a series of rules then we need to consider at what stage in the rules that we define and draw that pattern. For example, consider a wetland ecosystem. How are we going to describe and delineate the pattern(s) in that area? As the area changes (especially if it is a dynamic river ecosystem) so will the pattern. We can accept that because of our knowledge of river morphology through time. However, if we also take into account work by Weiher *et al.* (1998) then we find that in addition to these abiotic forces there are also the internal biotic ones which constrain the existence at any given time of a set of species. Although this probably does not matter at the biome level, as we gather more data through the use of, for example, geographic information systems (GIS), then an understanding of such constraints becomes vital for correct analysis.

The idea that each organism has a range of physical conditions within which it can function (the optimal range presenting the best of these conditions) is well known. This suggests that each species will have a spatial range (i.e. pattern) depending upon its adaptation to environmental conditions (e.g. see de los Santos *et al.* 2002): workers like Holdridge (see above) have based their work around such concepts. To understand this in more detail we need to consider two 'rules' which appear to work for a range of taxa. The first is the latitudinal gradient (Rosenzweig 1995) which states that the number of species declines as you go from the equator to the poles. Most groups of organisms follow this and the trend seems to work through time as well as space. The second is referred to as Rappoport's Rule which (again with exceptions) notes that the range of species declines from the poles to the equator (Krebs 2001). If we put the two together one could argue that fewer species would have more room to move around and have a large range but this ignores the actual areas involved. As they stand, both concepts are backed by data but a precise explanation has yet to be found. This assumes that species are evenly spread over their range and that there are no linkages beyond the obvious predator/prey relationships. There is some evidence to suggest that food webs are organised in a series of 'compartments', implying that species distribution may have more to do with other species (Krause *et al.* 2003).

The final aspect concerning the biological processes which underlie the patterns we see deals with the taxon cycle. Although this is seen as a relatively new and still controversial idea, it does get us to question the way in which species change through time and, therefore, distribution. Based on work on (mainly) bird species in the Lesser Antilles, research (Ricklefs and Bermingham 2002, Ricklefs and Miller 2000) has put forward an idea that taxa change with changes in ecological distribution. This would equate taxa with a 'life cycle' similar to that of 'species'. If this were found to be a widespread occurence, then it would add further complexity to our understanding of patterns.

6.4 Analysing patterns: environmental processes

The patterns we see are based partly upon the biological characteristics of the organisms involved.

There is a second set of factors which can influence the distribution of species on a broader scale. Physical features – geomorphology and geology – contribute considerably to global distributions. In this section we can outline the effects of plate tectonics, glaciation, barriers and islands on the distribution of species. Two plant species belonging to the Ericacae, *Daboecia cantabrica* (Plate 7) and *Erica ciliaris* (Plate 8) have geographical distributions which are of interest in this context. Both species have their most northerly and westerly occurrences in western Ireland, several hundreds of kilometres away from their main centres of occurrence in south-west Europe. Do their present-day occurrences in widely separate areas reflect the remnants of a formerly much more extensive range, the effects of glaciers, barriers or the peculiarities of island biogeography?

The distributions of land masses and oceans have changed throughout geological history. Although we did not understand all the details and even less about the mechanism during the nineteenth century, we were gathering evidence that suggested different arrangements than that currently seen. Fossil assemblages (e.g. Ernst 2000) contained similarities even though separated by oceans. Rock types were similar in South America, South Africa and Oceania. Early ideas of 'wandering continents' (as proposed by Alfred Wegener in the early twentieth century) did provide some explanation but increasing data made it less tenable. By the 1970s it was obvious that a new model was needed and so the first ideas surrounding plate tectonics were put forward. Here the earth's crust was composed of a series of plates. Upon most of these plates there was some less dense material forming the continents. Convection currents deep under the crust pushed molten material into weaknesses in the plates (usually the mid-ocean ridges). This would cause the plates to 'move', with the continents 'moving' relative to each other. Other forces would create tensions which in turn helped to shape the land masses. What is significant here is not so much the geological mechanisms but the implications for biogeography. One example that has tried to examine this is the Palaeomap project (Scotese 2000). By using a range of data it has been possible to combine continental locations with climatic belts and so produce an environmental history for most of organismic geological time. The climatic animation (http://www.scotese.com/paleocli.htm) provides an excellent review of current knowledge of this area. Of course, just because this work is being accepted it does not mean that there are no alternatives. For example, sea-floor spreading (where magma is forced along mid-ocean ridges) is the accepted version but there have been others, including an expanding earth theory suggesting that oceans were formed by a stretched earth. Whatever is finally agreed upon, it is worth noting the changing emphasis of this research area.

Plate tectonics shape forces over millions of years at the global scale. On a slightly smaller scale and shorter time span we have glaciation. The widespread movement of ice masses has had a considerable influence on the biogeography of Europe and North America. In terms of patterns we can focus on two aspects: the presence of ice and the alteration of the climate. During the last Ice Age (i.e. for the past 800,000+ years) the expansion of the northern ice caps caused considerable changes in the European and North American environments (Goudie 2001). The presence of ice covered the landscape and remodelled the location of soils and sediments. This had considerable effects upon the distribution of species as far away as the tropics. A common example is to look at the way in which alpine vegetation changed between glacial and interglacial periods. The alteration of climate is more than just changes around the main ice areas. Global depression of the sea level would cause changes in plant distributions around the world. Colder spells would lead to the contraction of tropical species' range (possibly into refugia). The corollary of this would be that a world heated under global warming would force arctic flora and fauna to retreat and significantly alter the distribution of glacial and periglacial features (Goudie 2001 – Figure 6.3). The Irish fringed

Figure 6.3 Glaciation and the English Lake District. This area has been extensively modified by glacial action, most notably by the creation of this lake caused by the valley being dammed by glacial debris. The impact of glaciers is not just felt here but far further south where periglacial action has altered landscapes significantly

sandwort (Plate 9) is a typical member of those primary colonisers in montane habitats which become established on recently available exposed, calcareous sites. This species is an endemic, with considerable morphological and genetic variability that appears to have arisen as a result of a long period of isolation in the single location where it currently occurs.

So far we have been looking at the way in which physical *events* have helped to shape the distribution of species. Distribution can also be influenced by physical *objects*. If we go back to first principles then the 'aim' of any species is to disperse so as to retain maximum influence over shaping the ecosystem. Anything which gets in the way of this could be considered a barrier. Barriers do not necessarily need to be physical (such as oceans) but can be physiological (we cannot live outside our tolerance limits), ecological (problems of competition and predation) or even psychological. All that is important is that they stop dispersal events. Of course, it is possible that some species find ways of overcoming barriers. For example, birds will fly over unsuitable areas and many marine shellfish have free-living larval stages that can travel considerable distances. If we also take organism size and ecological tolerance into account it follows that we have numerous 'barriers' specific to certain species or taxa all of which exercise some constraint.

Much of the work mentioned to this point has assumed large land masses and/or bodies of water. Some of the most interesting insights into pattern production have come from the study of islands. Islands represent ideal natural laboratories. They have a very definite boundary and their size makes detailed analysis possible. Interest in islands can be traced back to the work of Darwin and Wallace in the nineteenth century (Berry 2002) and from the study of islands such as Krakatau (especially after the explosive eruption of 1883 removed most of the vegetation from it). However, it is relatively recent work by MacArthur and Wilson in the 1960s (quoted in Krebs 2001) that has led to an upsurge of interest in the topic. They suggested that islands were in a dynamic equilibrium, i.e. they were subject to constant change but that the mean number of species stayed (more or less) constant (called the equilibrium theory of island biogeography – ETIB). Thus species numbers were enhanced by immigration and speciation and depleted by extinction and emigration. According to the Wilson–MacArthur model, each island would have an equilibrium number of species and that then larger islands would have relatively more per unit area than smaller islands. Even though this work is now almost 30 years old it continues to be controversial: workers continue to try to prove/disprove various parts of the concept. Whatever the eventual findings, its importance for us is that islands provide significant changes in the biodiversity of the surrounding area (which goes for land 'islands' such as mountains and conservation areas as well – see Box 6.3).

Biogeographic patterns are clearly influenced by the ecology and physiology of the organism (or rather their adaptation to specific environmental situations) and the nature of that physical environment. This leaves the final variable: time. Temporal changes are less easy to study given the obvious difficulty of researching one area for a

BOX 6.3

What is a place?

One of the cornerstones of the ETIB was the role of islands in the spread, or otherwise, of species. An island was taken as a piece of ground separated from others by some sort of barrier – in this case, water. Apart from that it is the same as any other area. That this is seen as reasonable is rarely questioned.

However, recent work by Walter (2004) questions that. He argues that an island is not just land but a unique area. In his argument, islands are similar because they share similar ideas of space which must affect the operation of ecological dynamics. In this sense they are different from mainland ecosystems (even nature reserves where the 'island' notion is prevalent). Mainland ecosystems have their own dynamics. Thus, ETIB fails because it does not compare like areas: we have two ecologies, not one. What makes this paper even more interesting is that it uses post-modern deconstruction of ideas rather than the more usual philosophies enjoyed in science. The points he makes should get us to evaluate carefully what we think we are looking at. It reinforces the need for biogeographers to take a holistic approach to the subject.

long enough period of time. One of the few exceptions to this is the Long Term Ecological Research program set up by the US National Science Foundation in 1980 (although many of the projects were started before this time). Of the 24 sites currently under study, Niwot Ridge in Colorado provides sufficient reason for the study of changes through time:

> Present-day environments cannot be completely understood without knowledge of their history since the last ice age. Palaeoecological studies show that the modern ecosystems did not spring full-blown onto the Rocky Mountain region within the last few centuries. Rather, they are the product of a massive reshuffling of species that was brought about by the last ice age and continues to this day. (Bowman and Seastedt 2001, p. 285)

It is the idea of 'reshuffling' that appeals. It suggests that environmental changes can occur faster than ecological response and that, as a consequence, organisms (mainly plants) are continually playing 'catch-up'. There is evidence for this in the Niwot Ridge study. Using a range of palaeoecological techniques (see section below), it has been shown that the tree range lagged behind environmental change by some 1,500–2,000 years (p. 293). The pattern, however, is not uniform. Some areas experienced greater change than others, suggesting that patterns can be very site-specific.

These studies of historical biogeography can provide considerable insights into species development and subsequent patterning. A relatively simpler and more common temporal change is succession. Although it could be argued that historical change and succession are the same thing, the two terms should really be seen as (largely) separate. Succession is used here as a development through time of an area resulting in a specific ecosystem. Historical change would be the replacement of one ecosystem with another where such replacement would not be seen as the normal development of the area.

As noted above, succession is difficult to study due to the time periods involved. One example comes from the eastern United States (Townsend *et al.* 2000). When farmers abandoned their fields and moved westwards the first plants to grow were annual weeds. These were replaced by herbaceous perennials which were, in turn, replaced by shrubs. Eastern red cedar replaced the shrubs and the local (secondary) sugar maple was finally established as the climax vegetation. Ice Ages are dramatic changes in distribution often over a restricted time period. Pollen diagrams (see below) are often used to show changes in vegetation, but other organisms have been studied and their previous and current distributions show considerable change. For example, a study of beetles found in late glacial deposits in Wales have present-day distributions as far away as Spain and northern Scandinavia (Lowe and Walker 1997).

Much of the work described above suggests some sort of conjunction between areas or species. What happens if species are separated in time as well as space? Isolation could prove to be one factor in the production of an area's biodiversity. Pole (2001) discusses the effect of New Zealand's isolation for 60 million years in terms of its biota: that the current pattern is a result of a complete turnover (i.e. replacement) of species rather than a more gradual replacement from a nearby land mass. Burnham and Graham (1999) suggest a similar problem with the neotropical realm in South America. Here, land bridges facilitated the spread of some species while at other times the area was isolated. The importance of this work is that it suggests a far more complex picture than one would normally expect, reinforcing the need for both contemporary and historical research.

6.5 Investigating patterns: methodologies

It is one thing to describe a pattern, but another to search behind the methodologies that made it possible. Gone are the days when all you needed to do to be 'ground-breaking' was to walk around an unknown area – these days it requires a sophisticated research program and often a good deal of laboratory analysis. Perhaps more importantly, it needs a critical eye on the sort of test used, the results obtained and the analysis deduced from that. Further, it is obvious from the work described above that a range of techniques are needed to find some sort of consensus.

Before starting this overview perhaps it is best to think about the basic material we are using. Two concepts are worth following in this respect. Firstly, we have groups of plants found together either as living, or previously living, together (biocoenoses). Here, it is obvious that the ecosystem can be inferred from the organisms found. Although there is more agreement on what is found it is still possible to disagree on its significance. Sometimes, groups of remains are found together which would not be seen in real life. These death assemblages (or thanatocoenoses) can bring together organisms from a range of ecosystems. This immediately raises the question of the sort of material found there. What has survived (and as important, what has not). The fossil record is notoriously selective about what becomes a fossil and what does not. Hard parts of animals are preserved but there is less chance of a jellyfish making an impression. This sort of situation makes palaeoecological reconstruction far more difficult. Of equal interest is the use of species as indicators of conditions. One of the most important ideas in late nineteenth-century science was that organisms had not changed: literally, the present is the key to the past (the principle of uniformitarianism). But what happens if it is not? What if species can adapt to some extent in the face of changing environmental pressures? This would make the reconstruction less accurate. This is not to say that either assemblages or uniformitarianism should not be used because of the problems inherent in them because, apart from anything else, that is all we have. However, it is equally foolhardy to accept results without equivocation. We do not have perfect information: we need to use it but be aware of its limitations.

The reconstruction of past distributions is crucial for our understanding of current patterns (see Niwot Ridge, above). Taking Quaternary environments as our case study we can appreciate the range of data that can be gathered. In their wide-ranging book, Lowe and Walker (1997) have divided techniques into four areas:

- **Geomorphological**. The production of maps showing key features (morphological mapping) can be supplemented with data on glacial erosion. This can be supplemented with remote sensing in terms of aerial photographs and satellite images. The aim here is to produce an accurate record of the surface showing both deposits and erosional features. This can be used to reconstruct both landscapes and the direction and timing of ice movements. Together this can make a geographical and geological chronology of events. For biogeography this is like producing the canvas onto which the plant and animal distributions are to be placed;
- **Lithological**. The mapping can give some indication of the materials present. However, by carefully collecting samples it is possible to carry out microanalysis on particle size, shape and composition which helps determine the direction and nature of the forces placed upon it. Data produced from this allow us to see where the material is coming from and what has happened to it. Much like the geomorphological information, the lithological data can tell us about the environments and their history. It is particularly important in considering the existence of glacial (cold) and interglacial (warm) periods;
- **Biological**. The use of uniformitarian principles means that we can gather biological samples and extrapolate from that to create likely ecosystems present in the area. Several biological materials are used. Perhaps the most well known is pollen. Due to its tough coating, pollen tends to withstand environmental pressures better than many other plant materials. It can be readily identified, which means that a core sample can produce a pollen-based chronology of events. A second biological method uses diatoms. Since their environmental preferences are fairly narrow it is possible to obtain good quality data from samples. In addition, they can be used to record water chemistry, sea levels and human disturbance. Both diatoms and pollen need very high magnification during the identification process. Macro-remains of plants or insects do not normally require this. Further, beetles can often provide accurate data due to their restricted environmental range. Their attachment to human habitation can also provide archaeologists with invaluable data about the living conditions of people and animals. Similarly, non-marine molluscs provide evidence about human living conditions and also about the nature of local ecosystems. Marine organisms represent a rich source of data. Molluscs, foraminifera and ostracods are among the more common animals producing data;
- **Temporal**. Of all the techniques used in palaeoecological reconstruction, the best known are the various physical dating methods. Radiocarbon dating can use a wide range of material to produce date estimates up to 30,000 years ago. The uranium series can also be used but the very long timescale makes it difficult. Tree-ring dating (or dendrochronology) uses the variations in annual growth to produce a pattern of ring widths which can be allied to specific dates. Similar annual layers can be seen in fine sediments called 'varves'.

Clearly many of these techniques are location/ material specific. Even allowing for this, the range of data is considerable. Many analyses combine two or more methods and use the comparison to produce a more accurate picture of events. Part of this accuracy is that it can allow for the errors and inadequacies inherent in any technique. For example, when we are looking at pollen diagrams and beetle analyses together we can appreciate that both have their errors. Pollen can spread easily and there is no guarantee that the pollen assemblage is from that direct area. Likewise, beetle assemblages are affected considerably by any post mortem change (from loss to alteration). The result is often a case of taking the best evidence available and finding a suitable compromise. For the biogeographer the most important lesson is that in producing patterns of distribution it is important to look at all the evidence, biological and non-biological, in order to come up with the best interpretation of events.

6.6 Summary

One of the most fundamental elements of biogeography is the creation of patterns of organisms. Due to the ease of construction, vegetation patterns are the most common. This is not to say that the matter is simple: there are practical and theoretical issues which need to be addressed before the pattern can be constructed. Since this work started over 100 years ago, there have been numerous attempts to divide the world into a series of regions based on a range of characteristics. Some have used biological features while others have relied on physical or environmental settings. Each has its own uses and merits, but at the end of the day we are trying to fit a distinct line around something that has no such distinct boundaries. The reasons behind this problem can be seen in the variables which control the distribution of a species – biological, physical and temporal.

Biological variables are those which suit the organism for the conditions in which it lives. Some of these relate to the characteristics of the species while others relate to the individual. Although it is possible to spend considerable time analysing each variable, at the end of the day the sum total fit/do not fit the individual for any given location. Much emphasis is placed on where an organism exists: it might be equally instructive to consider where it does not live! Also, despite the emphasis on the location of the individual it is important to realise that other organisms also play a part. The physical aspects are those linked to geology and geomorphology. These create conditions over geological time that have driven the existence and extinction of countless taxa. Temporal forces provide the third set of variables. We are aware of ideas like succession and uniformitarianism but the impact of time could be more widely realised. Species evolve at different rates creating the possibility of different assemblages (this could also impact upon uniformitarian principles).

Finally, we have the methodologies that can be used to gather data from which we make our distribution patterns. This part of the work can often be overlooked in the use of a new pattern but, like the small print of a contract, it provides us with the parameters within which we can operate. The example of Quaternary research is most instructive in this respect. It shows that we can use a wide range of physical and biological techniques but often come up with results which are contradictory. The skill comes in synthesising all the work and this cannot be done unless the methods are clearly understood.

APPLICATIONS – USING BIOGEOGRAPHY

Using ideas from this chapter in real-life situations

One of the first uses of island biogeography theory was the design of nature reserves and it remains a fundamental concept to this day. An allied concept is the notion of refugia – 'hotspots' of biodiversity that develop when all other areas around are changing. This idea has also been used to support the conservation of key areas especially in the tropics. In fact, the literature developed around refugia conservation has grown in the last five years. However, this is not universally accepted. For example, a paper by Wilf published in *Science* in 2003 has questioned some of the assumptions upon which refugia theory is based. Basically, the argument is that refugia are not needed to explain the great biodiversity of the region: that the area was species rich without special areas has been postulated. Although this might seem like a dry argument it is fundamental to biogeography on a number of levels. Firstly, it suggests that large ▶

parts of nature reserve design are based on false notions and that there might be other ways of creating reserves. Secondly, since many of these refugia are linked to biodiversity hotspots in the developing world which are themselves linked to increasing human pressure, it follows that we might be limiting human development in areas that could, after all, support it. Like so many of the examples given in these 'application' pieces, the work has implications beyond simple biogeographic theory: in this case, to human development in some of the world's poorest areas.

Review questions

1. Select one currently used distribution map from each of the four categories: organismic, environmental, holistic and conceptual. Selecting one taxon, compare and contrast the distribution maps.

2. What is the contribution of scale to biogeographical research?

3. Using one current distribution map, research its construction. What techniques were used, and how? What limitations could you find to each of the techniques?

4. Assess the validity of the taxon cycle to distribution maps.

5. Write a critique of the principle of uniformitarianism.

6. Describe the key changes in the distribution of tree species during the Quaternary period.

7. Select a research report detailing two or more methods used to reconstruct the vegetation patterns of the area. Analyse each technique, finding its strengths and weaknesses. How have these been handled by the report writers? Could an alternative viewpoint be supported by the evidence?

Selected readings

Whereas a wide range of texts have maps (e.g. Brown and Lomolino 1998), fewer go into the problems of making maps and comparing approaches. In this respect, despite its age, Dansereau (1957) is an excellent text covering material more modern texts leave out. The text by Bailey (1995) noted here is brief but a useful description of how a modern distribution map can be made. The best advanced works on distributions must be Rodwell (1992) and Dahl (1998).

References

Archibold OW. 1995. *Ecology of World Vegetation*. Chapman & Hall.

Bailey R. 1995. Description of the ecoregions of the United States. http://www.fs.fed.us/land/ecosysmgmt/ecoreg1_home.html.

Beeby A. 1993. *Applying Ecology*. Chapman & Hall.

Belyea LR and Lancaster J. 1999. Assembly rules within a contingent ecology. *Oikos*, **86**(3): 402–16.

Berry A. (ed.) 2002. *Infinite Tropics*. Verso.

Bowman WD and Seastedt TR. (eds) 2001. *Structure and Function of an Alpine Ecosystem*. Oxford University Press.

Brown JH and Lomolino MV. 1998. *Biogeography*, 2nd edn. Sinauer.

Bullock JM, Kenward RE and Hails RS. 2002. *Dispersal Ecology*. Blackwell.

Burnham RJ and Graham A. 1999. The history of neotropical vegetation: new developments and status. *Annals of the Missouri Botanical Garden*, **86**(2): 546–89.

Chase JM and Leibold MA. 2002. Spatial scale dictates the productivity–biodiversity relationship. *Nature*, **416**: 427–30.

Cox CB and Moore PD. 2000. *Biogeography – an Ecological and Evolutionary Approach*, 6th edn. Blackwell Science.

Dahl E. 1998. *The Phytogeography of Northern Europe.* Cambridge University Press.

Dansereau P. 1957. *Biogeography – an Ecological Perspective.* The Ronald Press Company.

Davies SJJF. 2002. *Ratites and Tinamous.* Oxford University Press.

De los Santos A, de Nicholás JP and Ferrer F. 2002. Habitat selection and assemblage structure of darkling beetles (*Col. Tenebrionidae*) along environmental gradients on the island of Tenerife (Canary Islands). *J. Arid Environments*, **52**(1): 63–85.

Duncan RP and Williams PA. 2002. Ecology: Darwin's naturalization hypothesis challenged. *Nature*, **417**: 608–9.

Ernst WG. (ed.) 2000. *Earth Systems: Processes and Issues.* Cambridge University Press.

Feminella JW. 2000. Correspondence between stream macroinvertebrate assemblages and 4 ecoregions of the southeastern USA. *Journal of the North American Benthological Society*. **19**(3): 442–61.

Goudie A. 2001. *The Nature of the Environment*, 4th edn. Blackwell.

Harding JS and Winterbourn MJ. 1997. An ecoregion classification of the South Island, New Zealand. *Journal of Environmental Management*, **51**(3): 275–87.

Hellberg ME. 1998. Sympatric sea shells along the sea's shore: the geography of speciation in the marine gastropod *Tegula. Evolution*, **52**(5): 1311–24.

Hessburg PF, Salter RB, Richmond MB and Smith BG. 2000. Ecological subregions of the interior Columbia Basin, USA. *Applied Vegetation Science*, **3**(2): 163–80.

Hraber PT and Milne BT. 1997. Community assembly in a model ecosystem. *Ecological Modelling*, **103**(2–3): 267–85.

Kennedy TA, Naeem S, Howe KM, Knops JMH, Tilman D and Reich P. 2002. Biodiversity as a barrier to ecological invasion. *Nature*, **417**: 636–8.

Krause AE, Frank KA, Mason DM, Ulanowicz RE and Taylor WW. 2003. Compartments revealed in food-web structure. *Nature*, **426**: 282–5.

Krebs CJ. 2001. *Ecology*, 5th edn. Benjamin Cummings.

Lees DC, Kremen C and Andriamampianina L. 1999. A null model for species richness gradients: bounded range overlap of butterflies and other rainforest endemics in Madagascar. *Biological Journal of the Linnean Society*, **67**(4), Aug: 529–84.

Li WKW. 2002. Macroecological patterns of phytoplankton in the north-western North Atlantic Ocean. *Nature*, **419**: 154–7.

Lonsdale WM. 1999. Global patterns of plant invasions and the concept of invasibility. *Ecology*, **80**(5), July: 1522–36.

Low T. 2002. *The New Nature.* Viking.

Lowe JJ and Walker MJC. 1997. *Reconstructing Quaternary Environments*, 2nd edn. Prentice Hall.

MacDonald G. 2003. *Biogeography – Introduction to Space, Time and Life.* Wiley.

Marchant R, Wells F and Newall P. 2000. Assessment of an ecoregion approach for classifying macroinvertebrate assemblages from streams in Victoria, Australia. *Journal of the North American Benthological Society*, **19**(3): 497–500.

Morin PJ. 1999. *Community Ecology.* Blackwell Science.

Ni J. 2001. A biome classification of China based on plant functional types and the BIOME3 model. *Folia Geobotanica and Phytotaxonomica*, **36**(2): 113–29.

Osbourne PL. 2000. *Tropical Ecosystems and Ecological Concepts.* Cambridge University Press.

Pakeman RJ, Hinsley SA and Bellamy PE. 2000. Do assembly rules for bird communities operate in small, fragmented woodlands in an agricultural landscape? *Community Ecology*, **1**(2): 171–9.

Pole MS. 2001. Can long-distance dispersal be inferred from the New Zealand plant fossil record? *Australian Journal of Botany*, **49**(3): 357–66.

Ricklefs RE and Bermingham E. 2002. The concept of the taxon cycle in biogeography. *Global Ecology and Biogeography*, **11**(5): 353.

Ricklefs RE and Miller GL. 2000. *Ecology*, 4th edn. Freeman.

Rodwell JS. (ed.) 1992. *British Plant Communities*, 5 vols. Cambridge University Press

Rosenzweig ML. 1995. *Species Diversity in Space and Time.* Cambridge University Press.

Rutherford MC and Westfall RH. 1994. Biomes of Southern Africa: an objective categorization.

Memoirs of the Botanical Survey of South Africa. (63). I-VII, 1–75.

Scotese C. 2000. Palaeomap project website. http://www.scotese.com.

Spellerberg IF and Sawyer JWD. 1999. *An Introduction to Applied Biogeography.* Cambridge University Press.

Stiling P. 1999. *Ecology: Theories and Applications,* 3rd edn. Prentice Hall.

Tester JR. 1995. *Minnesota's Natural Heritage.* University of Minnesota Press.

Thompson JN and Cunningham BM. 2002. Geographic structure and dynamics of co-evolutionary selection. *Nature,* **417**: 735–8.

Townsend CR, Harper JL and Begon M. 2000. *Essentials of Ecology.* Blackwell Science.

Walter HS. 2004. The mismeasure of islands: implications for biogeographical theory and the conservation of nature. *Journal of Biogeography,* **31**: 177–97.

Weiher E, Clarke GDP and Keddy PA. 1998. Community assembly rules, morphological dispersion, and the coexistence of plant species. *Oikos,* **81**(2): 309–22.

Wilf P *et al.* 2003. High plant diversity in Eocene South America: evidence from Patagonia. *Science,* **300**: 123–5.

Wright GR, Murray MP and Merrill T. 1998. Ecoregions as a level of ecological analysis. *Biological Conservation,* **86**(2), Nov: 207–13.

Yoshida T, Jones LE, Ellner SP, Fussmann GF and Hairston Jr. NG. 2003. Rapid evolution drives ecological dynamics in a predator–prey system. *Nature,* **424**: 303–5.

Zeigler SS, Pereira GM and Brown DA. 2004. Embedded scales in biogeography. In Sheppard E and McMaster RB (eds). *Scale and Geographic Inquiry: Nature, Society and Method.* Blackwell.

Websites

Palaeomaps. Examples of the way in which climate and land distribution has changed through geological time.
http://www.scotese.com/paleocli.htm

PART TWO

Biogeography in practice

CHAPTER 7

Studying and describing vegetation

Key points

- Vegetation description is fundamental to studies of succession and zonation and many other areas of biogeography;
- Vegetation is intimately linked with climate and geology, and increasingly, to human activities such as agriculture and forestry;
- It provides the groundwork on which applied biogeographical studies and increasingly, environmental assessments depend.

7.1 Introduction

The description of vegetation is one of the most important aspects of ecology and biogeography since it provides the foundation on which, for example, studies of succession or of animal distribution and abundance can be based. The biome concept relies significantly upon an accurate description and assessment of the distribution of vegetation types within a geoclimatic zone.

Vegetation types depend crucially upon climate and substrate, and quite small differences in the pedology, geology or hydrology of an area can cause major changes in the abundance or distribution of keystone species. Some species (or communities), because of their response to external factors, are extremely useful as indicators of particular local environmental conditions while others appear able to flourish in a wide range of habitats. In Europe, dwarf shrubs of the genus *Vaccinium* (bilberry, cowberry – Figures 7.1 (a) and (b)) show a consistent pattern of occurrence in relation to the duration of snow-lie in mountain regions. *V. myrtillus* and *V. uliginosum* (bilberries) are both deciduous and will commonly be found in exposed sites which are usually clear of snow, while *V. vitis-idaea* (which retains its small waxy leaves throughout the year) is more commonly found in sites where snow cover persists for a long time (McVean and Ratcliffe 1962).

The alpine catchfly *Lychnis (Viscaria) alpina* is frequently found in higher latitudes on unstable, mineral-rich soils subject to frost action (Figure 7.2). The occurrence of a subspecies in mountain areas which is able to tolerate the high levels of toxic metals such as nickel or chromium found in soils developing on outcrops of the ultrabasic mineral serpentine is interesting (Hultén 1950 and pers. comm.). The subspecies (*serpentinicola*) is known from Scotland and Norway and while it will grow in non-serpentine soils, it is rapidly outcompeted by more vigorous plant species. Biogeographers and ecologists look for species

Figure 7.1(a) The bilberry (blueberry, *Vaccinium myrtillus*) is a low-growing shrub, widespread in heaths and woodlands on acid soils throughout Europe. It tends to occur mainly on woodland margins and in larger glades in southern counties but becomes extremely common throughout much of the highland zone. The blue-black fruits are a useful food source for birds and small mammals (and ecologists). It is deciduous, like the bog bilberry (*Vaccinium uliginosum*, a more northerly plant), but the latter species has blue-green leaves and larger, more globular fruit. Both species are able to survive in exposed upland sites throughout the winter

Figure 7.1(b) *Vaccinium vitis-idaea* (cowberry) is found in birch and conifer woodlands and taiga throughout northern Europe, including parts of the British Isles. It is evergreen and the fruits are red when ripe; the plant is about the same size as the other species mentioned but is not able to survive in exposed sites. Where it occurs in mountainous areas, the site is one which is usually covered by some snow for most of the winter. In this photo, cowberry (in flower) is growing in association with dwarf birch and crowberry

Figure 7.2 *Lychnis alpina* is widespread throughout much of montane Scandinavia where it colonises open sites such as soil slips or recently disturbed ground. It has very poor competitive ability and is rapidly excluded from the vegetation succession except in areas where the soils have a low pH and contain appreciable quantities of heavy metals which cannot be tolerated by other normally more competitive species such as grasses

such as these to help in the investigation of a habitat or its environmental parameters and distribution. An even more dramatic example of species preference is shown by the lichen genus *Rhizocarpon*, in particular the yellow 'map' lichen, *R. geographicum*. This is widespread in upland areas throughout its global range, but only where the rock on which it is growing has a high proportion of silica (as in quartzite or granite). It is never found on calcareous or alkaline rocks such as limestone or basalt. Lichenologists have decided that the plant known as *R. geographicum* may be one of a group of morphologically similar species, differing very slightly in their chemical characteristics and habitat preferences.

7.2 The value of history

The earliest work on plants and their distribution was carried out as a part of medicine. Many early botanists and herbalists made excursions to find plants which had therapeutic properties and during the course of their travels often made notes of where they found these and other species. These lists were sometimes published and provide present-day botanists with historical evidence of the presence of species in places from which they may have subsequently disappeared as a result of changing land use. The herbalist and apothecary Thomas Johnson travelled extensively through southern and south-eastern England collecting and listing what he saw (Johnson 1629, 1632). Although the Deptford pink (*Dianthus armeria*) is no longer found in that part of south-east London, meadow clary (*Salvia pratensis*) still grows within the Greater London area, not far from where Johnson described gathering it.

Examples of vegetation description and plant geography can be traced back to the early nineteenth century. In 1819, Nathaniel Winch published a report on the geographical distribution of plants in northern England, while Macgillivray (1855) carried out one of the earliest ecological surveys in Scotland, noting in great detail all vegetation types along the River Dee from its source in the Cairngorm Mountains to the valleys. Towards the end of the nineteenth century, the description and, in particular, the mapping of French vegetation and plant communities was undertaken under the direction of Marcel Hardy at the University of Montpellier, alongside the mapping of the geology of the country. This was done primarily to provide information on land capability which would be of economic value. Hardy's work attracted the interest of two Scottish botanists, Robert and William Smith; Robert spent some time in the 1890s working with Hardy at Montpellier, and on his return to Scotland, started an ambitious programme of botanical surveys including vegetation mapping and description which would eventually cover the whole of the British Isles. Several coloured maps at a scale of 1:126,720 (half-inch to the mile) were completed by the brothers and published in the *Scottish Geographical Magazine* at the turn of the century. Typical of these is the map of northern Perthshire (Smith 1900 – Figure 7.3), which provided excellent information on the distribution of the

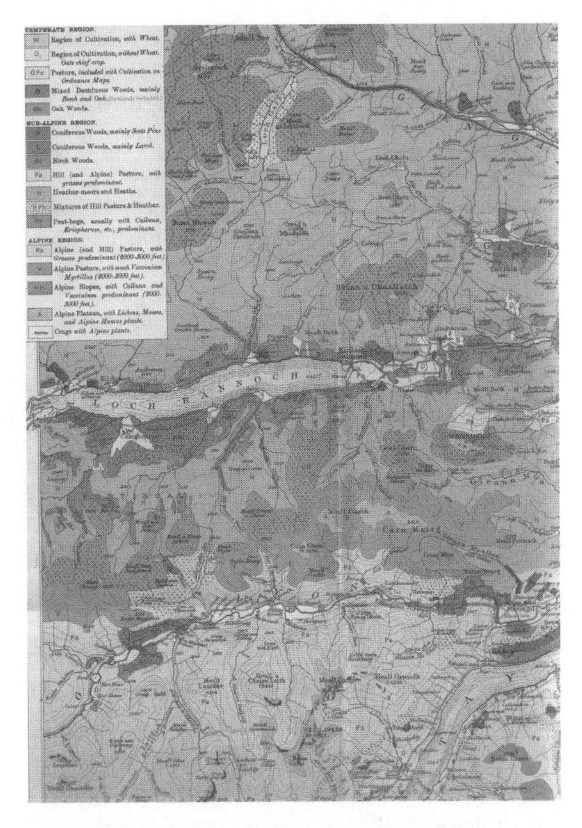

Figure 7.3 This is an extract from a 'half-inch' (1:126,720) map surveyed by Robert and William Smith and published over a century ago. While the exact boundaries of the upland and woodland vegetation types they recognise may have changed slightly as a result of successional and human influences, their present-day distribution is not much different

(Reprinted with kind permission of the Royal Scottish Geographical Society)

major vegetation types in this part of the Grampian Mountains. It says much for the quality of this work that it remains as relevant and useful today, over a century since it was produced.

Regrettably, Robert Smith died in 1900 at the age of 30 as a result of peritonitis and the project was taken over by his brother, William. Over the next decade, a few more maps were produced, including some of northern England, but the project had lost its momentum and by 1914, at the start of the Great War, it had ceased.

Some of the material was included in a volume, *Types of British Vegetation* (Tansley 1911) and repeated almost verbatim in the classic *British Islands and their Vegetation* (Tansley 1939) with scant acknowledgement. In its most recent form, vegetation description in the UK has been well served by the comprehensive National Vegetation Classification (NVC – see Box 7.1) which commenced in 1975 and was largely complete 14 years later. The five volumes of the survey covered over 30,000 samples and these were classified into a hierarchy of vegetation types using computer-aided analysis (see Rodwell 1992).

BOX 7.1

The NVC and phytosociology in the British Isles

The impetus behind this project was given by the realisation that as more and more ecological and environmental information was required for nature conservation (in particular, a systematic investigation and classification of habitats) and other processes, there was no up-to-date survey or classification of British vegetation along the lines that had been pursued elsewhere in Europe by phytosociologists. On a number of occasions work on British regions or habitat types had been published, beginning with Poore's application of phytosociological techniques to Scottish vegetation in 1955, followed by an extension of the methods by Poore and McVean (1957) and culminating in McVean and Ratcliffe's classic phytosociological study of the vegetation of the Highlands of Scotland in 1962 which began the process of relating the classification to other European traditions – in this case, Scandinavia. The development of a truly phytosociological approach to vegetation description in the UK was stifled by what many ecologists came to regard as an obsession with sampling strategies, objectivity and the respective values of classification and ordination. This was in marked contrast to traditional phytosociological description and analysis (including computer-aided analysis in the early 1960s) that was being carried out in Ireland under the influence of J. J. Moore. Luckily, some UK ecologists had decided that it was perfectly possible to integrate the use of analytical computer programs into phytosociological surveys and to use the computer to help in the display and interpretation of data, rather than use the analyses just as test beds for methodology – among the pioneers here were Ivimey-Cook and Proctor (1966). It was about this time that serious thought was given to setting up such a system which eventually produced government funding for the NVC which would provide standardised descriptions of named and systematically arranged types of vegetation within a classification framework. The brief did not include Northern Ireland and the work was to include major semi-natural and artificial habitats as well as natural habitats. This work is now complete, and since its publication it has had a dramatic effect upon many aspects of local and regional planning and conservation within the UK, since many planning and environmental impact assessments now require information on NVC habitat types within the designated area. The data have also been made available for other purposes such as the European Natural Vegetation Map project, a 1:2.5 million map of the major vegetation communities of Europe, coordinated by the Federal Agency for Nature Conservation (BfN) in Germany (Bohn *et al.* 2000).

7.3 The practice of vegetation description

Vegetation description, whether of individual species and their associates, or of communities of different plant species, is fundamental to ecology, plant geography and phytosociology.

Plant geography is defined by Dahl (1998) as the study of the geographical distribution of individual plant taxa, whereas phytosociology (plant sociology) involves the study of the geographical distribution of assemblages of plants (otherwise known as plant communities). Most modern vegetation description (including the NVC) follows the phytosociological model in which the basic material is provided by records consisting of standardised descriptions of vegetation plots (known variously as *relevés*, *aufnahmen* or quadrats) with lists of species and information about their relative quantities or abundances (Figure 7.4).

It is becoming quite common for such floristic information to be supplemented by target notes on the habitat and soil, although traditional phytosociologists tend not to record such detail. Individual vegetation records are taken in homogeneous stands of vegetation and can be combined together in tables ready for analysis. This may take the form of a vegetation classification in which similar records are grouped together, or by considering the vegetation table as a matrix of information within which relationships between species and plots can be elucidated, usually using computers. This is referred to as an ordination or indirect gradient analysis approach (Gauch 1982); the reference to gradient here is because the information in the table is considered as points in a multidimensional space which are rearranged by the analysis so that site records which are most similar to each other are placed close together and those which have least in common are at the opposite end of the arrangement. This is referred to as a gradient and, even with floristic data alone, provides some clues to possible controlling factors such as soil moisture or slope. More recent work in which environmental data, as well as floristic data, are subject to analysis at the same time can often yield clues as to the relationship between particular community groups and environmental factors such as soil pH or soil mineral concentrations. This is referred to as direct gradient analysis and can be a useful tool in environmental assessments. The use of the word 'direct' is appropriate here because of the use of habitat or environmental information in the analysis. The analytical technique may not identify the actual process(es) which influence the type of vegetation or the site, but it frequently enables the most likely factor(s) to be identified (see Coker 2000). A more detailed account of this process and more advanced aspects of data manipulation is given in Jongman *et al.* (1995). It should be stressed that the selection of an appropriate analytical procedure, whether manual or computerised, is vitally important, since an incorrect choice may produce erroneous results. It is equally true to say that even the most modern and sophisticated (or perhaps, currently fashionable?) analytical techniques may not provide ecologically meaningful results. Good quality vegetation description and interpreting the subtleties of the results of an analysis rely heavily upon good quality data and a degree of familiarity

Figure 7.4 This 50 cm square quadrat is divided up into 25 10 cm squares. In this application, the students are recording changes in plant species and cover along a transect line (shown by the tape)

with the habitat which cannot be achieved solely in a laboratory or computer suite.

There are a number of factors which will influence both the way in which vegetation description will be carried out and the purpose for which it is likely to be used (see Box 7.2). Any application or survey should have a rationale which is as clearly defined as possible. The collection of data should never be an end in itself but should be seen as providing the raw material for description and analysis and synthesis as part of the whole project.

BOX 7.2

Factors to be taken into account when devising a vegetation survey

- The collection of appropriate, valid and accurate information on the distribution of vegetation within the habitat under investigation;
- The examination, analysis and if appropriate, the application of these data using appropriate tools.

1. What is the purpose of the survey?
   - Reconnaissance (similar to the UK Phase 1)
   - Detailed (one-off inventory)?
   - Baseline survey for long-term ecological monitoring or management
   - Descriptive survey and classification of plant communities (phytosociology or NVC)
2. The scale of the survey?
   This means that the methods of description used may well vary according to the amount of ground being surveyed. Large numbers of detailed quadrats are not appropriate for surveying several thousand hectares of savannah – similarly, too few samples of a large habitat will tend to give a misleading impression of its biodiversity. There are no hard and fast rules determining exactly what the ratio of sampled to overall area should be.
3. The overall habitat type?
   It is important to match the technique to the type of habitat. For example, a method of sampling a temperate forest in Europe may not be totally appropriate for work in a East African savannah or Brazilian rainforest.
4. What constraints and resources?
   The major constraints here are usually climate and possible seasonality, time (as a window of opportunity), finance, manpower and equipment. To these could be added the extent of the area to be examined and the availability of expertise in identification of species.
5. The extent of the task?
   It is important to check if any published (or unpublished) work is available for the area. It can save a lot of time, particularly if it is of recent date and carried out by competent botanists (either professional or amateur). Such information can reduce the size of the task considerably, provided it is reliable.

Kent and Coker (1992) list a number of other questions which should be addressed:

- Is it necessary or relevant to identify all of the species present?
- If identification is required, do published floras or identification manuals already exist? If not, are local botanists available to instruct?
- Are environmental data to be collected and if so, what variables are to be measured?
- Are appropriate equipment and resources available to measure the variables with appropriate accuracy?

- How will the vegetation data be analysed? Are any particular analytical programs to be used? It is important to know this since some computer programs impose constraints on the form in which data have to be collected in the field.
- Has the dynamic nature of the vegetation been taken into account? This is where seasonality is often of great importance and information on the likely successional stages and climax state is valuable.
- There are almost certainly going to be issues over consistency of data gathering between workers. Errors may be introduced because of species identification lapses or differing perceptions of what constitutes a cover value of 75 per cent. This requires careful training before tackling the project.
- Speed versus accuracy is a critical aspect and the balance must be assessed for any project. Time available or factors such as seasonality of vegetation or climate can severely restrict the window of opportunity for a project as can a lack of resources. It usually comes down to collecting the best quality and quantity of data possible in the time available!

It is relevant to look at the question of species identification in more detail. There has been a major change in the way in which identification skills are taught, certainly in the UK, and most ecologists or biogeographers in the twenty-first century will have had little formal training and will need either to attend professional development courses for specific groups, or rely upon knowing someone who has such skills. In the UK, this is currently (2005) the focus of a major debate among professional organisations, the British Ecological Society and the universities, with some consultancies reporting difficulty in finding applicants for jobs with even elementary identification skills (see Box 7.3).

BOX 7.3

Why identification skills are important

Phytosociologists generally try to identify *all* species of plant present in their relevés – mosses, lichens, ferns as well as flowering plants. This is done because in many cases one cannot be certain which are the most important components of a vegetation type and missing out a species for whatever reason may cause difficulties later on. It tends to be time-consuming and very difficult if the operator is unfamiliar with the plants native to the area or lacks an identification manual or the help of an expert.

This detailed approach has much to commend it and it is the basis of the UK's National Vegetation Classification (NVC), as well as of descriptions of the plant communities of many other parts of the world.

But what if the survey required is merely to provide information of a general nature on the vegetation or habitats of an area? This is the aim of the Phase 1 habitat survey (Box 7.4) as used in the UK (JNCC 1990) as an aid for environmental audit. Vegetation types are identified simply (through a *hierarchy* of habitat types) and their extent colour-coded and plotted onto base maps. There are 10 Level 1 categories, A–J, covering all major habitat types, including cultivated and disturbed land.

Baseline surveys for monitoring purposes are often carried out on fixed sample areas which are recorded in detail, often accompanied by fixed point photography of the site. Regular resurvey at appropriate intervals, accompanied by photographs from fixed points, will combine to form a continuous record of vegetation development over time, allowing the investigation of variations in plant cover and changes in species over time. The number of species to be identified will vary according to the type of habitat and the purpose of the investigation, but in general terms a survey of this type ought to include all commonly occurring species in the area.

The detailed, descriptive survey of vegetation forms the basis of phytosociological studies, and of the NVC. It can be time-consuming and requires the careful identification of all species of plants growing within the sample area. Sample areas should be located in stands of visually homogeneous vegetation and the size is adjusted according to the type of habitat – thus, for woodlands, the NVC requires a sample area of 10 m square whereas grasslands are generally sampled with a much smaller frame, perhaps 2 m square. The resulting matrix of species occurrences in sites is often quite sparse, with less than a third of the possible matrix positions having any data. In areas in which there is a small-scale mosaic of habitats, care must be taken to avoid collecting data across habitat boundaries, which are usually quite distinct.

All the methods of vegetation description mentioned so far depend upon floristics, involving the identification and recording of species presence and often including their abundance. An alternative approach, often used for the classification of vegetation at small scales – i.e. the global distribution of a particular type,

BOX 7.4

Phase 1 habitat survey

The hierarchical approach comprises up to four levels, as in this simplified example:

Level 1	(A) Woodland and scrub	
Level 2	(1) Woodland	(2) Scrub
Level 3	(1) Broadleaved (2) Coniferous (3) Mixed	(1) Dense and continuous (2) Scattered
Level 4	(1) Semi-natural (2) Plantation	(1) Acidic (2) Basic (3) Neutral

Thus a plantation of oak trees would be classified as **A1.1.2**, and dense gorse (*Ulex*) scrub on a sandy soil would be **A2.1.1**.

The alphanumeric classification is often supplemented on maps by the use of appropriate colours.

such as temperate deciduous forest – relies upon the external morphology, life form, size and possible stratification of the species which are present. The advantage of this technique is that the species do not have to be identified in a particular taxon, and surveys using structural techniques can be undertaken quite rapidly over large areas. Floristic data are most usually applied at large scales, over relatively small areas, and while they produce good and highly detailed community-based data, the structural and physiognomic methods are excellent for rapid, broad habitat classification at the community level.

7.4 Vegetation description based on structure and physiognomy

While the most widely used methods of vegetation description rely upon detailed floristic studies, there are several techniques of vegetation description based on structure and appearance (physiognomy) which have been used. One of the commonest is the life form method devised by the Danish botanist Raunkiaer (1937 – see Box 7.5). This is described more fully in Kent and Coker (1992), but an outline of the method is presented for comparison with others.

The technique relies upon placing all plant species in the area into Raunkiaer groups, and presenting the proportions of each of the groups found within an area as a bar graph. Careful interpretation of the relative proportion will frequently demonstrate the close relationships which exist between life form and climate. A particularly good example of this is provided by Danin and Orshan (1990) in their study of life form spectra and climatic gradients in Israel (see also Figures 7.5(a)–(c) illustrating other aspects of this work).

An applied aspect of life form analysis is provided by work on anthropogenic environments such as derelict land and mineral spoil heaps, where the severity of the environment presents

BOX 7.5

Raunkiaer life form method

The method is based upon the height above (and occasionally below) ground of the parts of the plants which produce growth the next favourable season (perennating buds). It also assumes that species morphology is related to, and closely controlled by, climate. This means that humid, tropical environments tend to favour growth above ground while drier, less humid sites tend to favour species which can either survive the unfavourable season as bulbs underground, or as annual weeds which germinate rapidly in spring, flower and set seed by early summer and survive as seed through the hottest and winter months.

There are five main categories, each of which may be further subdivided.

1. Phanerophytes (species with buds emerging from aerial parts of the plant, more than 2 m above ground);
2. Chamaephytes (species with perennating buds borne less than 2 m above ground);
3. Hemicryptophytes (aerial parts of plants die back in unfavourable conditions, buds formed at ground level);
4. Cryptophytes (species with buds or shoot tips which survive unfavourable conditions below ground or under water);
5. Therophytes (annual species which survive the unfavourable season as seeds, growing only in the favourable spring and early summer months).

More information on the classification is provided in Mueller-Dombois and Ellenberg (1974).

Figure 7.5(a) Limestone pavement at Mullaghmore in the Burren National Park, Ireland. The development of deep, linear erosion lines (grikes) dissects this limestone area. The exposed limestone rock is a quite inhospitable environment whereas the deeper grikes often contain remnants of the woodland vegetation which once covered this area, including flowering plants, tree seedlings, ferns and bryophytes, flourishing in the sheltered and humid environment

Figure 7.5(b) The retreat of glaciers has provided areas for colonisation. Storbreen, in Norway, has an outer moraine (maximum extent) at the edge of the river, and dated at AD 1750. The current glacier snout is several kilometres up the valley. The intervening ground is covered with various types of vegetation – mainly dwarf shrubs in the older deglaciated area because the site (at 1,100 m) is above the local tree line. Sites nearer the glacier are characterised by grasses, mosses and lichens

Figure 7.5(c) Volcanic ejecta, such as cinders, form a very unstable and challenging environment. The photograph of Fujiyama in Japan shows colonisation and stabilisation by shrubs and dwarfed conifers (*krummholz*), mainly a species of larch (*Larix*). This particular situation develops just above the tree line

many challenges for plant growth and the more traditional floristic methods of assessment are unsatisfactory. Natural environments with equally severe problems such as volcanic areas or glacier forelands respond well to similar life form analysis, with a frequently wide range of life forms demonstrating the range and severity of environmental conditions on both a meso- and micro-scale. Conventional floristic analyses in such environments tend to fail because of the low levels of vegetation cover of most, if not all, species present.

In addition to the Raunkiaer life form technique, attempts have been made to use a combination of structure and physiognomy to provide classification schemes. The easiest to use in large areas is that proposed by Dansereau (1951, 1957) in which six sets of criteria are used – life form, size, cover, function (evergreen or deciduous), leaf size and shape and leaf texture (filmy/succulent/prickly). In this method, no detailed information on the species composition of the area is required, apart from the dominant species, which are described using symbols, shapes and shading to produce a profile diagram, which is then further embellished with letters which describe the size and coverage of the plants. The diagrams look a little bizarre, possibly because of the abstract appearance of some of the symbols, but do convey a great deal of information on the habitat. It is not a widely used technique in spite of its ease of use. A further structurally based system was devised by Fosberg (1961) as a means of describing vegetation with the IBP (International Biological Programme) which was established during the 1960s as a means of quantifying the energy budgets and primary production of the major world ecosystems (Peterken 1967). The structural approach was used since there is no way in which comparable floristic data (which by their very nature would be detailed and time-consuming to deal with) might be employed. The variations in geographical distribution of species would also render any floristic-based comparisons of little use – for example, the species composition of rainforest ecosystems in South America and West Africa differ greatly and floristic comparisons would be of little use. Structural and growth form comparisons would be much more informative.

The system is hierarchical, and starts off with three major vegetation categories – open, close or sparse. In each of these categories, there are 31 formation classes in which height and continuity are the dominant factors. Further subdivisions can be made on the basis of plant function, and finally by growth form and physical properties of the leaves, much as in the Dansereau method. This final (fourth) level enables field mapping to take place, with each category being known as a formation group. Apart from its use in the IBP, there have been few occasions when this technique has been used in the last 30 years. Goldsmith (1974) found the technique both rapid and satisfactory for conservation assessments of the vegetation on the island of Majorca. The major drawback was that it proved exceptionally difficult to produce valid vegetation maps since a lack of aerial photography or high vantage points meant that it was not possible to delimit accurately the boundaries of the Fosberg vegetation formations.

As with conventional floristic methods, the establishment of a suitable distribution of sampling points can be problematical. All but randomised placement tend to offer the chance of operator bias and subjectivity, an undesirable situation when one needs to establish a representative area in which to site the sampling points. The best solution is to describe a formation in a site which *appears* to be typical of that type of vegetation. This is not ideal, but in practice, it appears to work well and most vegetation surveyors tend to develop an appropriate level of expertise quite rapidly once in the field. The most successful structural studies appear to have been carried out in the tropics, as several papers from the Australian rainforests in Queensland have amply demonstrated (Webb *et al.* 1970, 1976, Webb 1978).

7.5 The value of taxonomy and identification strategies

The techniques for the description of vegetation using floristics rather than structure rely heavily upon the ability of the operator to identify the component species accurately. This is achieved most commonly by the use of an appropriate identification manual (or flora) for the area, and species whose names are uncertain when seen in the field may be collected and carefully preserved for later taxonomic work, having been assigned with a reference code for recording purposes. The herbaria of National Botanic Gardens can be of immense help in this respect but it is unlikely that instant identifications can be provided of several hundred dried specimens. It is important to realise that some areas may be so imperfectly known that no reliable identification manuals exist, and in such circumstances the effort involved in carrying out a survey, even with expert local botanical help, may outweigh any advantages of the floristic approach over a structural approach.

Another problem relates to the naming of plants. Conventionally, plants will have a local (vernacular) name and if they have been described by a taxonomist in a flora or similar publication, a scientific name as well. The vernacular name will differ according to the region, country or language but with luck, the scientific name, usually in Latin, will be the same worldwide. Unfortunately, this is not always the case. Taxonomists are keen to find out the earliest validly published name of a plant (or animal) and may, during the course of research through archives, come across one which was given before the first use of the current name for a species. This newly discovered name is then published and, with very rare exceptions, everyone has to use it in place of the name they were previously using. This has led, in the British Isles at least, to several changes of name for plants such as the dogwood, a shrub which started off as *Cornus sanguinea*, then changed its generic name to *Swida*, and for the present, is known as *Thelycrania sanguinea*. Even the name changes for the bluebell, a spring-flowering woodland bulb, promote confusion because early twentieth-century publications such as Salisbury's studies on the ecology of woodlands in Hertfordshire, UK (1916), refer to the bluebell as *Scilla nutans*, a name with which a twenty-first-century botanist would be entirely unfamiliar since it is now known as *Hyacinthoides non-scripta*, having in the intervening period been known as *Endymion non-scriptus*. Name changes are a source of considerable frustration and confusion; the situation is not helped when taking into account the preferred scientific name as used in the identification books of another country. Thus the alpine catch-fly, known in the UK as *Lychnis alpina*, is more usually known in Scandinavia as *Viscaria alpina*.

Figure 7.6 *Rhododendron yakushimanum*. Long-term isolation of plant or animal populations can lead to the development of new species or subspecies. Correct identification of endemic species such as this rhododendron from an island off the coast of Japan is important in aiding efforts to conserve the species and local biodiversity

Identification is an important aspect of any fieldwork and the ready availability of digital cameras means that if an unknown species is encountered in the field, the pressed dry specimen sent away for identification can be supplemented by a colour image of what it looked like when living (see Figure 7.6).

7.6 Patterns in vegetation

The plants which make up a typical piece of natural or even semi-natural grassland or woodland rarely grow in regular pattern such as would be seen in an orchard where fruit trees are planted in rows, an exact distance from their neighbours, or in a forestry plantation where a regular planting pattern makes it easier and less disruptive to extract timber or smaller trees before clear-felling takes place. The advancing rhizomes of sand sedge (*Carex arenaria*) produce buds at quite regular intervals which appear to 'stitch' their way across loose sand, but the pattern is linear and often radiates from a single plant in all directions. Pattern depends also upon the size of the sampling unit in relation to the size of the species under investigation and may repeat on a regular basis if the habitat is large enough.

Apart from this, plants of a particular species usually occur in either clustered or random distributions in most habitats, or an environmental factor such as wind or water carriage has produced linear strips or clumps of seedlings just where the seeds were deposited. Clustered distributions occur where the seeds are released and germinate near to the parent, or where the parent has produced offshoots which root and live an independent existence (see Figure 7.7). Clustered distributions can also arise when fruits or seeds are transported internally or externally by animals feeding on them. Seeds may pass through the gut and be deposited with the faeces, or attach to the exterior of the animal where they are removed during grooming. Some so-called 'random' distributions are not always random in the strict sense of the word. The distribution may appear to have no perceptible pattern but this might be because the scale of the overall pattern has not been appreciated. A further complication is caused by the influence of large- or even small-scale changes in the environment which will affect the ability of a plant to live in a particular habitat. Such a controlling factor might be the availability of water or the mineral composition of soil – there are many other possibilities too, including microclimatic variations.

Figure 7.7 *Sarracenia alata* (yellow pitcher plant) are a genus of insectivorous (pitcher) plant most commonly found in wet, woodland clearings and peaty areas (muskeg) in North America. This species was photographed in the Big Thicket National Preserve in Texas, USA, where thousands of plants were seen in a large woodland clearing. The flowers produce substantial amounts of seed which is shed close by and the dampness and low pH and nutrient status of the soil discourage competition from other species

The size of sampling unit (usually referred to as a quadrat) can be critical in the investigation of pattern. Too small a quadrat and it might fall within any pattern, while a larger quadrat may miss small-scale patterns. The scale of pattern in a particular habitat will depend upon the species present and their characteristics and it might be necessary to use different size quadrats for each species under investigation. This is obviously a non-starter, and either a compromise size is used or a series of nested quadrats, to suit the size of components within the ecosystem. As an example, one could use 1 m^2 for woodland floor species, 5 m^2 for the shrub and saplings layer and 10 m^2 or even 20 m^2 for the trees. The difference in sample size means that each set of data has to be analysed separately, and not as a combined set. While it is possible to get some sort of answer by doing a combined analysis, it is likely to be unreliable. Traditional phytosociological description and analysis tend not to be concerned with maintaining a constant size of sample area during the collection of data, but analysis by computer package assumes that the *same size* of sample area is used for all the data collection, otherwise the statistical validity of the results would be compromised.

7.7 Determination of quadrat size

The quadrat is used to define an area of vegetation which will be subject to survey, and its area is very important, the optimum varying from one vegetation type to another. The commonest method for determining the optimum size of quadrat for a particular vegetation type relies upon the minimal area technique and species–area curves originally described by Cain (1938) in which a small area, say 10 cm square, is examined and the occurrence of species noted. The area is doubled to 20 × 10 cm and the number of species counted again, followed by doubling to 20 × 20 cm, recording all species and more doubling. On each occasion, the number of species is plotted against the species area, and the resultant species–area curve usually shows a rapid rise in species count for the earlier sites, followed by a gradual decrease until a doubling of area no longer leads to a 10 per cent increase in the number of species recorded. This point is usually taken as the minimum area for the sample vegetation type, but the nearest larger standard size quadrat to this is selected. If, by doubling the sample area, several species not previously recorded are added, then it is quite likely that your sample is encroaching on a different vegetation type or an ecotone. It is important to understand that the minimal area technique only works properly in areas of homogeneous vegetation, but it is an extremely useful process for establishing a suitable sample size for vegetation studies in unfamiliar vegetation types. As a general rule, areas in which vegetation appears to change rapidly in terms of species should not be selected for a minimal area trial, but it is acceptable to use an area of vegetation which appears homogeneous. Typical quadrat sizes for a range of vegetation types are shown in Table 7.1. These sizes should be used for guidance, particularly in tropical environments.

Table 7.1

Vegetation type	Quadrat size (m)
Moss and lichen communities	0.5 × 0.5
Grasslands, heathland, dwarf shrubs <50 cm high	1 × 1 to 2 × 2
Tall grass, herbs >50 cm, shrubby heathland	2 × 2 to 4 × 4
Scrub and woodland shrub layer	5 × 5 to 10 × 10
Woodland canopies	20 × 20 to 50 × 50

7.8 Site selection

In floristic studies, the selection of a sample area for analysis may be done on a permanent or temporary basis. For a permanent quadrat, it is advisable to mark its position as accurately as possible, using surveying or a portable global positioning system (GPS) which can provide a location accurate to within a few metres. Permanent quadrat positions should be marked on maps or aerial photographs, and location photographs taken. If the area is secure, a marker should be set up to aid in refinding the site at a later time. Where this is not possible because of public access, a buried metallic marker is almost as satisfactory since it can be located by the use of a metal detector. Coker (in Coker and Kent 1998) used magnetised steel nails which were buried to mark the corners of a series of permanent sites for a trampling experiment. The location of the site was known quite accurately but the exact position of the permanent quadrats was fixed by the response of a pocket compass which was moved over the ground. The method was used satisfactorily over a period of 18 years.

Most vegetation description makes use of portable frames of varying sizes, according to the type of vegetation. The conventional quadrat is a square frame with sides about 0.5–1 m in length, but there is no reason why, for most purposes, a rectangular or even circular frame could not be used. The frame does not need to be rigid and many ecologists use lengths of thin cord with loops at appropriate intervals, which can be fixed in place using tent pegs or skewers. The advantage of this is that a large quadrat, say 10 m square, weighs a few hundred grams and is very little trouble to set out or carry around. Rigid metallic or wooden frames are heavy (particularly when they are more than 1 m square) and can be exceptionally difficult to manage in rugged or scrubby terrain.

Site selection can be achieved using randomly generated coordinates (produced either from tables of random numbers or by use of the random function available on pocket calculators). Pairs of random numbers are used to fix a position on a large-scale map, plan or aerial photograph of the area which has been gridded up. The process is repeated until enough sites have been identified to sample the area adequately. It may be necessary to put constraints upon the survey, either because of physical features such as lakes and rivers or sheer cliff faces where it would either be dangerous or impossible to survey, or because the area may contain other types of vegetation or attributes which it is not desired to sample at this time (such as areas of woodland within an area of grassland).

Other less commonly used constraints include stratified random sampling and restricted random sampling which are dealt with later in the chapter.

7.9 Recording of vegetation

It is becoming increasingly necessary for vegetation recording to be taxonomically and methodologically sound, particularly with the increasing emphasis on possible impacts on the environment which may follow establishment of housing, utility or other developments in or adjacent to ecologically sensitive areas.

In systematic studies of vegetation, it is vital to use a recording system which, while preserving as much information on species abundance as possible, is not too cumbersome to use. The 'percentage cover' method which is often used to describe the proportion of the sample area occupied by a particular species, can suffer from the drawback of spurious accuracy – it is simply not credible to say that a particular species has a cover of 4 per cent or even 84 per cent unless this is measured directly, and even with permanently marked sites, there will be daily and seasonal variations in cover which would render such 'accuracy' of little use, as will the impact of grazing herbivores. In most natural and semi-natural communities, substantial numbers of plant species will, individually, occupy less than 5 per cent of the area and many fewer species will occupy more than about a quarter of the area. Some workers use equal-sized classes with 5 or 10 per cent intervals which offer somewhat greater flexibility and ease of use. There is also the effect of vegetation stratification, where, for example, a layer of mosses or herbaceous plants lies under a low canopy of shrubs. In cases such as this, the overall cover will sometimes exceed 150–200 per cent (Coker 1988).

The only advantage to be gained by using a detailed percentage cover estimate is when the sample area is covered by a gridded quadrat with regular subdivisions of 5 or 10 cm side. A squared frame of this sort is quite easy to use, and for the usual 0.25 m^2 frame, either 25 or 100 smaller squares are available. The main drawback of this system is that it takes a long time and is unsuitable for any except small-scale studies.

Other quantitative methods for estimating cover include the use of weighted scales, such as those used by phytosociologists. Two such scales are in use of which the more frequently used is due to Braun-Blanquet. This is a five-point scale which is sometimes augmented with a five-point index of 'sociability'. The alternative to the Braun-Blanquet scale is the Domin scale, which has ten points, allowing greater discrimination at the lower end of the scale than the Braun-Blanquet; both Domin and Braun-Blanquet allow the presence of a species with less than 1 per cent cover or as a single small individual to be signified by a '+' sign. The differences between the two scales are shown in Table 7.2.

Cover estimates produced using the above techniques are usually subjective in nature and may

Table 7.2 The Braun–Blanquet and Domain cover scales

Value	Braun-Blanquet	Domin
+	Less than 1% cover	A single individual, less than 1% cover
1	1–5% cover	1–2 individuals, cover less than 1%
2	6–25% cover	Several individuals, but less than 1% cover
3	26–50% cover	1–4% cover
4	51–75% cover	5–10% cover
5	76–100% cover	11–25% cover
6		26–33% cover
7		34–50% cover
8		51–75% cover
9		76–90% cover
10		91–100% cover

be prone to error since there is a tendency to overestimate the cover of large, attractive, conspicuous and flowering individuals and to underestimate species which are not familiar. It should be understood that cover estimates from competent practitioners of either of the above scales will rarely differ by more than one class (Dahl, Barkman pers. comms.) when dealing with the same patch of vegetation.

In addition to visual numeric assessments of cover, some workers use the frequency symbolic approach, of which the DAFOR and ACFOR scales are examples, for rapid assessments of large habitats such as grasslands or heathlands when the use of conventional quadrats may be impossible or inappropriate on account of time or access limitations. The initial letters signify the following: **D**ominant, **A**bundant, **F**requent, **O**ccasional and **R**are, with the **C** in **ACFOR** signifying **C**ommon. There may be occasions when the frequency needs to be qualified, in which case, the prefixes 'visually', 'locally' or 'very' may be applied. The frequency symbolic approach is sometimes, erroneously, referred to as a 'semi-quantitative' technique – which is clearly not the case because no numbers are used.

The final method of recording is not subjective since it relies upon a species being present or absent from a sample area. Collection of presence/absence data is only desirable for reconnaissance studies or when time or other constraints do not permit a more accurate estimate to be made. There is no possibility of extracting any further information on species abundance from the bald statement.

In measuring species abundance, all the methods so far described rely upon the collection of qualitative (presence or absence) data or of quantitative data estimates. Kent and Coker (1992) make the distinction between subjective and objective data categories by inferring that objective methods of data collection are 'where more accurate and precise measures may be taken which should not vary from one recorder to another'. For the majority of cases this is probably true, but operator fatigue can lead to errors in recording and a recorder's unfamiliarity with the local plant life can provide similar opportunities for errors, particularly when seedlings or atypical growth forms are encountered.

7.10 Objective measures

Where it is not appropriate to use subjective measurements, no matter how carefully obtained, there are several methods which are objective enough for most purposes. Of these, frequency and density are the most commonly met with, although some workers make use of pin frames for an objective measurement of plant cover and others, particularly in agriculture and forestry, use biomass, yield and performance.

7.10.1 Frequency

This is the probability of finding a species in a given sample area. It is most commonly assessed by random throws of a quadrat in a designated area and noting the number of times that the species under consideration occurs in a large number of throws. A species occurring in 45 out of 60 throws has a frequency of three-quarters or 75 per cent. If a subdivided quadrat is available then the method can be refined by scoring presence/absence for the species in each of the subdivisions, and converting the number of presences to a percentage score. Kent and Coker (1992) point out that frequency also depends upon quadrat size, plant size and vegetation patterns. Frequency measurements are of most use when comparisons of vegetation are being made on a large scale.

7.10.2 Density

This is the number of *individuals* of a species found per unit area. While it is relatively easy to pick out individual upright growing plants and count them accurately within a sample area, species which spread by tillering or runners can cause problems owing to the difficulty in disentangling them. The process is rarely used on a large scale for whole communities because it is too time-consuming. It is particularly suited to work with tree saplings or populations of woodland or grassland herbs. Density estimates depend absolutely on quadrat size and this must remain constant for a particular survey; density is also affected by the scale of species distribution patterns in relation to the quadrat size and the size of the plants. Plotless sampling (Cottam and Curtis 1956) is a method of measuring density of tree species in woodlands and relies upon measurement to the closest target species from a randomly thrown pin. The pin is then thrown at random and another measurement taken. Tree density is related to the average distance between individuals, and a good estimate of tree density can usually be obtained which is within 5–10 per cent of the value obtained by direct counting. It is possible to use either the closest individual distance (tree to pin) or the nearest neighbour distance (tree closest to the pin to its nearest neighbour).

7.10.3 Cover estimation by point quadrat (pin frame)

This is a potentially very accurate and objective technique for measuring cover in low-growing vegetation. In its usual form, 10 long, thin pins are mounted in a frame which stands above the vegetation to be sampled. Each pin is withdrawn and lowered carefully through the frame into vegetation and the number of times it touches individual species is recorded. The frame is moved one-tenth of the length of the sample area and the whole procedure repeated nine times; the total of 100 pin samples and their 'hits' can be considered as a whole, in which case the cover of vegetation may exceed 100 per cent, or the number of hits per individual species can be extracted, and a series of percentage cover values derived.

Point quadrat work is time-consuming and difficult to use in scrub or tall grass, and its reliability is also affected by vegetation patterning. The pin used should be as thin as possible since thicker pins lead to overestimates of cover, and the best results are given by optical sampling using a frame in which the pins are replaced by a tube with cross-hairs. The observer looks through this and establishes intersections of plants with the cross-hairs which simulate a pin of negligible diameter. Goodall (1952) compared three pin sizes and found that percentage cover could be exaggerated by as much as 25–30 per cent if thick pins (4.75 mm diameter) are compared with the cross-hair estimates. For most purposes, pins with a diameter less than about 1.5 mm seem to yield acceptable results, and if results are being taken from a number of sites for comparison purposes, it is essential that the pin size remains constant.

Smartt *et al.* (1974, 1976) divided these and other abundance measures into three categories:

(a) **unbounded measures** – such as biomass or density which could have no fixed and final upper limit;

(b) **partially bounded measures** – such that there is a limit (where a quadrat area is being equivalent to 100 per cent but layers of vegetation can cause abundance values to be greater than this) as with percentage cover;
(c) **bounded measures** – these have a fixed upper limit, as with frequency records taken in a subdivided quadrat.

In the same work Smartt *et al.* recorded the same vegetation type using a wide range of different measures of species abundance. Their conclusion was that although the measure chosen for a given project should mainly depend upon the purpose for which the data are being collected, the practical differences between objective methods of abundance assessment (cover, density and frequency) and subjective methods such as visual cover estimates were very small. Subjective estimates of cover appear to work because the greatest number of species in a typical natural plant community occurs with low frequency and that few species occur with higher frequencies. If we take five equal percentage bands (0–20, 21–40, 41–60, 61–80 and 80–100), labelling them A to E and distributing the species occurrences among them according to frequency, the relationship that exists between them is that the numbers of species in A is greater than B which is greater than the number of species in group C; C and D vary slightly, there may be more or less species in C than D and sometimes equal numbers; group E tends to have higher numbers than group D. This phenomenon was first explained by Raunkaier in 1928 and expressed as a 'law of frequencies' which holds true for many ecosystems. If the numbers for each frequency class are plotted against numbers of species as a bar chart, a 'reverse J-shaped' distribution usually results.

7.11 Sampling strategies and design

One of the most frequently overlooked parts of vegetation description is the need to arrive at a sampling method which meets the requirements of the project and which is both statistically sound and within the time and resources budget. The sampling design should also take into account the proposed method of analysis and the likelihood of interpretation problems due to some form of autocorrelation between sites or (in the case of a long-term monitoring programme) over time. Spatial autocorrelation occurs between sample sites, because the vegetation of one site is likely to have an effect upon others near by. The effect might be one of shading or competition for nutrients. The same problem exists if comparative records are taken from one site over a period of time since the more recent sampling results depend upon the earlier sampling. It is difficult to be precise about such effects, but spatial autocorrelation can be reduced by not sampling too intensively in a small area.

Sampling techniques which are met with in vegetation studies include the following:

- stratified sampling
- random sampling
- restricted random sampling
- systematic sampling
- transects (line and belt)

Stratified random sampling of vegetation implies random assignment of sampling points to particular habitats, management regimes or age classes in the study area which have characteristics defined by the requirements of the scheme, and intended to ensure good balance of these within the whole sampling scheme.

Random sampling implies that every point within the survey area has an equal chance of being selected. The usual process is to set up a grid over the area, which can be defined by pegs and string, and to select sites using random numbers as described earlier in the chapter. Where it is not possible to set up a grid of coordinates, a randomised walk may be employed. The operator is assigned a random number between 1 and 360 (to give a compass bearing) and a second number to give a number of paces. The sample site is at the conclusion of the walk. Another random compass bearing followed by a random number of paces are generated and the process repeated. Purists or people with a lot of time on their hands may wish to increase the randomness by doing a random number of repeat pairs of compass bearing and paces, but for most purposes a single repetition will provide a sufficiently randomised sample for all but the most stringent work. It is worthwhile applying the words of the medieval philosopher, William of Occam, to the quest for randomisation: '*Entia non sint multiplicanda praeter necessitatem*', which translates approximately as 'matters should not be made more complex than necessary'. Good advice for any biogeographer or ecologist!

Restricted random sampling allows random assignment of sampling sites within an area to achieve a balance between, for example, differing habitats in terms of their extent or other characteristics. This ensures that no habitat is oversampled in relation to any other.

Systematic sampling implies the use of a series of regularly spaced sampling points which might form the intersections of a grid, or points along a line, a fixed distance apart. Careful observation before setting out the sampling network will usually show if the proposed sampling interval coincides with any underlying pattern in the vegetation, but normally, patterns tend to recur unevenly unless there is some regularly occurring environmental factor affecting them, such as old field drains in a pasture.

Kent and Coker (1992) make a distinction between vegetation description which tends to be inductive in its approach and rarely requiring a truly randomised sample, and the deductive approach which is necessary for the generation and testing of hypotheses using statistical probability, and which requires as rigorously randomised a series of samples as possible. This is to ensure that no sample is dependent upon another.

Transects are, in their simplest form, a line along which the occurrence of plants is noted. In most cases transects are deliberately set up to traverse areas of ground where there are likely to be changes in plant cover and a more or less evident environmental gradient such as soil moisture, change in soil type or exposure; an alternative approach is to run the transect line through an ecotone, such as a sand dune system, or a lakeside area where a range of communities from fully aquatic, emergent to terrestrial can be studied along a moisture gradient. Their locations are usually picked for a specific purpose and are non-random, being designed to demonstrate the maximum change or variation in vegetation over the shortest distance. A modification of the line transect method is where a series of quadrats are laid across a transect line which follows a particular feature such as a path. Examination of the vegetation patterns and occurrences in each of the transects will often provide useful information on the effects of, for example, trampling by pedestrians.

The belt transect may be thought of as a line transect along which quadrats are laid next to each other. Plant cover is recorded and tabulated for each quadrat. Changes in vegetation may be correlated with, for example, soil moisture, pH or humus content, measured from samples taken in each of the component quadrats.

7.12 Summary

In this chapter, we have looked at the history, theory and practice of a number of ways of investigating the structure, distribution and other parameters of a range of vegetation types using quadrat or point-based methods in addition to plotless techniques. The value and applicability of the techniques described demonstrate how important this aspect of biogeography will become in the light of increasing pressures for environmental impact assessments that take plants as well as animals into account.

APPLICATIONS – USING BIOGEOGRAPHY

Using ideas from this chapter in real-life situations

Most research programmes are relatively short term lasting only a few years. However, if we are going to note longer-term changes then the data recording has to be over a far longer timescale. There are a few long-term units dedicated to this line of work. The US Long-Term Ecological Research (LTER) Network (www.lternet.edu) looks after a series of 26 sites spread across the United States. The network was founded in 1980, although many sites (especially the classic Hubbard Brook ecosystem) have data going back to the 1950s. This particular research programme covers a diversity of ecosystems in one continental area (with some new ocean areas proposed). In contrast, the far newer Oxford Long-term Ecology Laboratory (www.geog.ox.ac.uk/research/biodiversity/lel/) was started in 2003 to gather data through time. Their research projects focus on archaeological and palaeoecological techniques to gather data covering a 2000-year timescale. By using core samples the researchers can see how the areas have changed especially under the influence of human activity. Although both approaches differ in that the US LTER gathers data over a longer lifespan and the Oxford unit takes material in a short time covering a far longer history, they are both committed to finding out more about changes and their causes. Given that geography is all about distributions (i.e. patterns) and what made them (i.e. processes) then this combines the two perfectly. Also, at a time when we need to know more about global warming potential, long-term records can be invaluable.

Review questions

1. What is the value of keeping records of vegetation surveys?

2. Why map vegetation?

3. Why is it necessary to match survey technique to habitat type?

4. How would you set about checking the identity of a potentially rare plant species from a single sighting during a survey?

5. What are the main differences between Phase 1 and NVC survey in the UK?

6. List the advantages and disadvantages of life form and physiognomic analysis. Which technique is likely to be the most useful for reconnaissance survey in an area of scrub and grassland?

7. How would you attempt to mark a permanent plot for future visits in an area that was subject to moderate human disturbance?

8. Describe how you might justify a sampling design to sample the plant communities of a mosaic of hedgerows in an intensively farmed landscape.

References

Bohn U, Gollub G and Hettwer C. 2000. *Karte der natürlichen Vegetation Europas* (Map of the natural vegetation of Europe). Bonn: Bundesamt für Naturschutz.

Cain SA. 1938. The species–area curve. *American Midland Naturalist*, **19**: 573–81.

Coker PD. 1988. Some aspects of the biogeography of the Høyfjellet with special reference to Høyrokampen, Bøverdal, Southern Norway. MPhil thesis, University College London.

Coker PD. 2000. Vegetation analysis, mapping and environmental relationships at a local scale, Jotunheimen, Southern Norway. In Alexander R and Millington AC. (eds) *Vegetation Mapping from Patch to Planet.* Chichester: John Wiley & Sons, Ltd.

Coker PD and Kent M. 1998. Long-term recovery responses of montane vegetation subject to recreational trampling. In *Vegetation Science in Retrospect and Prospect Studies in Plant Ecology*, **20**, 63. Uppsala.

Cottam G and Curtis JT 1956. The use of distance measures in phytosociological sampling. *Ecology*, **37**(3): 451–60.

Dahl E. 1998. *The Phytogeography of Northern Europe.* Cambridge: CUP.

Danin A and Orshan G. 1990. The distribution of Raunkiaer life forms in Israel in relation to the environment. *Journal of Vegetation Science*, **1**: 41–8.

Dansereau P. 1951. Description and recording of vegetation upon a structural basis. *Ecology*, **32**: 172–229.

Dansereau P. 1957. *Biogeography: An ecological perspective.* New York: Ronald Press.

Fosberg FR. 1961. A classification of vegetation for general purposes. *Tropical Ecology*, **2**: 1–28.

Gauch HC. 1982. *Multivariate Analysis in Community Ecology.* Cambridge: CUP.

Goldsmith FB. 1974. An assessment of the Fosberg and Ellenberg methods of classifying vegetation for conservation purposes. *Biological Conservation*, **6**: 3–6.

Goodall DW. 1952. Some considerations in the use of point quadrats for the analysis of vegetation. *Australian Journal of Scientific Research*, **5**: 1–41.

Hulten E. 1950. *Atlas of the Vascular Plants in N.W. Europe.* Stockholm.

Ivimey-Cook RB and Proctor MCF. 1966. Plant communities of the Burren, Co. Clare. *Proceedings of the Royal Irish Academy series B*, **64**(15): 201–311.

JNCC (1990) *Handbook for Phase 1 Habitat Survey – a Technique for Environmental Audit.* Peterborough: JNCC.

Johnson T. 1629. *Iter in agrum Cantianum.* (quoted in Gilmour J (1954) *Wild Flowers*, London, Collins New Naturalist)

Johnson T. 1632. *Descriptio Itineris in agrum Cantianum* (ibid.).

Jongman RHG, ter Braak CJF and van Tongeren OFR. (eds) 1995. *Data Analysis in Community and Landscape Ecology*. Cambridge: CUP.

Kent M and Coker P. 1992. *Vegetation Description and Analysis*. London: Belhaven Press.

Macgillivray W. 1855. *The Natural History of Dundee and Braemar*. London.

McVean DN and Ratcliffe DA. 1962. *Plant Communities of the Scottish Highlands.* London: HMSO.

Mueller-Dombois D and Ellenberg H. 1974. *Aims and Methods of Vegetation Ecology*. Wiley.

Peterken GP. 1967. *Guide to the Check Sheet for IBP Areas.* IBP Handbook 4. Oxford: Blackwell Scientific.

Poore MED. 1955. The use of phytosociological methods in ecological investigations. III. Practical applications. *Journal of Ecology*, **43**: 606–51.

Poore MED and McVean DN. 1957. A new approach to Scottish mountain vegetation. *Journal of Ecology*, **45**: 401–39.

Raunkiaer C. 1928. Domainsareal artstaethed of formationsdominanter. Konglike Danske Videnskaps Selskab, *Biologisk Meddeleser,* **7**, 1.

Raunkiaer C. 1937. *Plant Life Forms.* Oxford: Clarendon Press.

Rodwell J. (ed.) 1992. *British Plant Communities*, Cambridge: CUP.

Salisbury EJ. 1916. The oak–hornbeam woods of Hertfordshire, I, II. *Journal of Ecology*, **4**, 83–117.

Smartt PFM, Meacock SE and Mabert JM. 1974. Investigations into the properties of quantitative vegetational data. I. Pilot study. *Journal of Ecology*, **62**: 735–59.

Smartt PFM, Meacock SE and Lambert JM. 1976. Investigations into the properties of quantitative vegetational data. II. Further data type comparisons. *Journal of Ecology*, **64**: 41–78.

Smith R. 1900. Botanical Survey of Scotland II. North Perthshire District. *Scottish Geographical Magazine*, **16**: 441–67.

Tansley AG. 1911. *Types of British Vegetation.* Cambridge: CUP.

Tansley AG. 1939. *The British Islands and their Vegetation.* Cambridge, CUP

Webb LJ. 1978. A structural comparison of New Zealand and South-East Australian rain forests and their tropical affinities. *Australian Journal of Ecology*, **3**: 7–21.

Webb LJ, Tracey JG and Williams WT. 1976. The value of structural features in tropical forest typology. *Australian Journal of Ecology*, **1**: 3–28.

Webb LJ, Tracey JG, Williams WT and Lance GN. 1970. Studies in the numerical analysis of complex rain forest communities. V. A comparison of the properties of floristic and physiognomic–structural data. *Journal of Ecology*, **58**: 203–32.

Winch N. 1819. *An Essay on the Geographical Distribution of Plants throughout the Counties of Northumberland, Cumberland, and Durham.* Newcastle upon Tyne.

Websites

There are hundreds of references to vegetation description or vegetation analysis on the Internet and the sites which follow are a representative sample which appear to be relevant to this chapter.

A good summary of various vegetation analysis methods.
http://www.tuhsd.k12.az.us/Corona_del_Sol_HS/departments/Science/veganalysis.html

Vegetation analysis techniques used to 'read' saltmarsh vegetation.
http://www.britishecologicalsociety.org/epc/submission_bartlett

Useful site with summaries of exercises using a statistical package for vegetation analysis.
http://labdsv.nr.usu.edu/splus_R/

The web page of the Baltimore (USA) long-term ecological study. Worth browsing.
http://www.beslter.org/frame4-page_3b.html

An interesting application of vegetation analysis to animal territories.
http://www.nzes.org.nz/nzje/free_issues/NZJEcol3_44.pdf

CHAPTER 8

Studying animals and their distributions

Key points

- Variations in distribution can be seen as an interrelated set of long-, medium- and short-range factors;
- Short-range factors involve species and individuals and include ethology, adaptation, competition and food supply;
- Medium-range factors involve ecosystems and include environmental gradients, biotic dispersal and climate change;
- Long-range factors have a global focus and include time, geographical, geological and biological barriers;
- In mapping animals there are a number of problems to be faced including the mobility of the species and the theoretical paradigms used in the methods;
- An increasingly wide range of methods is used to sample animals for distribution study. These range from simple traps to satellite mapping;
- Studying individual cases in distribution helps highlight both progress and problems in the subject and is a vital part of the education in this field.

8.1 Introduction

Animals have always provided us with a paradox in biogeography. Their presence is obvious and their study popular: witness the numbers joining, for example, ornithological societies and the number of magazines devoted to birds. Set against this is the relative scarcity of texts in animal biogeography (or zoogeography as it is known) and the difficulty in studying their distribution due, in part, to their mobility. This popularity/scarcity dichotomy is an interesting one for students of the philosophy of science but does little for the biogeographer wanting to expand study in this field. Fortunately, with new tracking and recording techniques the matter of distribution is becoming clearer. This chapter seeks to give an overview of the topic, providing information on a range of tools which can be used but, at the same time, setting it in the context of both evolutionary changes and advances in methodology.

What do we *actually* mean when we talk about animal distribution? To say that x has a range y might help us when we draw a map but there again it might raise further questions. In reality, we need to be far more precise in our use of words. Two variables dominate our thinking – resolution and specificity:

- **Resolution** has to do with the scale being used (which is in turn a function of the method). For example, it is true to say that wood mice are found in Hampshire woodlands. This is the largest scale, local, and means that there are ecosystem patches where wood mice are present (and also large areas of non-wood where they are absent). It is also true at the regional level that these mice are found in southern England. Nationally and globally the answers become Europe and northern hemisphere. All of these statements are true but they highlight one issue – distribution depends on the scale of your study. This is particularly noticeable with the use in the UK of the national sampling grid where just one result could mean a 10 × 10 km square has a record!
- A similar challenge waits in terms of the specificity or taxonomic level used. On the way from species to kingdom the detail diminishes but the number of records, i.e. individual species, increases. The range of examples increases but contrasting ecologies might well mask subtle aspects of distribution (this does not happen with the single species but then you may not get an idea of the range). Although this is a theoretical aspect it does have practical applications. For example, when looking at the fieldwork methods available it is worth remembering that any real-life situation has trade-offs between detail and cost (time as well as finance) which need to be borne in mind.

8.2 What causes variation in animal distributions?

Species distributions are influenced by a series of three interrelated scalar factors. Short-, medium- and long-range factors combine to produce a seemingly complex pattern. To appreciate the advantages and limitations of the methods described in this chapter it is first necessary to outline these variations and show how they influence individual animals.

8.2.1 Short range

The first factors are those acting in a short range: ethology, competition and food supply. Ethology relates to the behaviour patterns of the animal and is crucial in getting meaningful results. Sampling procedures (see below) might be non-random and free from bias but it does not follow that the animal works in the same way! A student field study carried out by one of the authors provides a salutary point in this regard. The field vole is a small mammal of european grassland ecosystems. However, this does not mean the animal is evenly distributed. Behavioural patterns mean that they follow set 'runs' during feeding. If you place the mammal trap anywhere else, the chances are it will remain empty. Conversely, place the trap in the same position too long and the vole will become used to the free food (even with an enforced stay) and be a constant visitor (trap-happy). It does not take too much thought to appreciate the influence such behaviour has on numbers. In the first instance it would be possible to get a zero score while trap-happy voles suggest a very low population number (as no others are being caught!). Although this is a simple example, more complex research (Fischer *et al.* 1996) has shown that such variations can be crucial in understanding not just distribution but the population viability in fragmented landscapes. This example studied the behaviour of sheep in transferring plants and insects in their fleece. It demonstrated that the behaviour patterns of sheep influenced the types of material transported, thus affecting not just numbers of species but their distributions and even, in conservation settings, their survival. Another example comes from birds. Often there are problems with distribution analysis due to their long migrations. Research also reveals another ethological problem in that distribution appears to be sex-related with males flying further north than females but young males were found even further north which led researchers to propose a behavioural age barrier to migration patterns (Prescott and Middleton 1990).

Adaptation can alter the population sufficiently for there to be a noticeable difference without the creation of a new subspecies. For example, the banded snail *Cepaea* of Wicken Fen, Cambridgeshire, is grouped according to colour with the pale ones in the sunnier areas and darker ones in the shade of the fen plants. The argument that it is to reduce predation by more closely matching the physical environment might be compelling but is far from proven. Similar studies of the deer mouse in North America show colour variations probably related to soil colour (Brown and Lomolino 1998). Finally, competition and food supply can alter distribution and numbers. The kangaroo rat of the south-west United States includes several sub-species but their ranges do not overlap, probably suggesting competition. Food supply (and predator/prey relationships) might also produce variations. Studies on the introduction of predatory fish into the Great Lakes area (quoted in Brown and Lomolino 1998) have provided data on range limitation and thus changes in distribution.

8.2.2 Medium range

These short-range factors are influenced in turn by ecosystem-wide, medium-range ones. Most common in this regard are environmental gradients (see also Chapter 9), biotic dispersal and climate change. The environmental gradient is the change in a physical factor between two places. Thus aspect, altitude and latitude exhibit environmental gradients which can affect the distribution of habitats and their associated animal communities. Sometimes this can be quite large scale in nature, such as shown with the distribution of zooplankton and their relationship with marine hydrographic patterns (Sabates *et al.* 1989). There can also be a time element to this feature as demonstrated by the work of Caviedes and Iriarte (1989). In their study of rodents from pre-Quaternary onwards, distribution has been influenced by the environmental gradients of the Andean altiplano (or upland grassland). Here, movement to the south was restricted to the east slope of the Andes by the aridity of the Atacama Desert, while further movements were limited by the cold-arid conditions of the mountains.

Figure 8.1 Ground squirrel in Minnesota Zoo, USA. As habitats diversify through human activity so animal distributions will alter.

The concept of gradient can be expanded to cover both greater areas and a wider range of genera. Studies in China (Zhang and Zhang 1995) have demonstrated the non-random distribution of animals due to environmental conditions. Interestingly, this variation can still be seen if the scale of the analysis is altered, suggesting that it is a major factor from individual to community level. Long-term study of the Rhône valley in France (Resh *et al.* 1994) suggests that there are two sets of gradients working in the zoogeography of the area: a vertical gradient from interstitial to superficial habitats and a transverse one from main river to newly terrestrial sites. Given such diversity, selection and use of the best distribution analysis method is both desirable and fraught with difficulties.

Biotic dispersal refers to the ways and routes taken by species spreading out from their place of origin. If only certain routes are taken then it follows that there must also be barriers stopping movement. The key point is that barriers are species-specific, i.e. only certain species/genera can pass the barrier. It follows that any biotic dispersal will create changes in the community at both origin and destination. Some routes (corridors) permit the free passage of species. Filter routes are more specific in their requirements, while the 'lottery' or chance method depends on a complex series of chances coming together for successful movement. The key element for animal sampling is that habitats are not always strictly adhered to and that any barrier can alter species composition. Studies in the Lesser Sunda islands between Asia and Australia have worked out that the island chain itself is a filter permitting only limited exchange of species.

The final medium-range concept is that of climate change. Although this is often linked with such anthropogenic concepts as global warming (see Chapter 15) it can also have an effect on ecosystem distribution. Perhaps the most classic case in this regard is the recent Ice Age. Evidence from both pollen and animal remains has shown that distribution has been affected by the glacial/interglacial cycles. Such changes do not have to be confined to the large-scale event such as the Ice Age. There is increasing evidence of shorter-term change as well – see Box 8.1.

8.2.3 Long-range

Long-range changes are those affecting distribution on a global scale for a considerable portion of earth history – time, evolutionary changes, barriers and plate tectonics. So much is written about this that there is little need to repeat it here except to outline the key ideas and their impact on animal distributions:

- Time rates as one of the key forces. Often underplayed, time is a key concept in charting animal distributions. One of the earliest palaeontological concepts was the notion that the present was the key to the past (known as the concept of uniformitarianism). It argues that

BOX 8.1

The case of the evolving squirrel

The problem with evolution is that it is rare to see it in action due to the long timescales needed. However, research by Réale *et al.* (2002) suggests that shorter timescales may be possible which gives more evidence for species change (but also suggests a worrying trend for global warming if repeated elsewhere).

The red squirrel population of the Yukon, Canada, has been faced with an advancing of the timing of breeding of some 18 days in the past 10 years. Such a change could place great stress on the population and yet studies suggest that the population is still viable. If we assume that traits are variable both within and between populations/generations it follows that a large change should be met with a concomitant change in the squirrel. By using tagging and trapping over a 26-year period, the researchers found that there was a strong trend towards earlier breeders that could be seen as a phenotypic response within the population and a genetic response between the generations.

This study has utility beyond its subject. The rate of change of genetic and phenotypic variation is difficult to measure and so this long-term research gives valuable insights. It also highlights the way in which seemingly small changes in physical conditions can create larger changes in population structure. Whereas this squirrel can adapt it does not mean that this is universal. It is possible (although this research does not mention it) that in changing, the squirrel might eventually no longer fit into the community structure that is currently operating which might allow climate change, a far greater impact on ecology than presently recognised (see also Box 8.2).

BOX 8.2

Macaques and sika deer

The Japanese macaque, *Macaca fuscata*, is widespread in woodlands and forests throughout the Japanese archipelago, apart from Hokkaido in the north and the southernmost island of Okinawa (see Figures 8.2 and 8.3). Individuals of the mainland species (*M. f. fuscata*) are somewhat larger in size than those in island populations, for example at Yakushima, where the local subspecies *M. fuscata yakui* is stockier and also usually darker in colour. The mechanisms underlying this divergence into subspecies are almost certainly a response characteristic of long-term isolation and genetic drift.

According to Hill (http://www.biols.susx.ac.uk/home/David_Hill/index1.htm) the Yakushima subspecies is herbivorous and adapts its diet according to the time of year, generally eating a smaller range of fruit species than it does of shoots and leaves, and selecting leaves which have a low fibre content. Occasional shortages of food can lead to considerable mortality among the macaques; about 40 per cent of the population of about 4,000 died in the summer of 1998. Dead animals were examined post mortem and apart from appearing undernourished, several were found to have died from pneumonia.

Figure 8.2 *Macaca fuscata yakui*

Human activities at Yakushima, notably logging and plantations, adversely disturb the environment for the macaques, but when secondary forest develops towards maturity, this is an important habitat. The same can be said for the endemic subspecies of sika deer, *Cervus nippon yakushimae*, which shares many of the same habitats.

Compared to the mainland species, *C. nippon nippon*, the Yakushima deer is much smaller, standing about 55–60 cm at the shoulder compared with 90–95 cm for *C. n. nippon*. Several subspecies of sika deer are known from islands in

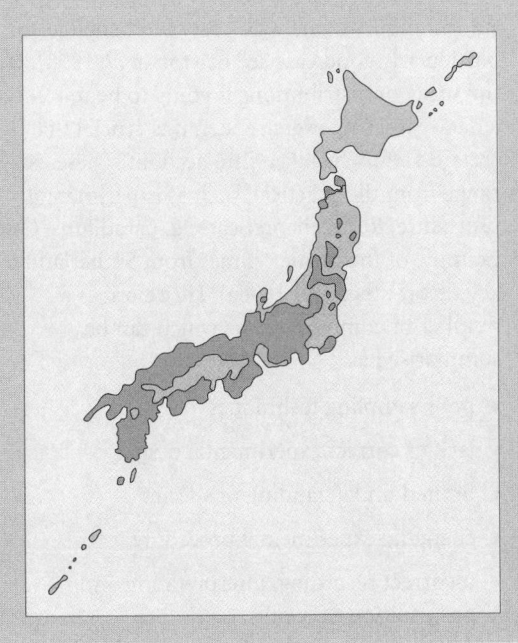

Figure 8.3 Distribution of *Macaca fuscata* in Japan. This species is found on all the larger islands of the archipelago (Honshu, Shikoku and Kyashu) except Hokkaido. An isolated population on Yakushima island, 40 km to the south of Kyushu, is recognised as an endemic subspecies (*yakui*)

this part of the Japanese archipelago of which *C. n. yakushimae* is the smallest, *C. n. keramae* is a little larger and *C. n. mageshimae* which is approximately the same size as the mainland subspecies (http://www.nzsika.co.nz/20296.htm). All these subspecies are grazers and browsers, and it would be interesting to see if food availability or some other factor was responsible for the variation in size.

Tsujino and Yumoto (2004) noted that high-density herbivore species such as sika deer often play an important role in forest regeneration in Yakushima, with a very high density (averaging 57 head km^{-2} estimated by faecal pellet count) in the western part of a lowland natural forest. Large-seeded tree species had significantly greater survivability in fenced quadrats than in unfenced quadrats. However, the survivability disagreed with feeding preferences. Sika deer activity increased seedling mortality of large-seeded species more than that of small-seeded species, and did not decrease much seedling survivability of non-preferred species. As might be expected, it was found that physical disturbance of the forest habitat by the high density of sika deer resulted in mortality for both preferred and non-preferred species, but that browsing of older preferred species was also an important factor.

the conditions in which animals are found today is the one their ancestors were adapted to. Note the changes in distribution and so one can infer climate changes.

- This can be suspect in many cases: it is equally likely that *evolutionary changes* have altered the physical requirements of the animal. In fact, evolutionary pressures can bring about changes in both distribution and physical characteristics. Such changes do not have to be long term. Studies in medieval shellfish (Ganderton 1981) have demonstrated that sub-specific differentiation could take place in as short a time span as two/three centuries.
- Finally, *geological barriers* and *plate tectonics* can cause considerable changes in animal distributions. Although such global changes might be seen as outside the scope of this chapter, the implications might be worth bearing in mind especially as some of the classic work in this regard was carried out by Darwin. Even today, there are questions as to the role played by these global forces in the local distribution of plants and animals. The Wallace Line (a line separating two areas in South-East Asia/Oceania either side of which distinctive species patterns exist) is just one example of the way in which ancient forces have helped shape modern distributions. It is only by appreciating these forces that we can hope to find appropriate methods to study their distributions.

8.3 How can we map these distributions?

All techniques come with costs and benefits (for a particularly good case see Lawton *et al.* 1998). If the study of distributions is going to be more than a data-gathering exercise then it is crucial that certain factors be taken into account. These can range from the practical, such as experimental procedure, to the theoretical, e.g. paradigms. One example of the former comes from Sutherland's key text (Sutherland 1996a). He describes a number of common errors which can be summarised as:

- poor sampling techniques
- lack of correct experimental design
- limited understanding of species
- changing experimental procedure
- incorrect recording, interpretation and presentation of results

Another case that researchers commonly meet (but rarely seem to acknowledge explicitly) is the idea of the trade-off. In the introduction above reference was made to trade-offs between scale of study and level of detail. The best situation is to have every place systematically checked for every species. Given the impossibility of ever achieving this, one must turn to sampling and use of techniques which permit results within the time and resource constraints given. While it can be argued that these aspects are a function of correct procedure there are other aspects which are equally important but less commonly seen in the literature, e.g. theory. Such concerns go to the heart of the subject, dealing as they do with the most basic assumptions.

Five examples can be used to illustrate concern in this area: paradigms, standardisation, comparative design, biocoenoses and extrapolation:

- A *paradigm* (Kuhn 1962) is a major body of knowledge accepted by practitioners in a subject as being the best available theoretical model. As such it changes only slowly. However, there are advantages in testing received wisdom, i.e. prevailing views. One study of historical biogeography (Zhao 1992) reveals weaknesses in the orthodox view of dispersal theory which are being replaced with vicariance studies. The argument presented is that the procedure that is commonly used is flawed in one or more respects. A different model will produce better results.
- Given the global nature of distributions it is not surprising that *standardisation* is necessary. In a study of moths (Odiyo 1987) it was felt crucial to develop a regional and national network for data gathering (similar ideas having been used routinely in botany, e.g. Rodwell 1992).
- Such requirements in animal ecology (taken for granted in studies such as meteorology) lead on to the third example – *comparative design*. The argument here is that science should test the tests and not just use them blithely. McClean *et al.* (1998) carried out a comparative study of statistical techniques to find the most appropriate one (i.e. best in terms of results) for wild turkeys in South Dakota. This gives an opportunity to rate various new designs.
- *Biocoenoses* (living assemblages) and their fossil equivalents thanatocoenoses (death assemblages) highlight some of the most fundamental tenets of the subject. Here the question is on the reliability of current data to represent past events. This carries with it some fundamental questions about our study, e.g. the way in which species might evolve/adapt in the long term which alters our perception of distribution (Andrews 1995).
- Finally, there is the use of data extrapolation to ‘fill in’ blank areas of knowledge. Although accepted practice, it does raise serious issues about reliability. Studies by Colwell and Coddington (1994) on global biodiversity stress the need for rapid survey techniques to fill in gaps in our knowledge. However, this comes at the price of accuracy although it might be valuable as a preliminary tool.

Despite the growing number of techniques available to study animals it comes down to two main categories from which to choose: evidence of existence or the collection of animals. The former relies on tracks, trails etc. to show that an animal was at a given place at some time; the latter requires that the animal itself be present. As the sophistication of techniques is growing it becomes important to see what other methods could be used/adapted. One of the most recent additions is the use of satellite tracking coupled with geographic information systems (Johnston 1998).

8.3.1 Track methodologies

Track methodologies rely on the animal leaving some visible sign of its location. These methods have the advantage that they can be carried out at any time (useful for nocturnal animals) and are useful where the animal is rare or very difficult to spot. The main disadvantage is that only a certain group of animals can be spotted this way: some sort of sign has to be left which obviously precludes smaller animals. Given these caveats there has been some extremely successful work carried out using the following techniques.

(a) Dung

Larger animals leave dung of a recognisable shape, colour, etc. By mapping dung patterns it is possible to work out the range of the animal. Excellent work on otter (*Lutra lutra*) distribution in the UK used dung (spraints) to map distribution of the species following conservation efforts in Wales. One example showing both the methodology and an evaluation is Mason and MacDonald (1987). The technique can be refined to the stage where each dung pellet is linked to an individual, allowing both individual and species distributions to be assessed (Komers and Brotherton 1997).

(b) Breeding sites/roosts

Many species will occupy a certain area for at least part of the year whether as part of a longer migration or as regular roost. Both birds and bats provide such sites. Furthermore, nests are usually distinctive of species. Thus a nest/roost count allows the distribution to be assessed. In a similar fashion some species will have a common display site where species can be assessed. Examples of both species and methods are recorded by Gibbons *et al.* (1996).

(c) Feeding signs

Some mammals, especially rodents, leave characteristic marks on their food sources, e.g. voles and mice can be tracked in woodlands by the tooth marks on nuts. This gives a useful indication of distribution in different ecosystems like closed woodland or where the species is nocturnal and/or rare (Cleave pers. comm.).

(d) Tracks/trails

Most animals leave a distinctive trail which, given the right ground conditions, can be preserved. There is no information as to the abundance of the species but at least some presence/absence data can be gathered. Similarly, other signs of presence such as hair or fur caught in hedgerows or on fences can be used (or even sampled using double-sided sticky tape to catch the hair as the animal passes – Sutherland 1996b).

8.3.2 Presence methodologies

The presence of an animal is the best indication of distribution. Most of these techniques involve capturing the species in some way and then releasing them unharmed (although this is not always the case with insects). Many of these methods require the animal to be marked or detained which brings in a whole range of questions related to the ethics of the experiment. Old systems where a toe was removed should by now be obsolete. Even marking an animal with paint might make it more visible to a predator which biases results. Young should not be handled to avoid rejection by their mothers and nursing animals should not be removed. Traps must be set correctly and visited frequently especially with small mammals whose need for constant food supply limits their capture times. In summary, all these techniques require a great deal of thought, training and in an increasing number of situations, legal permits. Sampling has come a long way since the days of Gilbert White of Selbourne who merely shot anything he wanted to identify! (White 1978): 'I procured one to be shot in the very fact; it proved to be the Sitta europaea (the Nuthatch)' – letter of 1768.

Not all of these techniques require the actual physical presence – birdsong is an excellent guide to species if the surveyor is trained. Common examples of presence methodologies are detailed below.

(a) Sound

Not all birds are easily spotted. Birdsong is an obvious way in which distribution maps can be made. This does require a great deal of study to recognise each birdsong but many ornithologists can distinguish several individual species. Other animals have distinctive sounds, e.g. whales, crickets. These can be used to record presence in an area although mimic species might create problems.

(b) Visual record

This might seem a simple technique but requires a great deal more than that (Lawton *et al.* 1998). Species should not be overly disturbed and whereas the technique is fine for a few large animals, for smaller animals with large populations, e.g. bats or birds, it becomes impossible to count without a permanent record, e.g. photograph.

(c) Trapping

This is by far the largest group of sampling techniques. Each one uses the characteristics of the animal to find a specific way of collecting specimens. Lawton *et al.* (1998) in a multi-species survey used netting for butterflies and flight interception traps for flying beetles. In situations where the animal is small and inaccessible, suction may be used such as that reported by Kennelly and Underwood (1985) for kelp forests or Macauley *et al.* (1988) for insects. Fish can be sampled using a box splitter (Winner and Mcmichael 1997), airlift sampler (Snow *et al.* 1987) or beam trawl (Rogers *et al.* 1998) among other methods. Very small animals can be sieve-sampled as shown by Schroeder *et al.'s* (1987) work on foraminifera. These techniques use the motion or location of the animal as a way of selecting the correct procedure. However, other aspects can be used: moths are usually caught in light traps although pheromone traps have also been used (Garrevoet 1997, Siliva *et al.* 1995).

(d) Fogging

It is possible to sample inaccessible areas such as trees by using a chemical fog. Insects will fall into prepared collection areas (Lawton *et al.* 1998).

(e) Aerial survey

Larger animals in remote areas can only be effectively analysed using some form of aerial survey. Photographs of the area can be taken quickly and effectively and counts/distribution research carried out elsewhere. In Australia, both camels (Short *et al.* 1988) and kangaroos (Hill and Kelly 1987) have been successfully studied in this way.

(f) Sonar/hydro-acoustics

Sonar can be useful in determining fish distribution although species analysis requires a great deal of knowledge (Perrow *et al.* 1996, Wilkins 1986).

(g) Banding/radio collars

Some animals have a large range (e.g. migratory birds) while others live in areas difficult to sample (e.g. dormouse). For both it is possible to attach radio-collars or bands to a few individuals. Banding is common for birds (Prescott and Middleton 1990), while radio collars have been used for large-scale study of dormice distribution (Cleave pers. comm.) and elks (Johnston *et al.* 1998).

(h) Extraction

This is a technique more common with insects where the aim is to create a situation disliked by the animal. Light/heat sensitivity can be magnified in a Tullgren funnel used to extract animals from soil and leaf litter. Alternatively, formaldehyde can be used to force insects from soil and leaf litter (Ausden 1996) although its toxicity may prelude its use in certain countries.

(i) Transects

Some insects might be more easily sampled in a transect line. To some extent, sonar is a form of transect sampling. Lawton (1998) used it for termite analysis.

(j) Questionnaires

These are not for animals but for locals when researchers are studying distribution. Such knowledge is vastly under-used but it can be seen to be a very cost-effective way of gathering information. Short *et al.* (1988) used it to supplement work on camel distribution in Australia, although they reported a considerable difference in results between aerial survey and questionnaire. There are increasing instances of indigenous knowledge being used especially where this is becoming more highly regarded, e.g. use of Aboriginal knowledge in Australia.

8.4 What can case studies show us about distribution patterns?

These three brief studies aim to evaluate critically the distribution and methodologies from which they are derived. Rather than act as definitive examples, they have been chosen to show a range of problems, distributions and species. Similar issues could be found with every species.

8.4.1 The Asian elephant (*Elephas maximus*)

> An object of worship, a target of hunters, a beast of burden, a burden to the people ... the Asian elephant once held sway over a vast region from the Tigris–Euphrates in West Asia through Persia into the Indian sub-continent. (Sukumar 1989, p. 1)

The Asian elephant has been a significant part of Asian ecology for millennia. Once it started to interact with the human population (over 2,000 years ago) the picture became less clear and the status of the elephant changed (mainly declined) until now it is part of the endangered species list. The reasons for its decline are important for our understanding of biogeography (see Chapter 15) but our main concern here is the gathering of data to work out distribution patterns. Despite its size, the elephant is known to be a reclusive animal that prefers woodland areas neither of which characteristics makes population estimation, or location, easy. In terms of our discussion above, its ecology and ethology make it a more complex case to start with. It might be thought that with 2,000 years of interaction we would know where every elephant is but each technique we have used has had problems, as detailed below.

(a) Early literature

One of the first records of the elephant was its use by various early nation states. There are records of elephants being bought and trained and there is evidence for domesticated elephants being used to find wild populations. Since elephants were a source of power (and military strategy) the records would only be good for domesticated stock. There would be far less concern about the precise area from which they were captured. After the invention of gunpowder the elephant became less important. Thus we have a record which is useful but upon which little reliability can be placed in terms of exact location.

(b) Statistical records

The start of British rule in India meant that records of hunts were kept as were data on ivory production. Both activities helped to reduce the number and distribution of elephants and although it told us less about the population left we could at least see where the range used to be. Trapping and other colonial records can be very precise but their job was to measure output not location.

(c) Questionnaire

According to Sukumar, the first comprehensive review of elephant numbers used a mixture of methods including literature search and questionnaires. The results were less than satisfactory because of the problems of accuracy from both sources. Later work (by the Asian Elephant Specialist Group of the IUCN) showed that these techniques tended to give a lower figure than seemed realistic. Despite this there are some advantages of this method: literature is relatively easy to locate (unless translation is needed) and the use of local knowledge is invaluable (a source recognised increasingly in conservation programmes such as found in Australia).

(d) Direct observation

Sukumar reports on a detailed study of the distribution in southern India. One of the key methods used was direct observation. Essentially this means recording each elephant as seen, although to improve this as a technique it would be possible to observe along a transect, or stay in one place for a longer time to see flow, or even travel by light plane or helicopter. Any one of these would give a fair idea of distribution (but not down to individual ecosystems) but is far less useful if numbers are required (Whitehouse *et al.* 2001).

(e) Tracks and trails

It is possible to detect elephant locations by the damage carried out but this would be mainly where numbers were very high and easy to see anyway (such as in a conservation programme). Dung is a useful marker but this leads in turn to considerations of how many animals were involved and when (Barnes and Dunn 2002, Laing *et al.* 2003).

(f) Genetics

This holds much hope in terms of finding out the development of the sub-species but less so when we need to know where rather than what (Eggert *et al.* 2003).

(g) Global positioning systems (GPS)

This has the advantage of being highly precise but also relating only to the elephants concerned. This is excellent in terms of finding ethology and ecology but less so in terms of distribution.

In conclusion one can see that the seemingly simple case of finding the world's largest land mammal is far less easy. Each technique has its problems: the point is not to reject every method (or else what would we use) but to see the limitations of each idea and adjust accordingly. We should also be aware of work done in fields similar to ours and adapt where possible (most of the references used here were for African elephants – there being no comparable work on Asian elephants).

8.4.2 Birds

> Documenting the distribution of a species is primarily a matter of determining where it is present and absent. (Brook and Birkhead 1991, p. 170)

Behind this simple fact lies a multitude of problems. Even the editors quoted here realised the enormity of the task especially when applied to birds, one of the most mobile of species. Perhaps the first question to be asked of birds is what *is* a distribution? See Figure 8.4. For plants, records of presence or absence are relatively easily constructed but what of the rare bird found well outside its usual range? Normally, if the distance is considerable (say 1000 km or more) then it would

Figure 8.4 Distribution is more than just a theoretical matter. Conservation relies on good distribution mapping. Bald-headed eagle, Minnesota, USA

be reasonable to refer to the individual as a vagrant – something blown into the area by strong winds perhaps. If the distance were in the order of a few hundred kilometres then it might be that the known range should be extended. To add to the possible inaccuracies of range we have a wide range of bird behaviour from the year-round resident to the global traveller. In addition, birds often have restricted areas even if the range is large, and many geographical factors can affect the distribution (Newton 2003). This gives a range of possible techniques which include the following.

(a) Field survey

The popularity of birds means that there is a considerable amount of personnel to help. This means that distribution maps can be made on quite ambitious scales although this is more normally a national activity. One of the most comprehensive of these is the UK British Trust for Ornithology's Breeding Birds in the Wider Countryside programme (BTO 2004) which is itself one of 34 different projects of varying sizes. The main part of the programme is to select 2,500 1 × 1 km squares at random throughout the British Isles. Volunteers visit a square three times each year, to record habitat and, in two transects, note bird species directly by sight or sound. Further details are recorded to help locate species (see http://www.bto.org/birdtrends/bbs.htm). The advantage of this scheme is that its sheer size means that the results are likely to be some of the most accurate obtainable. This can only really be justified with a very large unpaid workforce and so the amateur interest in ornithology is vital here. Despite the accuracy of the work there are problems. For example in 2001, many of the study areas were unvisited due to foot and mouth disease restrictions.

(b) Timed species counts

A simple technique usually involving noting every bird seen/heard within a time period. Pomeroy and Dranzoa (1997) compare this with a transect count, noting that the timed count is often as effective and usually more cost-effective to run. Providing the area is located accurately then some useful community-level data can be found. One obvious problem here is that nocturnal and marine species are less suitable.

(c) Spot mapping

The use of repeated surveys of measured plots. During each visit, the species under investigation is noted along with specific characteristics. This is regarded as one of the most accurate survey methods but it is more suited to finding individual territories for specific species. In a study of birds in India, Shankar-Raman (2003) compared line transects with spot maps, noting that the former is better for density but transects for distribution.

(d) Seasonal mapping

If the study involves a smaller area and the aim is to see distribution within/between ecosystems then a series of surveys needs to be considered on a longer-term, seasonal basis. A study of riparian areas in Australia by Woinarski *et al.* (2000) showed that distribution was highly correlated with ecosystem. This means that the common grid system used to show presence/absence might record the species in a particular 10 × 10 km square (such as is seen in the UK) but that this does not necessarily coincide with the actual location.

(e) Predictive models

Most of the methods mentioned above are based on the actual presence of the bird. If bird species are correlated with specific ecosystems then there is no reason why we could not predict the distribution by finding the right ecosystem. Although it sounds easy there are problems as Vernier *et al.* (2004) point out. The rise of GIS and the use of high-quality satellite data mean that this is one area likely to be pursued further.

Birds represent another example of distribution methods and associated problems. The ethology of birds contributes to the difficulties faced and it is only through the use of large-scale volunteer activity that much of the recording work is done.

8.4.3 Whales

Whales provide the third example. Like the Asian elephant they tend to be both elusive and rare; like birds they move in a medium and in locations that make global study difficult. Common methods seem to be the following.

(a) Sound surveillance

Whales give out distinctive calls. These can be tracked and, if using hydrophones (a series of microphones in an array) an idea of location can be gained. The University of Rhodes program (http://omp.gso.uri.edu/dosits/people/resrchxp/5.htm – accessed 17/3/04) is one such example. In addition, it is possible to obtain other data, for example swimming speed and respiration rate. The advantage is that it is possible to track individuals or small groups for weeks (as well as using the method for other species). However, groups are not the entire population and so this is an intensive activity.

(b) Opportunistic surveys

Whales are not predictable in the sense that their location is obvious. Many data can be gathered by being in the right place at the right time. This, coupled with data records such as photographs (many parts are unique to that individual), means that it is possible to track some whales for a considerable distance (even if the data only appear in random places where human and whale meet).

(c) Line transects

Although normally used in terrestrial vegetation analyses, a transect can be used with the difference being that the line is followed by a vessel and position is organised by a global positioning system. This more systematic way might yield better data but for a smaller area with greater cost (Reid *et al.* 2003). Aerial surveys follow the same procedure but can cover a wider area (Mobley 2001).

(d) Telemetry

It is possible to fit a radio tracking device but again, only on an individual basis. This gives data on routes, territories and behaviour (Culik 2003).

This final example shows another range of opportunities and problems associated with finding animal distribution. Readers should use this information to cast a critical eye over maps which suggest that distribution is an easy matter to elucidate.

8.5 Summary

The emphasis in biogeography has been on plants partly because of the ease of study and the ways in which this can be communicated to the public. Zoogeography, by virtue of the inherent mobility of its subject matter, creates more problems but since flora and fauna are intimately related it is important to understand the distributions of animals. One of the key problems with zoogeography is the defining of the concept of distribution. Although animal censuses are common it frequently takes little for an animal to be recorded in an area: often distribution ranges are little more than empty space through which an animal has been known to pass or near which the animal is supposed to live.

To this must be added the complex range of factors which influence animal distributions. Behaviour or ethology is a key factor in the short-range distribution: animals usually have a personal territory or set of runs determined by long-term useage patterns. In the larger scale, ecosystem factors will be important whereas at the largest scale, time and geological barriers combine to define the evolutionary potential of species.

Having accounted for the range of factors one needs to decide upon a sampling strategy and a mapping technique. As numerous studies have shown, this is far from a simple task with each method having its own advantages and disadvantages. The spatial and temporal range of animals is greater than that for plants and so we need to see which system will best record the actual distribution. We might divide this into track and presence methodologies, but as we see with both the overviews and the case studies there are difficulties to be overcome.

We leave this chapter with some understanding of the problems and prospects for animal biogeography and look towards a more detailed investigation of the physical factors that help create these patterns.

APPLICATIONS – USING BIOGEOGRAPHY

Using ideas from this chapter in real-life situations

Zoogeography is always going to be less easy than plant biogeography even if it is just because the subjects move around! However, there is a great need to find out about animal movements and none more so than the fishing industry. One such project is the Pacific Ocean Salmon Tracking (POST) research of the Census for Marine Life (http://www.coml.org/descrip/post.htm). The research programme uses the latest tracking technology to follow the salmon. Two different techniques can monitor the species on a long-term basis. Two tags are being trialled: a passive tag returning data when (if?) the fish is caught and an acoustic tag whose sound can be tracked continually. For certain groups of shelf-dwelling salmon this gives considerable detail about their life histories. Questions such as migration routes, migration behaviour, responses to changing ocean conditions and life-cycle statistics are just some of the data that are being gathered. The importance of this for biogeography is twofold. Firstly, ocean species are under-researched compared to terrestrial ones and this will help gather vital data (the system at full stretch will be able to track over 250,000 individuals!). Secondly, the data gathered will enable sustainable fishing programmes to be introduced and thus stop the excesses that have caused so many problems in the North Sea and North Atlantic.

Review questions

1. Outline the concept of distribution. To what extent does scale and resolution affect our understanding?

2. Describe the importance of ethology on distribution studies.

3. How can evolution alter our understanding of species distributions?

4. What is an environmental gradient and how does it affect distribution patterns?

5. Describe the types of barriers faced by migrating species. Give examples from a range of taxa.

6. Does the red squirrel alter our understanding of adaptation? How?

7. Analyse the problems associated with data gathering techniques.

8. To what extent can research trade-offs affect our results?

9. What is the value of standardisation?

10. For a range of taxa outline, and justify, the most appropriate distribution mapping methodologies.

11. For a named species critically evaluate, in detail, its distribution pattern including the ecology and ethology of the species and the data-gathering techniques used.

Selected readings

The aim of getting a wide review of animal distribution is rare because there have been few zoogeography texts published in the last 10 years. However, Newton's (2003) text does have a considerable amount of information but is restricted to birds. Sutherland (1996a) contains a great deal of discussion about sampling techniques. Although dated, Sukumar (1989) looks at distribution in a more classical sense with a wide-ranging discussion of changes in distribution through time.

References

Andrews P. 1995. Mammals as palaeoecological indicators. *Acta Zoologica Cracoviensis*, **38**(1): 59–72.

Ausden M. 1996. Invertebrates. In Sutherland WJ. (ed.) *Ecological Census Techniques*. Cambridge University Press.

Barnes RFW and Dunn A. 2002. Estimating forest density in Sapo National Park (Liberia) with a rainfall model. *African Journal of Ecology*, **40**(2): 159.

Brook M and Birkhead T. (eds)1991. *The Cambridge Encyclopaedia of Ornithology*. Cambridge University Press.

Brown JH and Lomolino MV. 1998. *Biogeography*, 2nd edn. Sinauer.

BTO. 2004. About surveys. http://www.bto.org/survey/index.htm (accessed 16/3/04).

Caviedes CN and Iriarte WA. 1989. Migration and distribution of rodents in central Chile since the Pleistocene: the palaeogeographic evidence. *Journal of Biogeography*, **16**: 181–7.

Colwell RK and Coddington JA. 1994. Estimating terrestrial biodiversity through extrapolation. *Philosophical Transactions of the Royal Society of London B Biological Sciences*, **345**(1311): 101–18.

Culik B. 2003. Review on small Cetaceans. http://www.wcmc.org.uk/cms/reports/small_cetaceans /index.htm (accessed 16/3/04).

Eggert LS, Eggert JA and Woodruff DS. 2003. Estimating population sizes for elusive animals: the forest elephants of Kakum National Park, Ghana. *Molecular Ecology*, **12**(6): 1389.

Fischer SF, Poschlod P and Beinlich B. 1996. Experimental studies on the dispersal of plants and animals on sheep in calcareous grasslands. *J. Applied Ecology*, **33**(5): 1206–22.

Ganderton P. 1981. Castle Acre Priory – environmental analysis. In Wilcox R. *Castle Acre Priory*, Norfolk Archaeological Society.

Garrevoet T. 1997. The investigation of clearwings using pheromones. *Levende Natuur*, **98**(5): 180–3.

Gibbons DW, Hill DA and Sutherland WJ. 1996. Birds. In Sutherland WJ. (ed.) *Ecological Census Techniques*. Cambridge University Press.

Hill GJE and Kelly GD. 1987. Habitat mapping by Landsat for aerial census of kangaroos. *Remote Sensing of Environment*, **21**(1): 53–60.

Johnston BK, Ager AA, Findholt SL, Wisdom MJ, Marx DB, Kern JW and Bryant LD. 1998. Mitigating spatial differences in observation rate of automated telemetry. *Journal of Wildlife Management*, **62**(3): 958–67.

Johnston CA. 1998. *Geographic Information Systems in Ecology*. Blackwell Science.

Kennelly SJ and Underwood AJ. 1985. Sampling of small invertebrates on natural hard strata in a sublittoral kelp forest. *Journal of Experimental Mariner Biology and Ecology*, **89**: 55–67.

Komers PE and Brotherton PNM. 1997. Dung pellets used to identify the distribution and density of dik-dik. *African Journal of Ecology*, **35**(2): 124–32.

Kuhn T. 1962. *The Structure of Scientific Revolutions*. University of Chicago Press.

Laing SE, Buckland ST, Burn RW, Lambie D and Amphlett A. 2003. Dung and nest surveys: estimating decay rates. *Journal of Applied Ecology*, **40**(6): 1102–11.

Lawton JH *et al.* 1998. Biodiversity inventories, indicator taxa and effects of habitat modification in tropical forest. *Nature*, **391**(6662): 72–6.

Macauley EDM, Satchel GM and Taylor LR. 1988. The Rothampstead Insect Survey 12m suction trap. *Bulletin of Entomological Research*, **78**(1): 121–30.

McClean SA, Rumble MA, King RM and Baker WL. 1998. Evaluation of resource selection methods with different definitions of availability. *Journal of Wildlife Management*, **62**(2): 793–801.

Mason CF and Macdonald SM. 1987. The use of sprajnts for surveying Otter *Lutra-lutra* populations: an evaluation. *Biological Conservation*, **41**(3): 167–78.

Mobley J Jr. 2001. Results of the 2001 aerial surveys of humpback whales. North of Kauai. http://npal.ucsd.edu/2001_Report/2001_NPAL_Report.html (accessed 16/3/04).

Newton I. 2003. *The Speciation and Biogeography of Birds*. Academic Press.

Odiyo PO. 1987. The use of biogeographical techniques in the study of migrant noctuid moths. *Insect Science and Its Applications*, **16**: 26–42.

Perrow MR, Côté IM and Evans M. 1996. Fish. In Sutherland WJ. (ed.) *Ecological Census Techniques*. Cambridge University Press.

Pomeroy D and Dranzoa C. 1997. Methods of studying the distribution, diversity and abundance of birds in East Africa – some quantitative approaches. *African Journal of Ecology*, **35**(2): 110.

Prescott DRC and Middleton ALA. 1990. Age and sex differences in winter distributions of American goldfinches in eastern North America. *Ornis Scandinavica*, **21**: 99.

Réale D, McAdam AG, Boutin S and Berteaux D. 2002. Genetic and plastic responses of a northern mammal to climate change. *Proceedings of the Royal Society B*. Online as DOI 10.1098/rspb.2002.2224.

Reid JB, Evans PGH and Northridge SP. 2003. *Atlas of Cetacean Distribution in North-West European Waters*. Joint Nature Conservation Committee.

Resh VH, Hildrew AG, Statzner B and Townsend CR. 1994. Theoretical habitat templets, species traits, and species richness: a synthesis of long-term ecological research on the Upper Rhône River in the context of concurrently developed ecological theory. *Freshwater Biology*, **31**(3): 539–54.

Rodwell JS. (ed.) 1992. *British Plant Communities*, 5 vols. Cambridge: CUP

Rogers SI, Rijnsdorp AD, Damm U and Vanhee W. 1998. Demersal fish populations in the coastal waters of the UK and continental NW Europe from beam trawl survey data collected from 1990 to 1995. *Journal of Sea Research*, **39**(1–2): 79–102.

Sabates A, Gili JM and Pages F. 1989. Relationship between zooplankton distribution, geographic characteristics and hydrographic patterns off the Catalan Coast. *Marine Biology*.

Schroeder CJ, Scott DB and Medioli FS. 1987. Can smaller Foraminifera be ignored in palaeoenvironmental analyses? *Journal of Foraminiferal Research*, **17**(2): 101–5.

Shankar-Raman TR. 2003. Assessment of census techniques for interspecific comparisons of tropical rainforest bird densities: a field evaluation in the Western Ghats, India. *Ibis*, **145**(1): 9.

Short J, Caughey G, Grice D and Brown B. 1988. The distribution and relative abundance of camels in Australia. *Journal of Arid Environments*, **15**(1): 91–8.

Siliva M, Tavareas J and Vieira V. 1995. Seasonal distribution and sex ratio of five noctuid moths (Insecta, Lepidiptera) captured in backlight traps on San Miguel – Azores. *Boletim do Museu Municipal do Funchal*, Suppl. 4: 681–91.

Snow NB, Cross WE, Green RH and Bunch JN. 1987. The biological setting of the Bios site at Cape Hatt Northwest Territories Canada including the sampling design methodology and baseline results for macrobenthos. *Arctic*, **40** (Suppl. 1): 80–99.

Sukumar R. 1989. *The Asian Elephant*. Cambridge University Press.

Sutherland WJ. (ed.) 1996a. *Ecological Census Techniques*. Cambridge University Press.

Sutherland WJ. 1996b. Mammals. In Sutherland WJ. (ed.) *Ecological Census Techniques*. Cambridge University Press.

Tsujino R and Yumoto T. 2004. Effects of sika deer on tree seedlings in a warm temperate forest on Yakushima Island, Japan. *Ecological Research*, **19**(3): 291.

Vernier LA, Pearce J, McKee JE, McKenny DW and Neimi GJ. 2004. Climate and satellite-derived land cover for predicting breeding bird distribution in the Great Lakes Basin. *Journal of Biogeography*, **31**(2): 315.

White G. 1978. *The Natural History of Selborne* (first published 1789). Shepheard-Walwyn.

Whitehouse AM, Hall-Martin AJ and Knight MH. 2001. A comparison of methods used to count the elephant population of the Addo elephant National Park, South Africa. *African Journal of Ecology*, **39**(2): 140.

Wilkins ME. 1986. Development and evaluation of methodologies for assessing and monitoring the abundance of widow rockfish *Sebates entomelas*. *US Marine Fisheries Service Bulletin*, **84**(2): 287–310.

Winner BL and Mcmichael RH Jnr. 1997. Evaluation of a new type of box splitter designed for subsampling estuarine ichthyofauna. *Transactions of the American Fisheries Societies*, **126**(6): 1041–7.

Woinarski JCZ, Brock C, Armstrong M, Hempel C, Cheal D and Brennan K. 2000. Bird distribution in riparian vegetation in the extensive natural landscape of Australia's tropical savannah: a broad-scale survey and analysis of a distributional data base. *Journal of Biogeography*, **27**(4): 843.

Zhang YZ and Zhang RZ. 1995. The prospective of zoogeographical study in China – a discussion on methodology. *Acta Zoologica Sinica*, **41**(1): 21–7.

Zhao T. 1992. Advances in biogeography. *Entomotaxonomia*, **14**(1): 35–47.

Websites

This can be a difficult area with some sites full of data and little discussion and vice versa. Most nations have a national bird census but the British Trust for Ornithology (http://www.bto.org) is one of the better places to start. In terms of distribution there are, as yet, no global organisations dealing exclusively with this, but the Animal Diversity Web (http://animal diversity.ummz.umich.edu/site/index.html) is a good place to start. For marine organisms, try the new IOBIS site (http://www.iobis.org/).

CHAPTER 9

Responding to conditions: environmental factors and gradients

Key points

- To understand the impact of environmental factors on species distribution we first need to appreciate the nature of the community;
- Environmental factors affect the distribution and abundance of all species;
- These factors are attributes which can vary in space and/or over periods of time;
- Although any physical factor can vary, in practice only a few key ones are significant in biogeography at the global and regional scale;
- Environmental factors rarely act singly and as a consequence, it is often difficult to distinguish which are the most important in any particular habitat;
- Environmental gradients occur in a wide range of situations in almost every type of habitat;
- Some gradients may be straightforward to interpret, such as changing light levels in woodlands or variations in soil moisture along a transect leading down to a pond;
- Other gradients may reflect more complex matters such as varying pH or humus content, or the presence of toxic materials in the soil;
- There is a range of techniques that can be used to analyse gradients;
- Human activities can modify the normal effects of environmental factors or gradients.

9.1 Introduction

Why is *that* plant *there*? A simple question seemingly, and yet one which lies at the heart of much of the work we do in mapping distributions. It is there because it is able to survive in such conditions. If it can also reproduce then we would consider that location to be part of the *range* (geographical area) of the species (see e.g. Breckle 2002, Archibold 1995). If we add time to this scenario then changes in either conditions or the species make this a far more variable setting. For example, conditions may change so much that the species cannot use that area or alternatively the species might change, making the conditions in the area unsuited for the 'new' populations. Our focus in this chapter is changing conditions, although to do this we must first appreciate the way in which species and communities respond.

Species distribution varies because physical environmental conditions vary. The aim of this chapter is to look at these changes in conditions.

The first element is the community and how it changes. Once we appreciate the biological responses we can see what the physical conditions can do. Next, we look at the key factors in gradients using examples to illustrate these ideas. Techniques used in gradient analysis are discussed and finally, human impact on gradients is outlined.

9.2 The nature of the community

Any investigation of organisms and their environmental relationships will have to concentrate on those factors which, as a result of careful observation, appear to define the composition of the plant or animal community. The local distribution of *Cassiope hypnoides* (Plate 10) appears to be controlled by one major factor, the availability of water from melting snow, since it is usually only found in the immediate vicinity of late-persisting snow patches throughout its geographical range. It is often easier to concentrate on the plant-environmental factors since plants tend to remain in one place whereas animals can move from one habitat to another if the former is less than ideal. One should also be aware of the problems caused by extrapolating from one set of measurements to account for the distribution of a species. The chosen site may not be typical in some way and this is where experience and observation are so important. Species show differing degrees of tolerance to factors such as soil pH or the availability of elements such as calcium throughout their geographical range.

Before looking at these factors in more detail it is important to consider how a community responds to environmental conditions. The change from one set of species to another is rarely simple because each species (and indeed each member of the population) will have its own responses to conditions. The important point here is that once we see how communities can respond then we are better able to distinguish between an environmental condition and a natural property of community interaction.

Although we are finding the situation more complex than originally thought, the fundamental concept for any organism is that of *tolerance*. Imagine an organism and its physical situation. For any given set of environmental conditions (e.g. light, temperature, exposure) there is a maximum and a minimum value that permits survival. Outside that range the organism will either die or fail to reproduce. It follows that there is a middle ground which represents the optimum conditions with a suboptimal area outside that. Multiply that by all the factors and you have ideal and suboptimal conditions producing a range within which the organisms can be found (although it is not that straightforward (see Box 9.1)).

BOX 9.1

Changes in tolerance

Black bearberry (*Arctostaphylos alpina*) shows a variable tolerance for calcium throughout its range (Figure 9.1). It is a trailing shrub found in mountainous areas throughout Europe where it grows well on thin (skeletal) soils developed over both calcareous and non-calcareous rocks, appearing to be indifferent as to the amount of calcium or humus in the soil. It also occurs quite widely in mountainous areas in the north-western part of Scotland, where it is restricted to exposed sites, growing on thin, acid humus or siliceous rock debris, and more widely in northern Scandinavia on similar sites with little or no available calcium. It is important to note that in Scandinavia, it frequently occurs on mountains ▶

Figure 9.1 *Arctostaphylos alpina*, Abisko, Sweden.

where the underlying rock may have exposed intrusions of basic rock containing some calcium. The bearberry grows well on the acidic substrates but poorly or not at all on the calcareous ones. The Scottish and northern Scandinavian sites represent the extreme of European distribution of this species. Another example is the daisy (*Bellis perennis*) which grows very commonly on almost any soil in lowland areas but which appears to show an increasing preference for calcium-rich soils as the altitude increases above 1,000 m. At such altitudes, daisies are never found on peaty or other substrates with a low pH even though the habitat seems otherwise quite suitable.

This situation is simple at the single-species population level but obviously becomes more complex as we add more species each with their own tolerances. This suggests that rather than ecosystems we see a mosaic of smaller units (e.g. Forman 1995). However, another school of thought has arisen which is of more use here. The assumption is that rather than have sharp boundaries between ecosystems we have a complex continuum of populations (Krebs 2001). Thus one population would grade into another (with the 'boundary' presumably being where the 50 per cent limit was reached). The value of this school of thought to us in this chapter is that it gave rise to *gradient analysis* – the study of vegetation changes in response to changing environmental conditions.

9.3 Gradient factors

An environmental factor is one of those physical or chemical attributes that has an effect on plant or animal life processes at a given place. The effect of a particular factor on living organisms may change according to variables such as time of day or season, to the physical situation of a site or its latitude (see e.g. Beniston 2000), or to the chemical and physical composition of the local environment. Environmental factors can operate at a variety of scales and their effects on plant or animal life can be quite specific. Many environmental factors are physical in origin, such as the amount of solar radiation which reaches a particular habitat, or the grain size of soil particles, while others arise as a result of the activities of living organisms, such as the amount of shade in woodland as a result of the development of the tree canopy. Others are chemical such as the availability of plant nutrients on a slope. Some factors are critical to the continuation of life in a particular habitat, for example water availability in dry (xeric) habitats or the amount and spectral composition of light in deep water habitats, while others only become significant when the organism is at the limits of its natural distribution. Finally, we must recognise that scale is also a key factor especially when species distribution can vary with scale (Potapova and Charles 2002).

In this section we look briefly at some of the key environmental factors – those that appear to be most significant in species distribution.

9.3.1 Lithosphere

As might be expected, rock type has a major effect upon the topography, and present-day landscapes reflect this dependence. Earth movements and erosive forces such as glaciation or wind erosion have produced a wide range of regional and local habitat types, each with particular climatic features. At the smallest scale, the presence of a rock or hummock will have a significant effect on its local (micro) climate by reducing wind speed and hence water loss from plants by transpiration, or by providing shelter from direct sunlight or rainfall. In extreme environments, such as arid or periglacial landscapes, the importance of such landscape elements will have considerable effects upon the local flora and fauna, permitting some species to colonise and survive in areas that would otherwise be too extreme (see Chernov 1985).

Local geology is also important, particularly as far as the effects of weathering on exposed rock surfaces and the development of soils are concerned. If the rock from which soil has developed contains significant amounts of bases such as calcium (as in the case of chalk or limestone), the plant communities which develop will usually have a greater biodiversity than those which have developed from the breakdown of rocks such as sandstones or granite. In areas where arable farming is possible, soils with a reasonable amount of calcium available tend to be preferred for crop growing whereas the poorer soils tend to be left as rough grazing or wasteland. It is sometimes possible to improve the quality of the grass in such areas by the application of calcium in the form of ground limestone, but the effects are often short-lived and uneconomic in the longer term as well as running the risk of destroying the ecosystem (see human impact, below).

In some parts of the world, outcrops of economically important minerals may influence the vegetation that grows in close proximity to them. The concentrations of essential elements (such as copper, magnesium or zinc) in soils is usually very low, of the order of a few micrograms per gram of soil, and in many cases excessive amounts may be toxic to the plants and, by bioaccumulation, to animals which feed upon them (see Figure 9.2). Other plants appear to be able to tolerate very high amounts of potentially toxic metals in the soil which inhibit the growth of non-tolerant species (see Box 9.2). This is particularly well shown in Kenya, East Africa, where the herb *Satureja abyssinica* occurs on soils in which the mineral chromite is found (Odhiambo and Howarth 2002). This association between plant species and particular heavy metals has been used as a means of identifying deposits for exploitation of metal ores, such as copper in Zambia or zinc in Europe where a species of pansy (*Viola calaminaria*) appears to have its distribution restricted to sites in which zinc concentrations are significantly higher than normal. The use of plants as indicators is a well-established part of geobotany and has been successfully used for a number of years in metal prospecting in Africa and North America. Some of these tolerant species accumulate the toxic metal in their tissues and animals that graze on them can suffer ill effects. If domesticated animals such as cattle or goats are involved, the toxic minerals are bioaccumulated and passed on to the consumers who may also show symptoms of intoxication.

Figure 9.2 *Minuartia verna* is frequently associated with sites where the soil contains significant quantities of heavy metals such as lead, nickel or chromium

Soils are controlled not just by subsurface geology (i.e. parent material) but also by slope angle and drainage. A typical example is the English chalk downlands such as those in Hampshire. Here, a clay cap on the hilltops gives a good clay rendzina; the slopes are a shallower undeveloped soil whereas the valley floor is often a richer alluvial soil possibly with poorer drainage. Such a sequence is referred to as a *catena*.

BOX 9.2

The impact of heavy metals on plant distribution

Copper is an essential element for many species of animal since it is important in the development of an effective immune system. Sheep that graze on pastures and moorland areas where soils contain higher than usual quantities of copper are at risk of copper poisoning because they accumulate copper very efficiently in their livers. Serious effects have been noted with levels of as little as 10 micrograms in grass, while cattle are much less sensitive, only displaying toxic effects with more than 100 micrograms per gram in their diet. Normal levels of copper in soils would be between 1 and 2 micrograms, but some soils around old mine workings can have more than 1,500–3,000 micrograms per gram of copper on slag heaps or mineral crushing areas. Copper is also important as a fungicide and has been used to control potato blight (*Phytophthora*) in potatoes, or powdery mildews on grapes. Regular spraying of copper fungicides in vineyards leads to elevated levels of copper in the soil and loss of ground cover vegetation. On slopes, pollution gradients can develop with potentially damaging results if copper leaches into water supplies and streams.

The concept of environmental gradients is well illustrated by the vegetation zonation associated with an exposure of serpentine (an ultrabasic rock) in the Jotunheimen region of Norway. Weathering of serpentine yields considerable quantities of the toxic metals chromium and nickel, both of which can have a significant and damaging effect on the plant and animal life of such areas. The exposed rock and thin surrounding soils bore very little vegetation apart from the alpine catchfly (*Lychnis alpina*) and some stunted grasses; drainage downslope was marked by an almost complete absence of plants apart from the grasses and a few lichens. Soil concentrations of chromium ranged from 430 micrograms per gram adjacent to the exposed rock, to between 380 and 210 in the downslope areas. Nickel concentrations were also extremely high at 275 micrograms in the soil, down to 124 in the downslope area. The localised nature of the effect was shown by analyses of soils along a horizontal transect. Soil chromium concentrations declined very rapidly over a distance of 1.75 m to less than 10 micrograms, and complete vegetation cover, mainly of crowberry, sedges and dwarf willow, was found within 2 m of the exposure, contrasting with the minimal plant cover on the metal-rich areas. *Lychnis* exhibits similar tolerance in a number of other metal-rich sites throughout its range, such as in the Lake District in England and the Grampian Highlands of Scotland. Rune (1953) has described such populations as being a separate subspecies (*serpentinicola*) of a plant which is quite common in mountainous sites throughout its range in Scandinavia and North America, growing on poor, acidic soils in exposed sites.

9.3.2 Atmosphere

Atmospheric gradients are more problematic than lithospheric ones primarily because of the nature of the medium. Many of the changes one can see are due to the hydrosphere or biosphere. For example, the decreasing amount of ground flora from the edge towards the centre of an area of dense forest is accounted for by the development of a thick canopy of branches and leaves which will restrict the amount of light reaching the forest floor. The relationship between light penetration and canopy or distance from the forest edge will vary according to tree density and species, but as a general rule, a dense broadleaf woodland canopy will let through less than 0.5 per cent of the solar radiation falling on the top of the canopy. The figure is higher in winter and spring, because of leaf fall, but dense branches and twigs can reduce light levels to 10 per cent or less.

However, day length and summer temperature (both controlled largely by atmospheric factors as well as location) can demonstrate clear gradients. A good example of this is the Australian Alps located on the Victoria/New South Wales border. Comprising less than 0.01 per cent of the land mass, it shows considerable variation in ecosystems towards the summits. Day length and maximum summer temperatures are commonly cited as key gradient factors controlling the tree line (maximum height above sea level for trees corresponding to a temperature of about 13 °C) and the growth of the alpine flora in the 'feldmark' of the high Alps (Specht and Specht 2000).

9.3.3 Hydrosphere

With both fresh water and oceans, the hydrosphere can highlight a number of key gradients. One current topic is the ocean halocline (distribution of salt concentrations) whose changes are being linked to global current patterns and thus air temperatures (Rahmstorf 2003).

An equally striking change is demonstrated in streams where two species of flatworm (*Polycelis felina* and *Crenobia alpina*) occur together. In lowland Britain, *Crenobia* occurs only in a few streams, fed by very cold, permanent and oxygen-rich spring water, while the much commoner *Polycelis* is less temperature-sensitive, occurs in most unpolluted streams and mixed *Polycelis* and *Crenobia* populations can develop around cold springs. As its specific name suggests, the environmental preference of *C. alpina* reflects its association with colder, alpine regions and it has been suggested that the relict populations of this species in lowland areas many represent a periglacial survival of some 8,000 years.

Zonation of animals and plants on a rocky seashore is another example of an environmental gradient which is determined primarily by the length of time that each section of the seashore is left high and dry by the tide. Species at the top of the shore, about high tide mark, will tend to be exposed for 20 or more hours a day, whereas those at the low tide mark may be left out of water from a few minutes or so. The former will have had to adapt to a dry regime whereas those on the lower part of the shore are often delicate species, readily damaged by drying out (Hayward 2004).

Finally, soil water content is one example of an important environmental factor whose availability can vary over time, as in seasonal fluctuations of water tables, or in space, such as the change in water content of soils along a transect across a coastal sand dune–dune slack system. Ranwell (1972) describes work done on sand dunes in the UK which demonstrates that soil moisture levels in these habitats tend to stabilise about 1 m below the surface for the stabilised slacks in between the sand dunes, but that the moisture content of all but the smallest dunes is dependent not upon the combined effects of rainfall and ground water, but on rainwater alone since the capillary rise of water in dune sand is rarely more than 40 cm above the water table. Open dune species such as the marram grass (*Ammophila arenaria*) can root down to about 1 m and dune stabilisation by this species can be seriously affected if rainfall amounts are insufficient to keep it alive.

9.3.4 Scale

The effects of environmental factors will vary according to the scale of the habitat that is under consideration. On a regional scale, the major environmental factors include climate and topography since these will influence the type of vegetation and thus the animal communities which can develop. Topography is particularly important for the distribution of rainfall, particularly in regions where there are hills or mountains. In much of western Europe, the prevailing wind direction is from the south-west and moisture-bearing winds from this direction tend to deposit much of their load on high ground with west-facing slopes, for example, the Western Highlands of Scotland, or the Cambrian Mountains of Wales (see Figure 9.3). Precipitation occurs when moisture-laden winds are forced to rise as a result of the hilly topography – cooling takes place and some of the contained moisture condenses into raindrops and falls to the ground. On the eastern side of the hills, amounts of precipitation are reduced for some distance inland, and the effect is referred to as a 'rain shadow'. Topography is also important on a more local scale. The temperature differential between north- and south-facing slopes has already been mentioned, but topography has a significant effect on surface and subsurface water movement. This in turn will influence water

Figure 9.3 Cwm Rheidol, Wales. Ancient valley woodland in the uplands of central Wales with abundant epiphytes (mosses, ferns and lichens), reflecting the high levels of precipitation on these hills and the resultant humid microclimate. Lower down the same valley, in early spring, showing the open nature of the birch–oak woodland on the slopes and the intrusive but abandoned mine workings

availability for both plants and animals. Steeper slopes tend to have faster drainage than shallow ones, and gravity will, together with temperature-related effects such as the freezing and thawing of soil moisture, move soil particles and rock debris downslope. Such instability is seen in the form of screes or of landslips – such habitats are quite precarious for all except specialised plant and animal communities which have adapted to an intermittently mobile existence.

9.4 Gradients – case studies

Gradients are a common feature in almost all habitats. In this context, the gradient implies that there is a change in the quantity or effect of a particular factor, usually in space rather than over a period of time. Thus, under calm weather conditions, air temperature decreases by about 0.6 °C for every 100 m increase in altitude. Similarly, the amount and spectral composition of light penetrating the oceans depend upon the depth of water (and any suspended materials that may cause turbidity), since light is scattered by the particles that cause turbidity and the longer (red, orange and yellow) wavelengths of sunlight are more readily absorbed by water than the shorter, green, blue and violet wavelengths.

Some environmental gradients can be very severe. This is particularly well demonstrated by the communities of animals which live adjacent to the hydrothermal vents which are found at great depths in many of the world's oceans. The effect of water pressure at such depths is to raise the temperature at which water boils, and the sea water adjacent to the superheated water and gas vents may be at a temperature of at least 300° C with no visible life. Within a few centimetres distance, the environment has become much cooler and a diverse ecosystem has evolved, dependent upon chemosynthetic bacteria that utilise the chemical output of the vents or 'smokers' which are rich in sulphides and other minerals. The zone of viability is very small, and animals which stray too close to the vents will be instantly killed by the high temperature (Wharton 2002).

The following cases represent a small fraction of the total and have been chosen to highlight applications and ideas in gradients.

9.4.1 North Sea

Less extreme gradients can be found in locations such as the Gulf of Bothnia, which lies between Sweden and Finland, and has access to the North Sea, by a narrow channel between Denmark and the south coasts of Sweden and Norway. Sea water usually has a concentration of about 35 parts of dissolved salts per thousand but in the Gulf, there is a strong gradient from south to north which is particularly noticeable in late spring and summer when the salinity at the northern end can fall to as low as 6 parts per thousand as a result of dilution by the huge volumes of fresh water derived from the melting of snow and ice which flow into that area through a number of rivers. The animal life associated with the less saline, brackish waters will differ considerably from that of the southern end of the Gulf where the salinity is six times greater. Very few species, apart from migratory fish such as eels or salmon, could survive such a change because of its effects on the salts/water balance and osmotic equilibrium of the body.

9.4.2 Diatoms in the USA (Potapova and Charles 2002)

Diatoms are unicellular algae, one of the key sources of fish food. They also have other uses such as bioindicators (if a good enough set can be generated). In this study, the aim was to determine the factors which gave rise to diatom distribution at the continental scale. One of the key issues was the *reason* for the distribution. For example, if species A is in place B but not place C, is that because C has unsuitable characteristics (i.e. A might have reached C but been unable to withstand conditions or compete successfully) or is it because A has not spread to C yet? Alternatively, is it only in B because it is the better place or because better-equipped species D has yet to reach there? With larger species this question might be easier to answer, but the size of diatoms coupled with their importance prompted this research. In addition to theoretical considerations there were also practical ones. For example, not all diatom species are fully recorded or described taxonomically which clearly limits the accuracy of the data gathering. Once data were gathered (which involved over 2,700 samples over 5 years), a form of gradient analysis, ordination (see below), was applied. The results showed a complex relationship but there was at least some agreement for suggesting three gradients:

- location in river system – upstream/downstream
- water chemistry
- temperature – especially controlled by altitude/latitude

Although the work was far more complex than shown here it does illustrate the significance of gradients and that there is often more than one gradient involved.

9.4.3 Large-scale patterns in relation to latitude and climate (Qian *et al.* 2003)

This study uses an arctic–tropic sweep of eastern Asia in determining the factors that control vegetation patterns. The scope of this study is broader in species and greater in area than the previous one: results here are applicable at the global level. Data were taken from a wide range of literature resources and printed databases (highlighting the value of even the smallest effort in distribution analysis and taxonomy), i.e. a desk study. Rather than attempt work at the species level, it was decided to use genera. From this, genera were divided into three groups (cosmopolitan, temperate and tropical) which were then further subdivided into geographical elements (giving 45 regions overall). Just over 2,800 genera were reported upon. Using correlation techniques (see Chapter 10) it was possible to demonstrate that latitude exerts a very strong gradient upon generic distribution patterns.

Plate 1 Multipurpose land use in an alpine valley, Les Diablerets, Switzerland. This Swiss mountain valley was formerly extensively forested but a great deal of the area was cleared for pasture. More recently, the development of winter sports facilities have contributed to further removal of trees which has increased the risk of avalanches on the steep slope.

Plate 2 Solifluction and frost-heaving. The extreme environmental conditions frequently associated with mountainous areas make for a challenging environment in which freezing and thawing play their part in the breakdown of rocks and formation of soils. The dynamic nature of this particular environment is such that few plant species apart from lichens and mosses can survive.

Plate 3 Oakwood at its altitudinal limit in England. Oak forest is the climatic climax vegetation over much of southern England (and western Europe) with mature trees in ancient woodland reaching 15–20 m. Black Tor Copse on Dartmoor in SW England represents a typical development of oakwood near the local tree line (450 m), with mature oak trees, many years old and no more than 2–2.5 m high, growing among huge granite boulders where sheep grazing as well as the relentless post-medieval use of available timber for charcoal burning (to supply the tin smelters down the valley), have otherwise deforested the area. Local environmental conditions are extremely severe and strong winds and late frosts in particular combine to keep the trees well trimmed.

Plate 4 A species-rich chalk grassland sward at Didling Hill, West Sussex. Chalk grassland habitats are arguably among the most species-rich ecosystems in Europe, provided the establishment of coarse grasses or scrub is prevented either by grazing or mowing. This orchid (*Herminium monorchis*) is no more than 30–40 mm high and is only found on well-grazed areas. It cannot survive competition from taller-growing species, which will eventually exclude it from the community.

Plate 5 Mountain avens. *Dryas octopetala* is an arctic alpine tundra and limestone grassland species which has the ability to fix nitrogen, not by the usual route of bacteria in root nodules but because of a symbiotic association with an actinomycete fungus, *Frankia*, which is associated with its roots. This input of nitrogen is extremely important in an environment where very few other sources of fixed nitrogen are available.

Plate 6 A glacier chronosequence in Norway. As ecosystems develop and mature, their biodiversity usually increases. This photograph of the Storbreen glacier foreland in southern Norway shows the outermost (1750) moraine and by the river and several later moraines within the foreland. Up by the snout of the retreating glacier, there is no vegetation at all but within a few years of freedom from ice cover, mosses, algae and grasses begin to colonise the moraine. Diversity increases over time and the presence of datable features (the lichen, *Rhizocarpon*) enables biogeographers to relate species diversity to time since deglaciation. After 250 years, the oldest part of the foreland supports a wide range of dwarf shrubs and willows, and many species of moss and lichen.

Plate 7 St Dabeoc's heath. Some plant and animal species have very restricted and often patchy distributions which may represent the remnants of a formerly much more widespread occurrence. *Daboecia cantabrica* is found as a native species only in Connemara, Ireland, SW France and northern Spain, a considerable disjunction – and the closely related but smaller *D. azorica* is an endemic, confined to the Azores.

Plate 8 Dorset heath (*Erica ciliaris*) in Ireland. This species occurs on heathlands in SW Europe (Brittany, Gascony, Portugal and northern Spain) and has a restricted distribution in the British Isles (Dorset, Cornwall and western Ireland) where its status is increasingly endangered due to human activities such as changing land use, and accidental fires.

Plate 9 Irish fringed sandwort. Some plants, such as sandworts, are extremely adaptable and appear to thrive in exposed or quite inhospitable habitats such as dry rocks or sandy places. One species in particular, the fringed sandwort (*Arenaria ciliata*), and closely related species such as *A. norvegica* are found on dry, calcareous rocks and soils in mountain areas in the northern hemisphere. A solitary occurrence of *A. ciliata* in Ireland is of interest because the plant, found on wet rocks on one mountain, is considered to be an endemic subspecies. This plant, *A. ciliata* ssp. *hibernica*, has a different number of chromosomes from other, closely related species (where the diploid number is 40). The Irish plant is a variable polyploid with 40, 80 or 120 chromosomes. It is also polymorphic (highly variable in appearance). The genetic and morphological peculiarities seem to support the idea of a relict species which has been isolated for a long period of time (perhaps from a formerly much more extensive distribution), and where mutations have occurred and survived through lack of competition.

Plate 10 Arctic moss heather. The distribution of some plant species may be controlled by the presence or absence of a particular environmental factor such as calcium or soil pH. *Cassiope hypnoides* is a tiny, prostrate shrub whose distribution throughout its global range appears to be controlled by the availability of snow melt water during the early part of its growing season. It is rarely found outside tundra habitats with late-persisting snow patches.

Plate 11 Japanese mangrove community. Mangrove swamps are a tropic–subtropical marine-fringing biome of considerable importance in reducing the influence of coastal erosion. This biome has a high level of biodiversity and the swamps can extend over many hectares. The ability of the propagules to drop from the parent plant and germinate and grow in sea water is particularly important as is the ability of the mature tree to transfer oxygen down to the roots in the de-oxygenated marine deposits. As the trees mature, they produce prop roots from the branches which assist in stabilising the tree and trap sediments. This photograph illustrates one of the most northerly mangrove swamps in Japan, composed of *Kandelia* sp. Conversion of coastal mangrove forest in the Indian subcontinent to provide breeding grounds for prawns to satisfy the frozen seafood trade is a major problem and could be the trigger for increased erosion on the coastline of Bangladesh.

Plate 12 What not to do in a National Park! The Burren National Park is one of the smallest in Ireland and is situated in the limestone country in Co. Clare. It is particularly noted for its karst geology and botany because of the unique combination of a mild and moist climate, limestone and human activity. Here, a number of plant species of widely differing climatic and environmental requirements seem to be able to coexist almost at sea level. In particular, the alpine spring gentian *Gentiana verna*, the arctic–alpine mountain avens *Dryas octopetala* and the Mediterranean orchid *Neotinea intacta* are to be found growing within a few centimetres of each other in the same mossy turf. A controversial proposal by the Irish government to set up an Interpretation Centre in the middle of the wilderness almost succeeded but was eventually stopped by the intervention of the European Court of Justice. Unfortunately, the area is prone to temporary settlement by travellers and tourists in camper vans, with a potential threat to the visual landscape and natural history interest of the area.

Plate 13 Cap Gris Nez, Côte d'Opale, France. European maritime grasslands form a plagioclimax/deflected succession which, in spite of its tolerance of wind-blown salt spray, is nevertheless quite fragile and susceptible to heavy human activity. This picture of Cap Gris Nez in NW France was the site of some of the German anti-invasion 'Atlantic Wall' fortifications in the Second World War. Apart from the remains of the control post and coastal artillery sites, it is possible on a fine day to get a view of Britain, 45 km away on the other side of the English Channel. The site is quite popular with walkers, many people visit it for its military links and it now has the status of a protected landscape. Unfortunately, the geology of the area is such that substantial cliff falls have removed significant parts of the vegetation, paths, etc.

Plate 14 An abandoned mine. The Cwm Ystwyth copper mine site (near Aberystwyth in Wales) is one of a large number of small mines which were exploited during the eighteenth to the early twentieth century after which it became uneconomic to extract copper or other minerals. The toxic nature of the mine waste is clear from the different types of vegetation seen growing on the spoil heaps. The initial pollution-tolerant grasses and herbs growing on the spoil are succeeded in time by birch, and eventually oak woodland results after at least 150 years.

Plate 15 A national treasure, nearly lost! Bradfield Wood is a National Nature Reserve that nearly disappeared under the plough, when the UK Ministry of Agriculture was giving cash grants to farmers to convert what was then considered by the Ministry's advisers as being in 'old and poor condition, therefore useless, non-productive woodland' into fields to grow excess amounts of barley or sugar beet in the 1960s.

About 50 per cent of the woodland was lost before the Suffolk Nature Conservation Trust managed to buy the remainder and it is now properly managed on behalf of English Nature by a full-time forester, contractors and voluntary labour. It is undoubtedly an ancient woodland (more than 400 years old) since thirteenth-century records exist of timber provided for the construction of the medieval cathedral at Bury St Edmunds, which is near by. The wood is currently managed on a sustainable basis, with rotational coppicing and sale of woodland products such as fencing or charcoal.

The photo shows a typical cut section shortly after coppicing has taken place. In the wood, there are coppice stools, several metres in diameter which, when dated, turn out to be 500 or 600 years old. It is astonishingly rich in plant and animal species, some of which are very scarce outside this area.

It was totally inexcusable for the UK government's Ministry of Agriculture to connive, possibly through ignorance or worse, at the near destruction of such an important environmental resource by failing to check with the Nature Conservancy who were at that time the government's conservation advisers! **A nation that destroys its heritage has no future!**

Plate 16 An upland blanket bog. Bogs are repositories of much that happens in their vicinity – pollen, dead vegetation, twigs, fallen trees, dead animals and dead people (sometimes). Blanket bog is a vegetation type largely confined to temperate and cool temperate countries or areas adjacent to the Atlantic Ocean – so there are blanket bogs in north and west England, Wales, Ireland and over much of the Scottish highlands and islands (and the Falkland Islands in the South Atlantic), all acting as local time capsules. Sampling of the buried peat and examination of plant remains and associated pollen and spores can reveal a lot about the local environment, vegetation and land use in previous centuries. Information of this sort is of considerable interest to archaeologists and biogeographers and palaeoecologists.

The fact that the bog is continuously wet means that decay rates are slow in an anaerobic environment and even the most delicate structure can be retained and the sample identified. It is possible to work out the age of the material by carbon dating, and to devise an approximation of the vegetation type at a particular point in time – either *in situ* or close by.

The photograph shows a typical domestic turf cut from which the sods are piled up to dry for several months before use.

Within the British Isles, peat or 'turf' has been the fuel of many people for thousands of years because wood was unavailable, difficult to obtain or too expensive. More recently, the exploitation of milled peat for power station fuel has been tried by the Irish Electricity Generating Board (Bord na Mona) – a bad move since they have now drained and destroyed most of the best bog sites and eliminated the habitat for many rare plants. In the UK, and many parts of Europe, use of peat in gardening and horticulture has been an environmental disaster with many lowland bogs being destroyed over the past century. This is industrial-scale exploitation of a barely renewable resource. Use as a fuel by country people has caused some, but not serious, environmental damage.

Equally unwelcome was the draining of large areas of blanket bog in remote and wildlife-rich parts of the UK, for the development of forestry as a method of tax planning by some sections of society. The performance of many such plantations is quite poor and does not merit the destruction of unique wetlands in internationally recognised wildlife sites.

Figure 9.5(b) A major problem with abandoned mines concerns acid drainage water which contains heavy metals and solids in suspension. This example from Cwm Rheidol shows a precipitate of iron hydroxides (ochre): the water has a pH of about 2.5 which is far too acid for freshwater organisms to survive

Figure 9.5(c) The Rheidol mine drainage is diverted into settling tanks to get rid of suspended material prior to de-acidification by limestone chippings before reaching the river

9.5 Gradient analysis

From the cases mentioned above we can see that the responses to gradients can be seen at all scales in all species (to a greater or lesser extent). Often the results are complex so a simple statistical test or graph is unlikely to provide the answer. This section focuses on those methods which seek to make sense of the data.

Trying to establish which factors are likely to be of importance in any particular habitat or environment is often quite difficult. Measuring them – or even knowing how to measure them – is even more of a challenge (see Figure 9.6 (a) and (b)). From a theoretical standpoint, there is an almost infinite number of potential factors which could possibly be of importance in any particular habitat. The ideal state of affairs would be if all these factors could be measured or monitored on a continuous basis, but this would lead to an excessive amount of data and the expenditure of a lot of time and effort for a limited outcome. While it is desirable to have continuous data available, time and equipment constraints mean that most data are collected either as samples over short periods of time using automatic data recorders, or more commonly, as 'snapshots' at a particular time of day or season. While such data are not ideal, they do give some indication of the local environment and will enable conclusions to be drawn regarding species tolerances.

Figure 9.6(a) Ambersham Common in southern England. The transition from dry, ling heather (*Calluna*)-dominated heathland sloping down through wet heath and bog is reversed as the topography changes. The gradient here is related to the water content of the soils down the catena

Figure 9.6(b) Wet heath zonation. A *Sphagnum* moss, sundew (*Drosera rotundifolia*) and cross-leaved heath (*Erica tetralix*) wet heath community, also at Ambersham Common

Figure 9.7 A conspicuous and common lichen of boreal taiga and tundra, *Nephroma arcticum* would be expected to occur frequently in similar habitats in Scotland, but it is extremely rare even though the range of habitats look identical. Its global distribution is circumpolar and as far south as northern Japan. The most logical explanation for its rarity is that the precise microclimatic requirements for its establishment and long-term persistence are rarely encountered in Scotland or elsewhere in the UK

Before moving on to examine some of the techniques used it is worth considering the following parameters to our study of gradients:

- **Measurement**. Some factors can be measured or estimated in the field with a fair degree of accuracy, such as temperature, wind speed or soil depth, while others may be impracticable to measure in the field, such as the concentration of carbon dioxide in the soil atmosphere or the concentrations of particular nutrients in the soil, except as part of a larger programme;
- **Variation.** Some factors vary widely over small distances, depending upon the topography, geology or hydrology of a site, and there are often quite significant changes from one time of day to another according to the direction in which the site is facing. The environment of south-facing slopes in the northern hemisphere is almost always warmer and less humid than that of nearby north-facing slopes of a similar gradient and vegetation type (consider the impact this has on one example – *Nephroma arcticum*, Figure 9.7);
- **Interaction of factors.** Environmental factors can be very difficult to study individually under field conditions since they tend to interact in a multiplicity of ways. For example, Dahl (1951) has shown that in Scandinavia, the lower altitudinal limits of occurrence of many mountain plant species are closely correlated with the maximum summer temperature isotherms, and the critical temperature varies over a range of at least 7 °C for different montane species. The clear implication is that the lower limits of distribution of some species are directly controlled by one environmental factor, namely the highest summer temperature. Following on from this, in a climate which might be marginal for the development of a mountain flora, it is likely that some species would become rarer and more likely to be confined to higher-altitude sites than others which are more tolerant of higher temperatures;

- **Statistical requirements.** Although this is strictly part of analysis and not gradients it is worth mentioning at this point. Dale (1999) in his detailed examination of spatial analysis emphasises that methodological decisions will impact upon results and, therefore, interpretation. Among others, he notes the importance of sampling, quadrat size and line orientation (i.e. direction of sampling) when analysing gradients.

What techniques are available to us? Some have been mentioned above (and will be described further in Chapter 10). However, to demonstrate something of the range, we can summarise three of the more common ideas (taken from Dale 1999):

- **Continuous presence/absence.** Perhaps the simplest technique, this uses the location of species along a line measured either continuously or at preset point sampling areas. The value of this technique is its simplicity and the fact that it can be graphed giving a clear visual representation;
- **Overlap.** This takes the previous technique one stage further, looking at the number of species whose location overlaps. This can produce a mathematical representation of overlap which can be subject to significance testing. From this stage we can use increasingly sophisticated techniques to look at aspects of the nature of the boundary between species and any gaps in distribution;
- **Ordination.** This is one of a range of techniques very commonly used in many areas of ecology (and geography). The basic idea behind ordination is to plot a series of data points using more than two axes. Where there is a better fit for data, the axes are more significant (see Websites, below).

9.6 Human impact on gradients

Human activity is responsible for the accentuation of many changes in environmental factors and might be described as an overriding factor in many parts of the world. Human activities and population growth are responsible for so many environmental problems whose consequences threaten the future of the human race. (One example of this is the 'ecological footprint' – the area needed to support human existence – see Sanderson *et al.* 2002.) These include the effects of global warming which have meant that many species of plants and animals which are dependent upon cold mountain habitats for survival at the fringes of their current distribution will now be lost as global temperatures rise. Similarly, distributional limits of some species will move northwards and this may have consequences for agriculture and pest control (for pests, see also Fobil 2002).

9.7 Summary

The study of environmental factors and gradients is both fascinating and challenging, partly because of the interactions which occur and also because it provides valuable insights into the adaptability of many species of plants and animals to life under a wide range of conditions. From this chapter you will have been able to appreciate that because of the potential interactions, it is rarely possible to identify which one or two factors most characterise a particular type of environment. The best way to study environmental factors and gradients is through long-term observations and analysis of the behaviour or abundance of particular species in different areas. Species which have a low level of adaptability to changing environments will invariably be lost from habitats as they undergo successional changes, or as a result of climatic change or land management.

APPLICATIONS – USING BIOGEOGRAPHY

Using ideas from this chapter in real-life situations

A gradient is a change in conditions between two places. If these differences can be discovered and mapped then it is possible to find out more about how the earth works. There have been numerous studies, but one area which has been drawing more attention of late is the distribution of physical and chemical features in the ocean. One example of this is thermohaline circulation. Differences in salinity create differences in density which in turn can drive the vast ocean currents. The most significant broad-scale current is called the ocean conveyor belt – a link of currents, both surface and deep-water, which drives nutrient and energy distribution. One example of this can be seen in the Antarctic Ocean where deep upwelling of the belt leads to an abundance of food supplies for the krill upon which the marine ecosystem depends. Recent research has shown that this pattern is not fixed. Changes in ocean temperature have been recorded and there is talk that this could alter the position of major currents, for example the Gulf Stream which keeps the UK in relatively mild winters. If this should happen then there are predictions of major climate changes along with widespread loss of species. The bottom line here is that by studying such changes we can better inform ourselves about the consequences of our actions.

Review questions

1. Describe how you might attempt to investigate the factors which control the distribution and abundance of an invasive species such as the seaweed *Sargassum* which is beginning to assume pest proportions on some parts of the European coastline.

2. Modern conservation practice suggests linking fragmented habitats with wildlife corridors. What is the rationale behind this idea and what environmental factors should be considered in the planning of such corridors?

Selected readings

There are a range of texts dealing with physical factors but there are very few focusing specifically upon gradients. One of the better new ones is Breckle (2002), while Archibold (1995) is a comprehensive look at the global situation.

References

Archibold OW. 1995. *Ecology of World Vegetation*. Chapman & Hall.

Beniston M. 2000. *Environmental Change in Mountains and Uplands*. Arnold. (Chapter 2, Characterisation of mountain environments).

Bradshaw AD, Humphreys MO and Johnson MS. 1978. The value of heavy metal tolerance in the re-vegetation of metalliferous mine wastes. In Goodman GT and Chadwick MJ. *Environmental Management of Mineral Wastes*, pp. 311–34. Alphen aan den Rijn, The Netherlands: Sijthoff and Nordhoff.

Breckle S-W. 2002. *Walter's Vegetation of the Earth*, 4th edn. Stuttgart: Springer.

Chernov YI. 1985. *The Living Tundra*. Cambridge: CUP.

Dahl E. 1951. On the relation between summer temperature and the distribution of vascular plants in the lowlands of Fennoscandia. *Oikos*, **3**: 22–52.

Dale MRT. 1999. *Spatial Pattern Analysis in Plant Ecology*. Cambridge: CUP.

Fobil JN. 2002. Remediation of the environmental impacts of the Akosombo and Kpong dams. Horizons Solution site. http://www.solutions-site.org/artman/publish/article_53.shtml (Accessed 5/8/04).

Forman R. 1995. *Land Mosaics*. Cambridge: CUP.

Funk DW, Noel LE and Freedman AH. 2004. Environmental gradients, plant distribution and species richness in arctic salt marsh near Prudhoe Bay, Canada. *Wetlands Ecology and Management*, **12**: 212–33.

Grime JP. 1979. *Plant Strategies and Vegetation Processes*. Chichester: John Wiley & Sons, Ltd.

Handley JF. 1986. Landscape under stress. In Bradshaw AD, Goode DA and Thorp E, *Ecology and Design in Landscape*. Blackwell Scientific Publications.

Hayward PJ. 2004. *Seashore*. London: Collins.

Krebs CJ. 2001. *Ecology: The Experimental Analysis of Distribution and Abundance*, 5th edn. HarperCollins College Publishers.

Luo T, Pan Y, Ouyang H, Shi P, Luo J, Yu Z and Lu Q. 2004. Leaf area index and net primary productivity along sub-tropical to alpine gradients in the Tibetan plateau. *Global Ecology and Biogeography*, **13**: 345–58.

Odhiambo BO and Howarth PJ. 2002. Chromium concentrations in vegetation samples from West Pokot District, Kenya. *Environmental Geochemistry and Health*, **24**: 111–22.

Potapova MG and Charles DF. 2002. Benthic diatoms in USA rivers: distributions along spatial and environmental gradients. *Journal of Biogeography*, **29:** 167–87.

Qian H, Song JS, Krestov P, Guo Q-F, Wu Z-M, Shen X-S and Guo X-S. 2003. Large scale phytogeographical patterns in East Asia in relation to latitudinal and climatic gradients. *Journal of Biogeography*, **30**: 129–41.

Rahmstorf S. 2003. Thermohaline circulation: the current climate. *Nature*, **421**: 699–701.

Ranwell DS. 1972. *Ecology of Saltmarshes and Sand Dunes*. Chapman & Hall.

Rune O. 1953. *Serpentine Vegetation*. Acta Phytogeographica Suecica No. 31, Stockholm.

Sanderson EW *et al.* 2002. The human footprint or the last of the wild. *Biosciences*, **52**(10): 891–904.

Specht RL and Specht A. 2000. *Australian Plant Communities*, Oxford: OUP.

Wharton DA. 2002. *Life at the Limits*. Cambridge University Press.

Websites

Gradients are not a significant area in biogeography in terms of websites. To understand more about the use of ordination techniques (one of the key statistical ideas) try here: http://www.okstate.edu/artsci/botany/ordinate/ which gives a vast amount of detail. One good study of gradients can be seen at http://www.lincsbap.org.uk/Rivers/springs1.htm. A larger example of research is the Sevilleta LTER (long-term ecological research) project at: http://sev.lternet.edu/ while the LTER project is at http://lternet.edu/.

CHAPTER 10

Quantitative and statistical methods – an introduction

'If you torture the data long enough, Nature confesses.' Ronald Coase (Nobel Laureate)

Key points

- For statistical data, to be useful, it must be collected in an appropriate manner for the type of test that is going to be used;
- Before setting out to solve a problem, make sure that all the necessary information and techniques are available and that their use and limitations are understood;
- Planning is very important as is the use of an appropriate test;
- Where an hypothesis is to be tested, it is very important to make sure that it is the correct one, and that the methodology to be used is robust;
- The use of a computer program for calculating statistics is often a time saver, provided the data are reliable and the program has been validated;
- Remember the acronym and adage 'GIGO' – garbage in, garbage out;
- Do not rely upon complex programs to massage your data to produce impressive tables and graphics – if the sampling scheme or data are poor or deficient, the results will be useless!

10.1 Introduction

The purpose of this chapter is to guide readers towards a range of resources for statistical and other analyses, both printed and available through computer packages or via the Internet. We felt that it was neither appropriate nor sensible to produce yet another chapter on how to do means, standard deviations and other statistics, but instead to show how the more quantitative aspects of a survey can be dealt with and how to choose the most appropriate statistical method for a particular application.

10.2 Planning

During the planning stages of any ecological survey, there are five major points which have to be taken into consideration:

1. The purpose of the survey and its timescale;
2. Hypothesis formulation and prediction from this hypothesis;
3. How many data need to be collected;
4. The type of data which needs to be collected since different statistical tests may need categorical or nominal (presence/absence), ordinal (rankable) or interval (continuous data);
5. The statistical test(s) which may be needed to analyse the data (which may feed back into a consideration of the type of data required).

Although it might seem pedantic to say so, before any survey takes place, the reason for it must be properly established, together with a clear idea of the timescale during which it will take place. If the organisms or habitat to be studied are seasonal in their occurrence or activity, it would a waste of time and resources to arrive either too late or early, or to have to leave halfway through the investigation because of lack of supplies or other potential calamity. Any organisation, particularly if the survey is to be undertaken in an area with which the surveyor(s) are not familiar, should include local contacts who are familiar with the habitat. The same remarks must apply to the question of the hypothesis. It is not a good idea to go to an area without having first considered what the purpose of the survey is, if it was to be a reconnaissance or more detailed study, and most importantly, if it was being carried out to test the validity of a hypothesis.

One of the commonest problems in an ecological investigation concerns point (3), the establishment of the number of samples that need to be taken or surveyed. Henderson (2003) provides an excellent introduction to this aspect, using estimates of the mean and variance or frequencies of the population based upon preliminary surveys. This is an aspect of project planning which tends to be done by intuition rather than as a result of careful consideration. An experienced worker may be able to design and establish a valid scheme without the need to do more than a preliminary survey, but where time allows, the necessary calculations should be carried out to ensure that nothing of importance is missed through inadequate sampling. In some cases, the sampling technique will probably involve a whole population (for example, a threatened species in a particular habitat).

10.3 Which test to use?

The statistical tests that can be applied to data which have been gathered by sampling fall into two broad categories:

- **Descriptive** – involving the collection, organisation, summarising and presentation of data;
- **Inferential** – including the making of inferences, testing of hypotheses, determination of relationships and prediction making.

For our purposes, data are collected from observed variables describing some sort of event. Data exist as values or measurements that have been assumed by the variables which describe the event, as in the pH of a series of soil samples that have been collected along a transect line, or the heights of trees in a woodland sample. Variables are broadly classified either as qualitative (non-numeric) or quantitative (numeric variables are either discrete or continuous). Discrete variables assume values that can be counted – as the number of animals in a particular field at a specific time, whereas continuous variables can assume any value between two limiting values – for example, the amount by which a plant has grown during the last week. A collection of data values is the data set and each value is a data value.

Data may be collected from a whole population, or if time or other constraints apply, it may be necessary to take a smaller but representative part, which is a sample. The sample is a subset of the population, but a census is a sample of an entire population, for example, in a particular habitat.

Both populations and samples have particular characteristics associated with them, namely parameters and statistics; a *parameter* is a fact or characteristic of a population, such as a birth or death rate, whereas a *statistic* is a characteristic or fact regarding a sample. An example of the latter might be the number of eggs laid in a clutch compared with the mean clutch size for that particular species.

Data types vary in their quantitative nature and the most quantitative is the *interval* type – real measurements such as height, girth, mass and pH. With interval data the amount of difference between, say, oak saplings of 25, 50 and 100 cm height is such that the tallest is four times the height of the shortest and the difference between the first two is half the difference in height between the second and third.

Ordinal data measurements are such that they can be ranked with respect to each other even though the exact differences may not be easily quantifiable, such as the amount of weed infestation in a series of fields or the degree of damage caused to habitats by human activity.

Categorical data are, as their name suggests, data which can only belong to one or other particular category – such as presence or absence, infected or not infected and damaged or undamaged.

In some cases, a hypothesis might require the collection of two sets of data – which could/might be collected as matched pairs, such as when it is desired to investigate a possible relationship between a plant and a particular environmental factor such as soil calcium content. An example of the use of unmatched pairs could be to study the differences in numbers of snails in two habitats – on different soil types, where there is no particular need to pair the observations.

Many people using statistical tests will have only the most general idea of the topic and the following sources are recommended: Jaisingh's *Statistics for the Utterly Confused* (2000) and the Internet-based online statistical primer *Hyperstat* by David Lane (1999). For many people, one of the most difficult aspects of ecology or biogeography is the selection of an appropriate statistical method to apply to their particular project. During the planning stage, it is assumed that an appropriate sampling plan will have been drawn up in order to investigate a suitable number of sites, or to take a satisfactory number of samples, and that an appropriate method of analysis will have been arrived at.

The choice of test to be used is made potentially less difficult for the non-specialist since there are a number of published 'decision trees' (e.g. Chalmers and Parker 1989) which lead the investigator through a series of alternatives and which, with care, will produce a valid choice. These 'trees' do not deal with every conceivable test but will suffice for the majority of investigations. A slightly different approach is taken by www.graphpad.com in which a choice can be made online. The tree which follows is divided for convenience into two parts – which assumes that you know if you are looking for relationships and associations between data items or for similarities and differences (see Figure 10.1).

It cannot be emphasised enough that even with good quality data, the choice of an inappropriate test yields outcomes which are of no or limited use. Equally, poor quality or inadequate data cannot be processed or manipulated by statistics to produce a valid result.

10.4 Data types and information content

Any measurement which involves numerical aspects such as height, mass or number is part of a continuous sequence since a value can be expressed exactly, rather than fitted to a small number of discontinuous classes. This does not preclude the possibility of putting continuous data into a ranked order or a series of classes, in which case some of the precise information content will be lost. Conversely, ranked or presence/absence data cannot be converted into validly continuous data since the necessary information is not present.

It is possible to interpret the content of data as a hierarchy in which the greatest amount of information is possessed by numerical data such as interval data, followed by ranked data and classifications and finally, the lowest amount by categorical, presence/absence data.

10.5 Types of statistical tests

10.5.1 Parametric tests

Parametric statistics are based upon population parameters such as means and standard deviations which assume that the distributions of the variables being assessed have certain characteristics, such as a normal distribution or

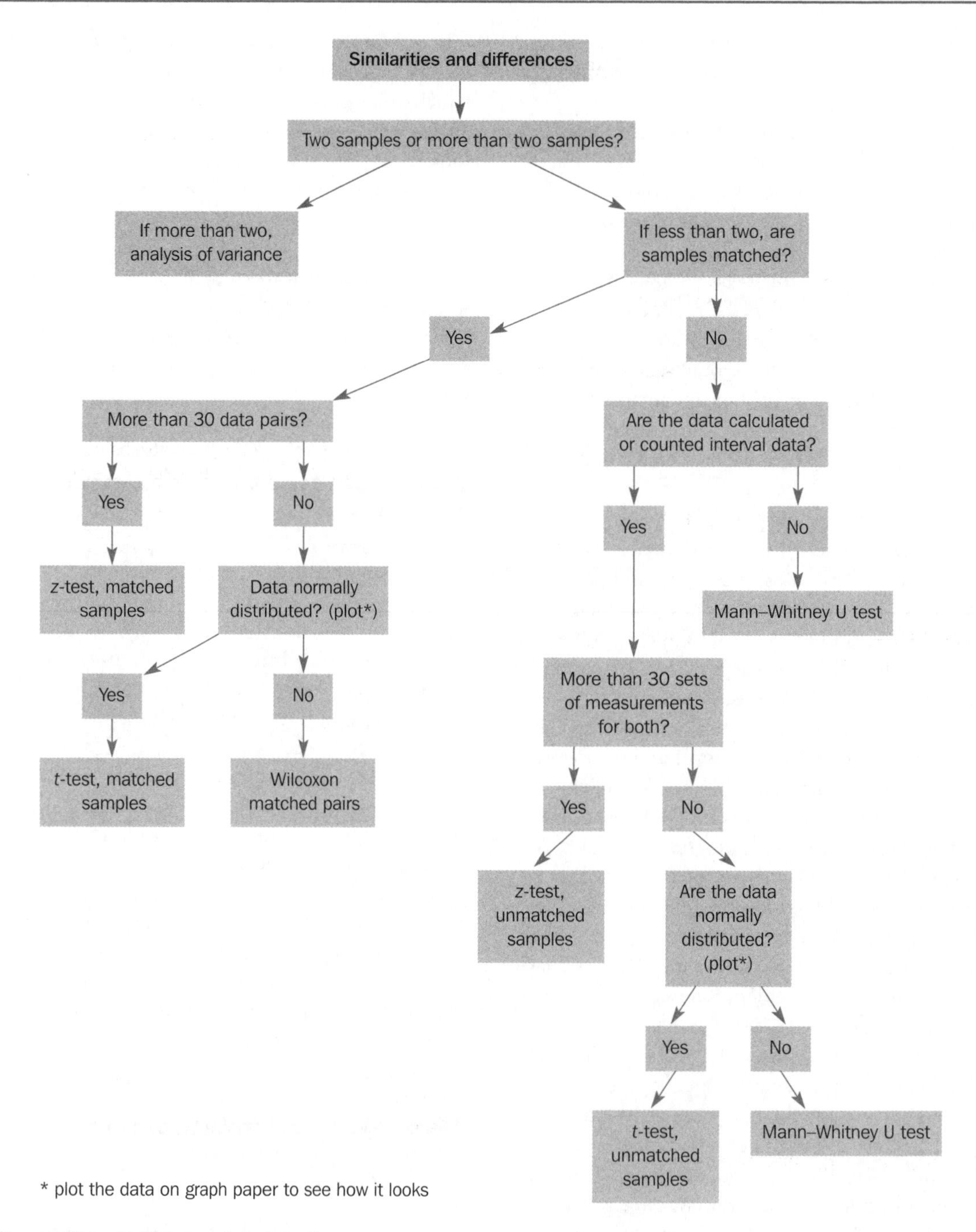

Figure 10.1 Statistical test decision trees

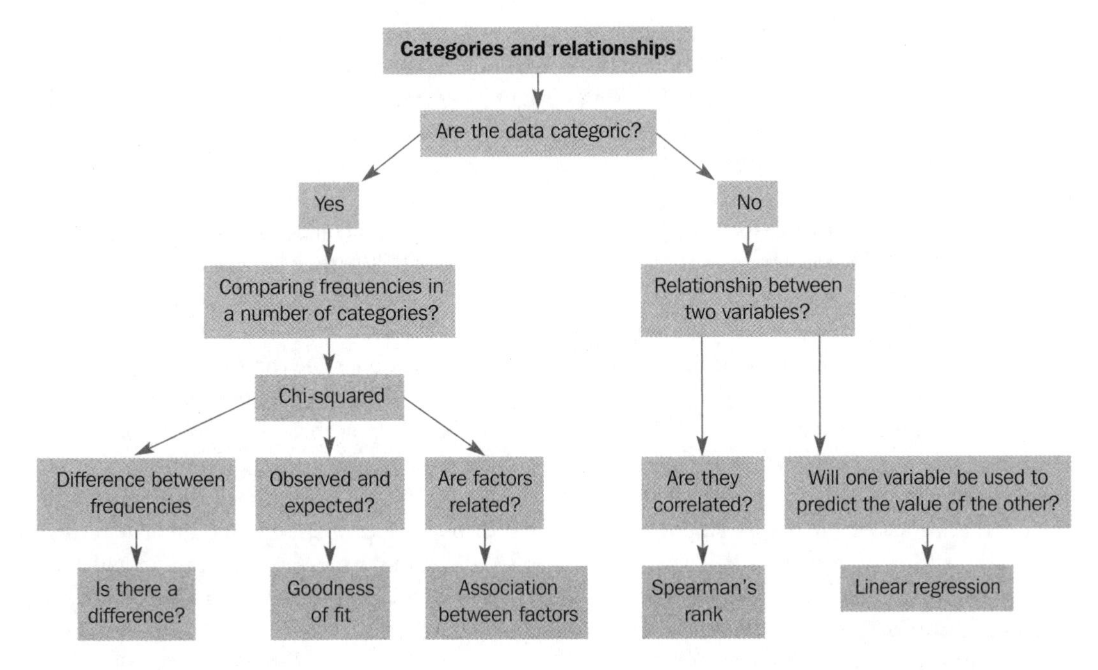

Figure 10.1 Continued

equal variances. The data are applied to appropriate equations which adjust a general purpose model to fit the particular data set. They are generally very robust and utilise the full information content of the original data, and they often retain considerable power to detect differences or similarities even when these distributional assumptions are violated – some variables violate the assumptions so markedly that a *non-parametric* alternative is more likely to detect a difference or similarity.

Typical parametric methods include t-testing, analysis of variance (ANOVA) and least squares regression.

10.5.2 Non-parametric tests

These types of statistical tests have much less rigorous assumptions concerning the distributions of the variables and the variances of comparison groups. It is important to note that these techniques tend to rely upon the rank of the individual observations rather than their absolute numeric values. This implies that non-parametric statistical tests have less stringent assumptions, and the computations are generally easier to understand. However, they do have a number of disadvantages – the metric and numerical values of the results are lost since only the ranks, not the numerical values, are used and statistical power (as compared with using parametric statistics) is lost.

It is worth comparing a few tests:

Parametric test	Analogous non-parametric test
Student's *t*-test	Wilcoxon rank sum test
ANOVA	Kruskal–Wallis test
Paired *t*-test	Wilcoxon signed (rank) test *or* the sign test

The Wilcoxon rank sum test is used in cases where independent data are available for the comparison of data for two different groups (e.g. litter size in a prey and predator relationship). The Kruskal–Wallis test is used in instances where you want to compare results for more than two groups (e.g. fish size in ponds, lakes and rivers). The sign test and the Wilcoxon signed rank test are broadly similar: the sign test is used for paired data (such as in a pre- and post-treatment study), and involves counting up the number of cases for which the second value is greater than the first (positive signs) to see if the number of positives are greater than expected by chance alone. The Wilcoxon signed test is similar to the sign test but also ranks results by the magnitude of the difference of the paired values.

Having selected the test and collected appropriate data, how will the data be analysed? There are a number of statistical sites available on the Internet which enable users to input data and receive an appropriate analysis (such as www.graphpad.com), and the Excel spreadsheet (Microsoft Office) has a small range of tests which can be augmented (e.g. with XLstat).

A more detailed treatment of statistical methods will be found in Chapter 4 of Kent and Coker (1992), while among others, Cadogan and Sutton (1994) is a useful introduction – particularly for those who are less confident with statistics – using worked examples from a range of habitats and analysis types.

10.6 Multivariate analysis

Multivariate analytical techniques are extremely powerful methods for dealing with species distribution, and abundance or environmental data; matrices of data can be investigated for patterns, gradients or similarities. While it is possible to use some multivariate techniques on very small data sets, using a pocket calculator (if you have a lot of time and patience), most data are currently processed using a computer. The techniques fall naturally into one of two categories – *classification*, in which, for example, species or sites are grouped together, and *ordination*, which arranges sites or species in order, relating possibly to an environmental gradient or other influence such as grazing or other management. Computer programs which do one or other of these tasks have been around for more than 40 years but their widespread availability, and potential use in some aspects of modelling, are features of the last 15 or so years since the personal computer revolution of the 1980s. For a detailed explanation of these processes and how they are used in ecological and biogeographical applications, see Chapters 5–8 in Kent and Coker (1992).

Classification can be carried out using a divisive criterion, splitting the data matrix up into smaller groups based upon the amount of similarity between individuals, or agglomeratively in which the data set is split up and then rearranged and joined into larger and larger groups. The results are often displayed using a dendrogram in which the levels of division or of agglomeration are displayed in a hierarchy.

10.7 Multivariate analysis software

There is a reasonable amount of software available at present, and most of it is designed to run on IBM-compatible PCs, with Mac software being a rare item (Syntax 2000 – see below – is one such package). Most modern Mac computers will run PC software under 'emulation' conditions, but this *can* be a slow process since the software emulation uses a lot of processor power.

Up until the mid 1980s, most multivariate analysis was done on mainframe or mini-computers but the development of the IBM PC and clones meant that the mainframe programs could be revised so that they would work on the desktop machines. Initial limitations of computer memory were overcome by ingenious programming, and then by hardware development so that present-day computers have more than enough memory and processing power to run appropriate applications.

Initially, the programs were non-graphical and derived from the mainframe versions by very simple conversion. Data could be entered directly into the program via the keyboard, or read in from text files, but a mistake in data entry would terminate the analysis and lead to frustration. Over the past few years, more and more of the programs have been rewritten from first principles to work with Microsoft Windows© operating system (generally 98 or later versions) which has better graphics possibilities and allows several programs to run at the same time. There has also been a change from the single-method analysis program to a package in which a set of data can be readily analysed with several different programs. These programs are generally much more user-friendly and program output is much more easily comprehensible, particularly if presented with the aid of good colour graphics.

While the older PC versions of many analysis programs are still available and will run on modern PCs (Kent and Coker 1992), their use is not recommended except as a penance! The unfriendly nature of the original DOS interface will not commend itself to users of Windows. Many packages are available, some more expensive than others, and all have their advocates. In our experience, three PC-based packages are currently available which will do the majority of analyses that are likely to be required, at a reasonable price. There are a number of others which may be available on institutional computer systems or as downloads but the three that follow are particularly recommended.

The first of these is *MVSP version 3.1* for Windows by Warren Kovach, which is available as a free download for evaluation. MVSP performs several types of ordinations including principal components analysis (PCA) and correspondence or detrended correspondence analysis (CA/DCA). It also does canonical correspondence analysis (CCA), a technique which is currently popular in ecological studies. It can also perform classification using cluster analysis, with a wide range of distance or similarity measures and clustering strategies. A useful adjunct of the package is its ability to calculate several diversity indices on ecological data, including Simpson's, Shannon's and Brillouin's indices. It has a good user manual.

Once the data have been analysed, the results can be directly plotted. For example, by selecting the ordination axes that are required, scattergrams will be drawn. Joint plots of both variables and cases can be drawn for CA results. Euclidean biplots of PCA results (with variables as vectors) can be produced, as can biplots of the environmental variables in CCA.

MVSP is particularly useful because it has a number of data manipulation features, such as transformation, merging of two or more data files, and conversion to formats such as range-through. It will also import data from and export to a variety of formats. Its other advantage is that it will run on a very modest older computer system with a small amount of memory, but it does require at least Windows 98.

The second package, *PC-Ord*, by MjM Software has been available for some years and in its current version (4) is extremely well featured, with software for many types of classification, ordination and species diversity applications. The range of features available in this version is similar to those offered by MVSP. The user manual is comprehensive but a related book, *Analysis of Ecological Communities* (McCune *et al.* 2002) offers a rationale and guidance for selecting appropriate and effective analytical methods from PC-Ord which will be appropriate in all aspects of community ecology and biogeography. The combination of the package and book is very powerful, and worth purchasing. A free demonstration download for evaluation is available.

The third package is *CAP* – one of several available from Pisces Conservation. The Community Analysis Package is very well featured, has a well-written manual which can be downloaded before purchase, which is something neither MVSP nor PC-Ord currently offer. There is also a free demonstration download. As might be expected, this package offers virtually all the same analyses as the others but adds some, such as NMDS (non-metric multidimensional scaling) which some ecologists prefer to PCA since it frequently, but not invariably, produces a better ordination than standard PCA. A recently published book (Henderson 2003) is a very useful companion and is well recommended, particularly for animal population studies.

Which one is best? All three are very good, represent excellent value for money and are used on a daily basis by many ecologists and biogeographers worldwide. Manuals are good and they are easy to use once you get used to them. Try before buying and visit the appropriate websites for further information.

The first of two packages which should be mentioned for the sake of completeness is *Canoco* version 4.5. The original PC version of this was issued in 1987 and ran under DOS, and version 4 is the first Windows version. This package only does ordinations and statistical analyses based on ordinations, but it has good graphics and a comprehensive manual. It is good but *very* expensive and most of its functionality is available in the previously mentioned packages. A time-limited trial version is currently available on request from the agent noted below in Websites.

The second package is *Syntax 2000*, a modular program which is very well featured and possibly more suited to advanced studies. Five modules are planned, of which three, dealing with ordination, hierarchical and non-hierarchical clustering are available, with two more, as yet unreleased, which will deal with evaluative strategies and matrix and ranking. No further details for the two unreleased modules are available. The original program was limited to various forms of cluster analysis and was DOS-based, but the present one is available in both Windows and Mac versions, with a manual. Unfortunately, no trial version is available and it is an expensive piece of software. The program was originally available in 2000, and it is a matter for some regret that the missing modules are not yet available. The stockist's website gives further information on the package's capabilities, together with some screenshots.

10.8 Which program?

The choice of analytical technique – whether to use a classificatory or an ordination program – depends upon the desired outcome. For the sort of data that is likely to be collected and analysed using multivariate techniques in an ecological or biogeographical context (species occurrences in sites, possibly with some environmental data for each site), an ordination approach often provides an insight into spatial relationships between species or sites by arranging species in order along a series of computer-generated axes (based upon eigenvalues) which can then be plotted as a series of scatter diagrams. Most ordination programs restrict plotting to three or four, in pairs, but technically, there are as many of these derived axes as sites or species. If labels are provided for each species or site, then the scatter plot will show species or site groups which will probably bear some relationship to the situation from which the data were collected. The axes do not necessarily indicate a *particular* environmental gradient, but an inspection of the spatial relationships of species or sites on the scatter plots compared with their location and companion species in the field might show up a moisture, soil pH or management gradient or pattern – thus giving an indirect gradient analysis. The technique known as CCA allows the simultaneous ordination of biological and environmental data which leads to the derivation of a direct gradient analysis (ter Braak 1987) from the data. The ability to relate, for example, vegetation distribution and local environmental factors is a very powerful one (see e.g. Coker 2000).

In choosing which ordination method to use, it is as well to note that the type of analysis chosen (PCA, CA, DCA or CCA) will produce ordination plots which will not necessarily correspond very closely because of the technique used, but it should be possible to identify similar groups in each case if the data are sound. More advanced work may need to take account of mathematical problems caused by using the first axis ordination as the starting line for the second and subsequent axes. This can lead to a nonlinear response curve or arch effect which is sometimes seen in analyses carried out using PCA or CA, and is least problematical in DCA (the detrended version of CA). Another aspect which sometimes causes problems in more advanced work is one of autocorrelation which means that it is possible to predict the value of a variable at a given point in space by knowing its value at other locations. Positive autocorrelation is the usual form in biogeography and implies that the value of a variable, such as species density, will be similar to that of a similar point close by. For example, if a series of samples are taken down a slope, the measurements will have a certain degree of spatial autocorrelation because of their proximity to one another. Temporal (time-based) autocorrelation occurs if successive samples are taken from a site over a period of time. At present, the effects of autocorrelation are generally ignored by biogeographers and ecologists because the work that they do is more descriptive and inductive. Deductive studies require the use of inferential statistics and are outside the scope of this book.

Certainly for most applications, the use of ordination to explore the data, followed by an application of classification (such as cluster analysis) seems to work well. TWINSPAN (Two-Way Indicator Species ANalysis) (Hill 1979) has been used to classify the large number (>30,000) records for the UK NVC (National Vegetation Classification). This program initially produces an ordination and then uses this as a basis for classification of a species and sites data matrix into a hierarchy of groups which are used as the basis for the community classification.

10.9 Summary

This chapter has attempted to show how the type of mathematical analysis of data to be used depends upon the desired outcome of the initial survey which should have been settled at the planning stage. Any application of statistical or other analytical software or procedures is liable to give misleading or incorrect results if the method used is not applicable. It also shows how a decision tree approach to test selection can be helpful for less experienced workers, and gives an overview of some of the more appropriate software and sources for statistical help. The final section on multivariate methods shows the background and some of the strengths and weaknesses of the approach and how to select a package and approach which will produce a reliable and appropriate analysis.

APPLICATIONS – USING BIOGEOGRAPHY

Using ideas from this chapter in real-life situations

It might seem almost unnecessary to consider the application of statistics in the modern use of biogeography. Despite their obvious value there is still a great deal we need to know. One application of statistics is the modern web-based database which can be used to gather basic statistics and other data about a vast range of subjects. Two examples demonstrate the value of statistics. The UK Department for Environment, Farming and Rural Affairs (DEFRA) is a major producer of statistics. One of their sites, the *e-digest of environmental statistics* (http://www.defra.gov.uk/environment/statistics/index.htm) aims to give people access to a range of basic information about the environment ranging from air quality to wildlife. This, coupled with a range of data from elsewhere, gives the researcher a chance to analyse changes in the environment. At present, this site gives only limited data. A far more ambitious scheme is that of the World Resources Institute's *Earthtrends* data portal. Here, the aim is to provide a considerable range of global data (http://earthrends.wri.org) which can be gathered and analysed. The advantage of this system is its accuracy and depth of data recording both of which make the site, and its companion biennial (World Resources), one of the most cited reference texts on the environment. The increasing use of online data sites (especially those of the UN) means a greater access to key information. It follows that any biogeographical group adding to this is helping to provide baseline data against which global change can be measured.

Review questions

1. What is the purpose of an hypothesis?

2. For each of the following data types, give three practical examples to illustrate the meaning of the term to someone who is unfamiliar with the practicalities of data collection: categorical (nominal), ordinal and interval.

3. Why do some statistical tests use population data and others use sample data?

4. What is the distinction between parametric and non-parametric tests?

5. What is an emulator?

6. What, apart from the way in which they display data, is the difference between a dendrogram and a scattergram?

7. Using the diversity data from the website, use a computer-based method to determine the Shannon and Simpson's indices.

References

Cadogan A and Sutton R. 1994. *Maths for Advanced Biology*. Cheltenham: Nelson.

Chalmers N and Parker P. 1989. *The OU Project Guide*, 2nd edn. Montford Bridge, Shrewsbury: Field Studies Council.

Coker P. 2000. Vegetation analysis, mapping and environmental relationships at a local scale, Jotunheimen, Southern Norway. In Alexander R and Millington AC. (eds) *Vegetation Mapping from Patch to Planet*, pp. 135–58. Chichester: Wiley.

Henderson PA. 2003. *Practical Methods in Ecology*. Oxford: Blackwell Publishing.

Hill MO. 1979. *TWINSPAN, a FORTRAN Program, for Arranging Multivariate Data in an Ordered Two-way Table by Classification of Individuals and Attributes*. Ithaca, NY: Cornell University.

Jaisingh L. 2000. *Statistics for the Utterly Confused*. McGraw-Hill (also available as an electronic download).

Kent M and Coker P. 1992. *Vegetation Description and Analysis*. London: Belhaven Press.

Lane D. 1999. *Hyperstat*, 2nd edn. Atomic Dog Publishing.

McCune B, Grace JB and Urban DL. 2002. *Analysis of Ecological Communities*. Glendale Beach, OR.: MjM Software Design.

Ter Braak CJF. 1987. *CANOCO, a FORTRAN Program for Canonical Components Analysis by Partial Detrended Canonical Correspondence Analysis, Principal Components Analysis and Redundancy Analysis*. Wageningen, The Netherlands: TNO Institute of Applied Computer Science.

Townend J. 2002. *Practical Statistics for Environmental and Biological Scientists*. Chichester: Wiley.

Waite S. 2000. *Statistical Ecology in Practice*. Harlow: Prentice Hall.

Websites

CANOCO.
http://www.microcomputerpower.com/

CAP.
http://www.pisces-conservation.com/

MVSP.
http://www.kovcomp.co.uk/mvsp/

PC-Ord 4.
http://home.centurytel.net/~mjm/

Syntax 2000.
http://www.exetersoftware.com/cat/syntax/syntax.html

Hyperstat.
http://davidmlane.com/hyperstat/index.html

XLstat (Excel spreadsheet add-on).
http://www.xlstat.com

Graphpad.
http://www.graphpad.com/quickcalcs/index.cfm

Graphpad resources.
http://www.graphpad.com/index.cfm?cmd=library.category&categoryID=2

CHAPTER 11

Modelling

Key points

- A model is an attempt to describe a system;
- Models can rarely describe the precise functioning of even a simple ecosystem because of the interactions which occur with other nearby systems;
- Island biogeography is an example of a relatively successful attempt to explain the particular situation which isolated ecosystems face;
- Most current work on models is concerned with climate and ecosystem interactions.

11.1 What is a model?

A model is an attempt to describe a system, theory or phenomenon that accounts for its known or inferred properties and may be used for further study of its characteristics. In ecological and biogeographical terms, models have frequently been used to examine changes in natural or anthropogenic vegetation under the influence of climatic change, or to investigate the optimum way in which to manage a renewable resource such as fish or timber stocks to obtain the maximum sustainable yield. A model is not, despite what some modellers think, reality!

11.2 Types of models and their limitations

Models can be simple or complex, depending upon the application. For example, a simple graphical model based upon annual records of animal pelts purchased by the Hudson Bay Trading Company in Canada in the late nineteenth century were used as the basis for a model of lynx and snowshoe hare population predator–prey interactions. This showed, as might be expected, the effects of a decline in prey numbers on the predator – not immediate, but with a delay before the onset of predator decline. It also showed that when prey numbers began to increase as a result of lower predator numbers, there was a delay (lag) before predator populations began to rise. Figure 11.1 (see Box 11.1) illustrates a typical set of observations from 1900 to 1920.

Some early ecological modelling attempts were not always very successful, partly because of the lack of processing power in the computers that were used, but also because of poor planning (or possibly lack of expertise) on the part of the modellers who initially failed to take into account the vitally important role of decomposers and detritivores in recycling essential nutrients and minerals from dead material and wastes in a woodland ecosystem simulation. The failure to realise this point would mean that available

BOX 11.1

Lynx and snowshoe hare in the Canadian North-West Territory (NWT)

The primary food of the lynx is the snowshoe hare and therefore the population cycles of these two species are closely linked (see Figure 11.1). When hares are plentiful, lynx eat little else, taking about two hares every three days. When hares are scarce, lynx also prey upon mice, voles, squirrels, grouse and ptarmigan, and they will also eat carrion. However, these food sources often do not meet the lynx's nutritional needs. Some lynx cannot maintain their body fat reserves, and become more vulnerable to starvation or predation. Other lynx manage to remain healthy by using alternative prey and food sources when the hare numbers are low. When snowshoe hares are scarce, many lynx leave their home range in search of food.

Across most of the boreal forest, hare populations experience dramatic fluctuations in a cycle that lasts 8–11 years. At the peak of the cycle, snowshoe hares can reach a density of up to 1,500 hares per km^2. The habitat cannot support this many animals, and as predation increases and starvation sets in, the population starts to decline. Continued predation accelerates the hare population decline, since lynx and other predators are at a population high. When the hare population reaches a low level, it stabilises, for several years. The food plants slowly recover and the hare population starts to increase again. Since hares have several litters each year, the hare population increases rapidly. After a year or two at high densities, the hare cycle repeats itself.

The lynx population decline follows the snowshoe hare population crash after a lag of

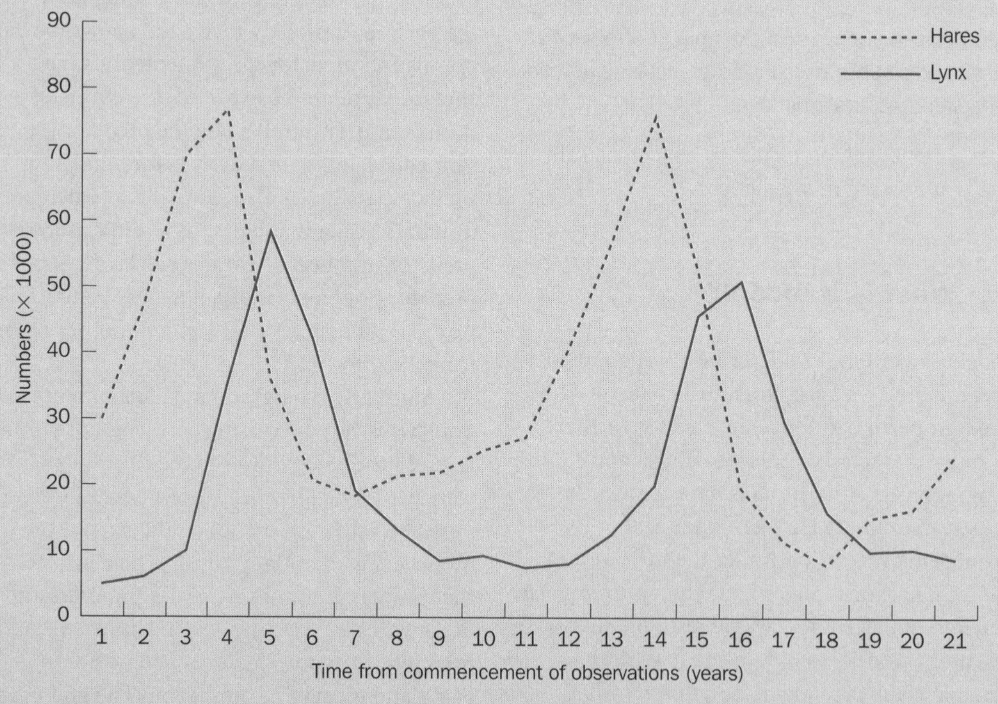

Figure 11.1 Lynx and snowshoe hare interaction, 1900–1920

1–2 years. As hare numbers start to decline, lynx continue to eat well because they easily catch the starving hares. When hares become scarce, the lynx numbers decline as well. Their lack of fat reserves makes them less able to live through starvation and cold temperatures. Food shortages also cause behavioural changes such as increased roaming and loss of caution. This increases their vulnerability to predation. Malnourishment has the most significant effect upon lynx reproduction and population levels. When females are in poor condition, fewer will breed and not all of these offspring will produce litters. Litters will be smaller, and most, if not all, of the few kittens born will die soon after birth. This means that for a period of 3–5 years, few or no kittens survive to adulthood. Studies have shown that the level of kittens in a lynx population may be zero at the population low, and as high as 60 per cent when their numbers are increasing. Low lynx population levels last for 3 or 4 years. When hares become plentiful again, the lynx population begins to increase as well.

The highs and lows of the lynx population cycle do not occur at the same time across the NWT. For example, in the early 1990s, lynx numbers peaked two years later in the north-western NWT than in the south-western NWT.

See: http://www.nwtwildlife.rwed.gov.nt.ca/NWTwildlife/lynx/lynxharecycle.htm

nutrients for plant growth, such as nitrogen and phosphorus, would become limiting, and there would be a growing excess of dead plant material and dead animals which were not decaying. The system would eventually shut down.

There have been attempts, particularly in the 1950s and 1960s, to predict the functioning of ecosystems using a model which relies largely on the basis of energy flows, in which very detailed snapshot measurements of energy flows for ponds or small areas of grassland were supposed to show how the system functioned. What was never fully admitted to is the simple fact that almost every ecosystem is a complex web or network of food and energy flows, varying in amount from hour to hour or day to day. More recent attempts at modelling ecosystem function have used the prediction from competition theory that the greatest biodiversity occurs in ecosystems where the twin stimuli of stress and disturbance are intermediate, rather than high or low. The humpback model, as it is known, predicts that the number of species in an ecosystem will rise, for example, according to increases in nutrient availability up to a point where this is at a moderate level. Further additions of nutrients lead to increasingly more eutrophic conditions which usually result in a decline in biodiversity, followed by a complete collapse of the system. The humpback model describes exactly what Lachavanne (1985) found in nutrient-stressed oligotrophic lakes in Switzerland where productivity – a useful indicator of plant performance – was highest when the nutrient level was mesotrophic.

The idea of habitat models has been extremely important in the development of ecology since the mid twentieth century, particularly since Whittaker's work on species response to environmental gradients (Whittaker 1956). McCune points out that despite the central importance of species response functions (SRFs), our understanding of them is surprisingly primitive, particularly when considering more than one factor at once (McCune 2004). He goes on to identify the problem as being one that conventional statistical models cannot represent SRFs very well, and proposes the use of a more flexible process using non-parametric multiplicative regression (NPMR) as a tool to represent SRFs as species response functions in a multidimensional space.

It is important to realise the limitations of all forms of modelling as management tools. For example, Beissinger (1998) provides an interesting overview of simple to complex population simulation models, focusing on the extreme, and often excessive, data requirements for complex models; the conclusion from this is not to build a model for which you do not have the supporting data. Mills *et al.* (1996) worked on a grizzly bear model and in presenting a real-life example of species management, they also showed the importance of using modelling only as a tool and not something on which to base important management decisions. Similarly a paper by Starfield (1997) contrasts two views of modelling – the model as a representation of 'truth' and the model as a problem-solving tool. It outlines how to use a model usefully and efficiently. The paper was written in order to encourage wildlife managers and scientists to view models as problem-solving tools to be used as routinely as they collect data and analyse them and not as representations of reality as was so often the case.

CASE STUDY

Island biogeography as a model?

In an earlier chapter, we looked at island biogeography – a theory developed by MacArthur and Wilson (1967) which used a quantitative mathematical modelling approach to explain the number of plant and animal species found on an island as a function of, or balance between, immigration and extinction. For biogeographers, this is arguably one of the most interesting and thought-provoking models, in spite of recent work which tends to throw doubt on the role of equilibrium states.

The number of species found on an island is a function of:

- the area of the island
- its topography and elevation
- its climate
- its distance from the source region
- the species richness of the source region
- the state of equilibrium between rates of colonisation and extinction.

This implies that these are factors which must be included in our model. There are several additional factors which must be considered for inclusion in our model:

- The *rate of colonisation* will initially be high because most species will be new and because the island can be reached by those with good powers of dispersal. Over time, immigrants will increasingly belong to species that have already reached the island, and in consequence, the rate of new species colonisation declines.
- Additionally, the *rate of immigration* will be affected by the location of the island, because those islands close to the source region will have higher rates of immigration than those which are further from the source region, assuming that the islands have comparable climates and landscape.
- The *rate of extinction*, on the other hand, will usually start at a low level but gradually rise because the greater the number of species present, the greater is the risk of extinction as populations of individuals become smaller due to competition for food or space. Small populations are certainly at greater risk of extinction than larger ones.
- Initially, the few species present can occupy a greater variety of available ecological niches than would be possible on the mainland, where competition with other species exists. However, as the island is occupied by new species, competition takes place and may result in either the species becoming extinct or, as a result of separation in space or time, the species undergoes divergence and the resultant organisms become genetically different from each other.

- As new species arrive, smaller populations can be supported because there is a constant amount of food available which would not be enough for larger ones. As already indicated, smaller populations have a greater chance for extinction, and as new species arrive, the rate of extinction must rise. Eventually, the system settles down and the number of species present will be the result of a balance between the rate of immigration and the rate of extinction.

MacArthur and Wilson's theory has, in the past, provided a useful framework for a more structured approach to biogeographical modelling and has since been applied to studies conducted on mountain peaks, cave environments, individual plant species isolated within larger populations of other plant species, and relict areas, but recent work has shown that the theory itself is not as tenable as first thought in that different levels of equilibrium were reached on particular islands, depending on the amount of human and other disturbance (such as volcanic eruption or glacial activity) that the island has sustained. This finding tends to support the idea that a *dynamic equilibrium* may exist on islands which enables a small level of positive or negative change to exist. It is the precise operation of feedback loops which keeps such systems in balance.

These disturbances may be very long-lasting, such as climate change or a rise in sea level. As a result of this, populations of islands are likely to be continuously fluctuating and adjusting. For example, some bird populations on Pacific islands could best be interpreted as declining towards a new, lower balance because rising sea levels and erosion processes, possibly as a result of global climate change, are constantly diminishing the available areas of the islands. Small fluctuations and adjustments are a feature of systems which are reasonably stable and represent the effects of positive and negative feedback acting to preserve a state of relative equilibrium. Less stable systems may have an excess of positive feedback which unbalances the system, for example, by one species breeding more rapidly than another and monopolising food and territory. Negative feedback tends to stabilise a system by damping down change. For example, stress caused by overcrowding in some animal species frequently depresses birth rates as well as increasing death rates and as a result, the population declines and overcrowding decreases.

11.3 Issues in modelling

Some earlier attempts at modelling the behaviour of ecosystems tended to make use of techniques used in community ecology. These allowed ordinations (computer analyses which arrange matrices of sites and species in the order of their response along axes of potential environmental gradients) of the different response variables (species) and, reciprocally, that of sites (relevés) which were then interpreted in environmental terms. Since these were indirect gradient ordinations, inferences concerning the relationships between environmental variables and the ordinations were made after the ordination procedure either by inspection of the results or through further analyses such as regression. However, the process of ordination can now be given a stricter environmental basis by methods of canonical ordination (e.g. CCA which is a direct gradient analysis) which enable measured environmental data to be simultaneously ordinated with vegetation data. Furthermore, the axes of canonical ordinations can be tested for their significance by such means as the Monte Carlo test by comparison with a random data set generated from the observed response and explanatory variables. The Monte Carlo method makes use of randomness. A Monte Carlo test is a simulation in which many independent trials are run independently to gather statistics.

11.3.1 Scales of modelling and vegetation description

The choice of technique for vegetation description is dependent on the scale at which the model is required to work. At a small scale it is possible to describe vegetation using species names. At a larger scale, description in terms of a generalised vegetation type such as heather moorland or birch forest, general physiognomic or environmental features, has been often used (e.g. biome).

Biogeographical models based upon correlation were criticised by the IPCC (the Intergovernmental Panel on Climatic Change) Impacts Study because they did not show causal relationships between climate and plant physiology, and as a result, their presumed ability was to predict only the current distribution of biomes. In defence of the modellers, it must be said that relatively little relevant physiological information was available, and most of that was derived from laboratory studies – hardly ideal. In later mechanistic-based models such as BIOME 3.0, the representational sophistication of the vegetation increased further to describe vegetation not in terms of biomes but in terms of plant functional types (PFTs). The notion of PFTs is not a particularly new concept in vegetation classification and is related to physiognomic approach, such as that introduced by Warming (1909) and Raunkiaer (1934), demonstrating that particular growth forms such as dwarf shrubs were considered to characterise particular habitats which have particular environmental characteristics such as temperature range or rainfall. Experience has shown that for many purposes, the growth-form approach to modelling of ecosystem impacts and feedback loops on a large scale works well and is superior to approaches based at the genus or species level.

If dynamic vegetation change is to be part of the ecological description of sites for this type of model then some method of defining the seral (successional) gradients which exist within them is essential. The application of ordination techniques to data for different sites should delineate the conditions and give pointers towards possible developments of those sites. For instance, a number of sites in a woodland could be ordinated using two (biological) indicators of extremes of prevailing environmental conditions, such as humus depth and light availability – in this case the presence of humus-loving or light-requiring species. Along the gradient it could be reasonably expected that regeneration of trees would take place somewhere between the occurrence of these two species types. The occurrence of different plants on different sites could define the site conditions and their gradients. Categorisation of sites would give a basis for this. There would then exist a number of possible directions for site development depending upon initial site conditions and/or category of site (assuming that there are some well-defined site categories). If the survey area is large enough, it should be possible to find each stage of site development within the existing survey data (or at least to be able to infer what it should be). The range of possible site developments can be specified in general terms in a model topology.

11.4 Current modelling methods and concerns

Much of the current work on biogeographical modelling has tended to concentrate on problems allied to climate change and its actual or potential effects on natural and anthropogenic vegetation. This is of great importance as far as the likely effects of global warming on both short- and long-term agricultural trends. It is not simply a case of higher temperatures, but changes in patterns of precipitation and weather system tracks. In this context, modelling has helped in our understanding of the complex relationships between physical, ecological and biogeochemical aspects of interactions between vegetation and climate on a range of spatial and temporal scales from individual plants to global simulations

(IPCC 1995). Some global impact studies of biosphere–atmosphere interactions have been performed with biogeographical and/or biochemical models (VEMAP 1995).

Biogeography models attempt to simulate global equilibrium distributions of vegetation under, for example, current and future climates. Most of these models are correlative in nature and rely on observed associations between vegetation and climate. The majority have, for reasons of simplicity, been developed using a rigid environmental envelope methodology in which a few climatic constraints determine the pattern of vegetation cover. Models such as BIOME-1 (Prentice *et al.* 1992), MAPSS (Neilson *et al.* 1992) and BIOME 3.0 (Haxeltine and Prentice 1996), focus primarily on modelling potential vegetation types and have combined aspects of correlative and mechanistic approaches. However, despite recent advances in our level of ability to understand the factors that control vegetation distribution, current models are not yet able to simulate the distribution of different types of vegetation uniformly well.

Current modelling initiatives in biogeography tend to stress the importance of developing reliable global dynamic models of vegetation for the prediction of realistic, rather than conjectured, impacts of future climate changes on vegetation and hence the variety and value of future vegetation types. Some approaches look at such aspects as leaf area index and productivity (Woodward *et al.* 1995), while others stress the link between vegetation form and dynamics with local climate – thus for a given climatic type, it should be possible to identify a plant functional type which is able to survive local soil and climate and produce the maximum possible productivity (Foley *et al.* 1996). In Box 11.2 two examples of modelling programs are mentioned.

BOX 11.2

Examples in modelling

There are a number of appropriate modelling packages of which Simile is a typical example. Simile provides an efficient visual modelling environment, allowing a user to draw the elements of a model, and the relationships between them, with the minimum of effort. The package provides a range of tools for visualising the behaviour of the system to be modelled, using graphs, tables or animations. Models are stored as a set of declarations about the model structure, rather like a blueprint for its design which helps users in trying out ways in which the computer can assist the modelling process, beyond simulating the behaviour of the system. http://simulistics.com/.

A recent paper (McCune 2004) describes the theory behind a new method of habitat modelling using a form of multiplicative regression. A modelling package (Hyperniche) is available for purchase or trial from MjM Software and provides a very useful introduction to the topic. http://home.centurytel.net/~mjm/nicheoverview.htm.

McCune describes a habitat model as representing a relationship between a species and factors which control its existence. Species performance can be measured in many ways, such as abundance or demographic rates, while habitat factors are essentially variables that describe a particular location or habitat. In a broader sense, habitat factors include interactions such as those which occur with other species, climate, history of disturbance or even the time of day and these are usually easier to work with in a practical fashion, using appropriate statistical tools.

11.5 Current global-scale climate–vegetation relationship analyses

It is widely accepted that climatic factors such as ambient air temperature, incident solar radiation and water availability play an often crucial role in the distribution and functioning of vegetation. Climatic factors such as these are quite complex and can be included in a vegetation model in a number of ways:

- They can be used as threshold constraints in phenology (the study of the annual cycles of plants and animals and how they respond to seasonal changes in their environment);
- They can be external drivers of physiological functions such as transpiration, photosynthesis and respiration;
- They can be used as scalars (or levels of magnitude) in representing the likelihood of different events in the life of vegetation such as the probability of a disturbance;
- It is also possible to use climatic variables as a factor in vegetation classification itself – for example, mist forest or snow bed vegetation.

It is only comparatively recently that a probabilistic (what if?) approach has been used with any degree of success to elucidate the relationships between global vegetation and climate (Shevliakova 1996). This computationally efficient approach is applicable to different types of vegetation and to groups of physiologically important climatic variables. The probabilistic approach to the analysis of vegetation distribution was originally proposed in the 1960s by Goodall (1970), but made little progress at the time because there was both insufficient global climate and environmental information available and a lack of computational power (David Goodall, pers. comm.). This approach has been used on a small scale to simulate potential (future) distribution patterns of plant species in Swiss and other central European mountain forests (Kienast *et al.* 1996).

As always, the choice of variables and identification of potential feedback loops, positive or negative, are very important if a valid model is to be produced. Appropriate factors have to be investigated and accepted or rejected, depending upon their importance to the model function; this may mean detailed investigations of, for example, soil nutrient status as well as its mechanical composition or the minimum temperature which the vegetation can survive in winter. In modelling, one has to take into account the *most likely* factors and be aware that sometimes these cannot be readily quantified. It is possible to measure quite reliably a variety of environmental parameters with modern instrumentation, but if the vegetation type is intolerant of burial under snow for more than a day or two, an extended period of snow will probably damage it just as surely as overgrazing or overheating during an unusually warm summer, regardless of the sophistication of the apparatus. Soil particle size distribution can be worked out and moisture or humus contents measured, but our measurements are almost always snapshots. We can measure many physiological factors about our vegetation, but just how much is actually important in deciding whether or not that plant will grow under apparently similar environmental conditions several hundred kilometres away. Similarly plant species differ considerably in their tolerance of environmental change, with the most tolerant species found in more places than the least tolerant.

11.6 How are models likely to affect our perceptions of biogeography?

Modelling is undoubtedly a technique with many applications, on both large and small scales, which will be of increasing importance in ecological and biogeographical research and practice. While acknowledging the importance of modelling and its value as a predictive tool in such diverse areas as resource management or climate change, it is vital not to lose sight of the fact that, almost without exception, our attempts at modelling parts of the environment are still relatively simplistic. It is relatively straightforward to devise the mathematical relationships which define the *major* parameters of a model but a lot less easy to produce a model which reliably simulates, for example, nutrient flow in a woodland ecosystem. This state of affairs arises simply because we do not yet understand enough about the many interacting factors which control life on earth to be able to put them into anything more than a relatively crude model. We should take the analogy of the feedback loop very seriously and as our research improves the 'fit' between the model and what it purports to simulate, so we should refine the model as a process of feedback. This refining process should not only include more climatic or environmental information which is relevant to the situation, but also a more detailed knowledge of the ecology, distribution (and environmental interactions) of keystone species. Successful modelling is not merely a question of using very powerful computers to churn through masses of data, but of having the right sort of information to start with, and a wide-ranging appreciation both of the complexity of ecosystems and of the innate variability of environmental response of individual plant and animal species. Providing good data like this is often time-consuming and not particularly glamorous or well funded, but it needs to be done. This is why it is important for anyone interested in ecology and biogeography to get down to learning not only about the latest trends in modelling and computing but also how to identify and observe species accurately in the field and to see how they interact with their environment.

APPLICATIONS – USING BIOGEOGRAPHY

Using ideas from this chapter in real-life situations

One of the greatest expansion areas in environmental science is the use of models to simulate situations and test hypotheses. Biogeography is well suited to be a major part of this research activity especially in the fields of climate change and biodiversity. One area that seems to be attracting a great deal of attention is the use of geographic information systems (GIS). This type of application combines database management functionality with locational information and high quality graphical display and can produce powerful data sets and analyses that can then be placed onto custom-built maps. This gives researchers the ability to query the data and suggest ways forward. One of the many examples of this can be found in the Biogeography Program of the US government's National Oceanic and Atmospheric Administration (http://biogeo.nos.noaa.gov/). As part of their research they devised an add-on module for a standard GIS program. This has enabled them to construct far more accurate images for a range of ecosystems (one example being coral reefs). Another case has been the Channel Islands National Marine Sanctuary (http://biogeo.nos.noaa.gov/projects/assess/canms/cinms/). The object of the exercise is to create a GIS map of the area and construct effective cost–benefit scenarios for a range of conservation options. There is also the aim of using the project to further develop GIS tools. In carrying out this project it is also possible for the research team to see if the data currently gathered are sufficient, if distribution patterns can be found and how they might compare with proposed conservation boundaries.

11.7 Summary

Modelling is a process which can help us to understand how an ecosystem or process functions. Models are rarely completely representative of the system or process but attempt as far as is possible to emulate the major phases or parts. Generally, most ecosystems can be modelled once the extent and activity of the controlling factors are known and allowances made for the dynamic nature of the interaction – the lynx–snowshoe hare model is a good introductory example. Island biogeography is another, rather more complex model, and although modern work has cast some doubt on its universal applicability, it works quite well in most cases. Currently, much emphasis in modelling is centred on climate change and biosphere interactions. Modelling is now more widely recognised as a predictor of what *might* happen rather than what *will*, provided it is used as a means to an end, rather than as an end in itself.

Review questions

1. Why do scientists, economists and engineers use models?

2. Why, in particular, are positive feedback loops dangerous in the real world?

3. What is the effect of a negative feedback on the functioning of a system?

4. Investigation of climate change has been a major focus of modelling effort. What purpose does the modelling serve in the 'real' world?

References

Beissinger SR. 1998. On the use of demographic models of population viability in endangered species management. *Journal of Wildlife Management*, **62**: pp. 821–41.

Foley JA, Prentice IC, Ramankutty N, Levis S, Pollard D, Sitch S and Haxeltine A. 1996. An integrated biosphere model of land surface processes, terrestrial carbon balance, and vegetation dynamics. *Global Biogeochemical Cycles*, **10**(4): 603–28.

Goodall DW. 1970. Statistical plant ecology. *Annual Review of Ecology and Systematics*, **1**: 99–124.

Haxeltine A and Prentice IC. 1996. BIOME3: an equilibrium terrestrial biosphere model based on ecophysiological constraints, resource availability, and competition among plant functional types. *Global Biogeochemical Cycles*, **10**(4): 693–709.

IPCC 1995. Climate change 1995: Impacts, adaptations and mitigation of climate change: scientific–technical analyses. Contribution of Working Group II to the Second Assessment Report of the Intergovernmental Panel on Climate Change (IPCC). Cambridge University Press.

Kienast F, Brzeziecki B and Wildi O. 1996. Long-term adaptation potential of Central European mountain forests to climate change: a GIS-assisted sensitivity assessment. *Forest Ecology and Management*, **80**: 133–53.

Lachavanne JB. 1985. The influence of accelerated eutrophication on the macrophytes of Swiss lakes: abundance and distribution. *Verhandlungen der Internationalen Vereinigung für theoretische und angewandte Limnologie*, **22**: 2950–5.

MacArthur RH and Wilson EO. 1967. *The Theory of Island Biogeography*. Princeton University Press.

McCune B. 2004. Nonparametric multiplicative regression for habitat modelling. http://www.pcord.com/NPMRintro.pdf

Mills LS, Hayes SG, Baldwin C, Wisdom MJ, Citta J, Mattson DJ and Murphy K. 1996. Factors leading to different viability predictions for a Grizzly bear data set. *Conservation Biology*, **10**: 863–73.

Neilson RP, King GA and Koerper G. 1992. Toward a rule-based model. *Landscape Ecology*, **7**: 27–43.

Prentice IC, Cramer W, Harrison SP, Leemans R, Monserud RA and Solomon A. 1992. A global biome model based on plant physiology and dominance, soil properties and climate. *Journal of Biogeography*, **19**: 117–34.

Raunkiaer C. 1934. *The Life Forms of Plants and Statistical Plant Geography*. Oxford: Clarendon Press.

Shevliakova E. 1996. *Application of Statistical Methods for Modeling Impacts of Climate Change on Terrestrial Distribution of Vegetation*. Pittsburgh: Carnegie Mellon University.

Starfield AM. 1997. A pragmatic approach to modeling for wildlife management. *J. Wildlife Management*, **61**(2): 261–70.

VEMAP 1995. Vegetation/Ecosystem Modeling and Analysis Project: comparing biogeography and biogeochemistry models in a continental-scale study of terrestrial ecosystem responses to climate change and CO_2 doubling. *Global Biogeochemical Cycles*, **9**(4): 407–37.

Warming E. 1909. *The Œcology of Plants*. Oxford: Clarendon Press.

Whittaker RH. 1956. Vegetation of the Great Smoky Mountains. *Ecological Monographs*, **26**: 1–80.

Woodward FI, Smith TM and Emanuel WR. 1995. A global land primary productivity and phytogeography model. *Global Biogeochemical Cycles*, **9**(4): 471–90.

Websites

In addition to those mentioned in the text, further information on ecological modelling can be accessed via the WWW-server for Ecological Modelling which maintains a register of models and hosts the ECOBAS project which deals with mathematical descriptions of ecological processes:
http://dino.wiz.uni-kassel.de/ecobas.html

An Internet search using Google will provide many hundreds of useful references and selective use of the advanced search feature will enable you to highlight references and sites of interest.

PART THREE

The human dimension in biogeography

CHAPTER 12

Environments under threat – the biogeography of change

Key points

- Ecosystems are dynamic: they are subject to constant change;
- Anything which causes or leads to change in an ecosystem can be regarded as a threat to the survival of that particular ecosystem;
- Threats come from both natural and human sources;
- Natural changes to ecosystems arise from long-term evolution through medium-term successional and zonational changes through to short-term catastrophic changes such as volcanic activity;
- Threats from human sources include the creation of ecosystems (e.g. from agriculture), from modification (e.g. grasslands) and destruction (e.g. rainforests) as well as short-term catastrophic events such as warfare;
- One of our most important practical tasks is the measurement and assessment of threat levels;
- A key current use of biogeography is to find and implement solutions to threats facing ecosystems.

12.1 Introduction

Ecosystems are in a state of continual change and yet such dynamism is rarely acknowledged in a biogeography whose maps tend to accentuate the static. This, however, can work to our advantage in the study of change. Assume that the *status quo* is the unchanging nature of a given area and its associated ecosystem. Anything that disturbs that situation could be seen as a threat to it. It is entirely possible (and within ecological theory) that the ecosystem will try to outcompete or otherwise nullify the change. If the change is halted then the ecosystem wins and the *status quo* is maintained – in ecological terms the system is stressed but recovers. However, if the stressor is too great then the ecosystem loses and the change is permanent, i.e. the ecosystem has been unable to overcome the threat. The boundary between win and lose depends upon the nature of the threat, its extent and the nature of the ecosystem. It is possible that some ecosystems are in a more vulnerable position than others. We tend to refer to these as vulnerable or fragile ecosystems (or as threatened/endangered species for individuals).

This places change at the heart of a dynamic system and the concept of threat as a force promoting change. Such a point is crucial: it means that we can study various changes in the ecosystem and recognise common factors and traits which lead (or do not) to change. In this chapter we examine a range of threats both natural and human and assess the impact that they can make. A new, but rapidly expanding, field is the study of threats and the changes that can occur. Under the umbrella title of 'Environmental Impact

Assessment' these techniques help us to quantify or at least identify the nature and magnitude of threat. It is appropriate that we investigate these techniques because, apart from their value, they link threat with the final section, the solutions to threat. Since the widespread recognition of threats and the need to reduce or remove them, a number of strategies have been employed from international law to local politics (see Figure 12.1). Do we need to bother? Consider this:

> Amazonian rainforests are some of the most species-rich tree communities on Earth. Over the past two decades, forests in a central Amazonian landscape have experienced highly non-random changes in dynamics and composition. Our analyses are based on a network of 18 permanent plots unaffected by detectable disturbance. (Laurence *et al.* 2004, p. 171)

Figure 12.1 *Phyllodoce caerulea* is one of more than 60 species of plants that are protected (in the UK) by legislation because of its rarity. While relatively common in suitable habitats in northern North America, Siberia, Iceland and Scandinavia, outside this main area of distribution it occurs only in very small quantity in three sites in Scotland (and one relict site in the French Pyrenees). The plant has been known in small quantity in one of its Scottish sites for 170 years but was only discovered in two more sites in the last 30 years. It has probably never been at all common or widespread in Scotland and traditional UK *Calluna*-moorland management practice (sheep grazing and rotational burning to provide mosaics of mixed-age heather for red grouse (*Lagopus scoticus*)) will restrict it still further because of its poor competitive ability at the limits of its geographic range

Although it uses some highly cautious language the actual point is that rainforests are changing even though there is no obvious human activity. This implies that change may have a longer-range impact than just the area it occurs in. It follows that anything which might cause change needs to be considered: this study suggests we cast our net further than previously.

12.2 Causes of threat

Much of the literature seems to assume that threat is a human-centred term, i.e. people cause a threat to the environment and it changes. This can be seen as overly simplistic. If the object of any ecosystem is to survive then anything that promotes change must be seen as a threat to that system. If, because of threat, the ecosystem changes then it matters little if the change is brought about by natural or human agencies. The inclusion of natural agencies is important because it links the natural dynamic with human activities like conservation which seeks, in effect, to maintain the *status quo* even in the face of natural changes through succession. This picture is further complicated when one considers that, in addition to the various individual causes of threat, there are factors such as time and scale to take into account and that each of these will operate alongside (or nested with) the others. This produces a large range of possibilities, leading some writers to conclude that the landscape is better served by focusing on the idea of mosaics rather than large homogeneous areas (Forman 1995) – see Figure 12.2.

12.2.1 Natural threats

Natural threats to ecosystems can be divided into three main areas – long, medium and short term depending on the duration of the threat. The area covered by such forces can be as small as a few islands or the entire globe; they operate at all scales:

Figure 12.2 An extinct volcano, Victoria, Australia. The unique wetland ecosystem built here has been subject to both natural and human forces. Volcanoes are relatively uncommon in Australia and this extant cone has supported a unique system. Natural changes such as drought and flood are added to by human activity of which the most recent is conservation. Careful monitoring has allowed koala numbers to increase considerably (and probably further than the 'natural' system might allow). What impact does each of these have on the area?

(a) Long-term threats

These are brought about by the development of species, i.e. evolution. This seeks to replace one species with another better suited to the conditions at the time of change. How or why this happens is still a matter for dispute but the biogeographical implications are clear. In terms of geological timescales, the plants and animals in an area are likely to change completely (but not all at the same time, or rate or area, making palaeoecological reconstructions difficult). There are significant problems with this approach. Uniformitarianism, one of the key tenets of geology (*the present is the key to the past*), stumbles when it reaches evolutionary scales. An extinct species might look like a present day one, but have a different ecological range. New species might evolve and others die out all without any obvious pattern, suggesting that the further one goes back in time the more fraught are one's assumptions in biogeographical analysis. Despite these problems, there are some examples which can be used to illustrate this theme; one of the better known are the Galapagos finches first brought to our attention by the works of Charles Darwin. Darwin noticed that the individual islands of the Galapagos appeared to have distinct species of finch. He argued that with the islands being so isolated (about 1,000 km from the Ecuadorian coast) it was unlikely that several species of finch arrived to colonise specific islands. It is more likely that the area was colonised by one set of finches and that they spread to the islands in the group and evolved to suit the conditions peculiar to that island. Despite the presumed ability to do so, Galapagos finches do not appear to have interbred or even just moved islands. It is not only finches which have demonstrated this evolutionary divergence of form. Darwin also noticed the changes in body shape among the Galapagos tortoises related, it is thought, to differences in food supplies. The biogeographical problem arises when one tries to map such distributions. When do the differences become great enough to constitute two species rather than two varieties of the same species?

(b) Medium-term threats

A similar situation arises when one considers the changes that occur through time (succession) or space (zonation). The timescale for this depends on the area and the nature of the change: it can be anything between one year and thousands. Both are natural responses to immediate changes in environment and involve the substitution of one or more species with others. The effect may be so profound that an entirely new ecosystem is created. The chalk downland areas of southern England provide a suitable example. The early chalk rock colonisers gave way to grassland, scrub and eventually forest. Human activity removed the trees and created a semi-permanent grassland (see also Figure 12.3). Despite the seeming simplicity of this it should also be borne in mind that research shows considerable complexity in responses between species, e.g. in shading between tree species (Gilbert *et al.* 2001).

Figure 12.3 Despite the idea that this is a natural scene, farmland near Abergavenny, South Wales, shows the impact of thousands of years of modification

A more instructive case is given by Stiling (1999) who describes successional changes in Alaskan glacier vegetation. The first colonisers of the glacial moraine are algae, lichens and bryophytes. These plants are then invaded by fireweed (*Epilobium*) which is in turn replaced by alder trees. Ultimately, spruce trees occupy the role of climax forest vegetation. The interest in this case lies in the interactions between the various stages (or seres). The old idea of succession was that each preceding plant association left the environment sufficiently altered for a new species to invade. This would be repeated until the climax vegetation was reached. The process (called facilitation) appeared consistent with the evidence until more evidence was brought to light. Research demonstrated that facilitation only operated at one time – at other times it was competition between species. Lichens and mosses colonised the bare rock and soil. With little nitrogen, these plants could not rely on the soil but collected nutrients from the air and water. The next stage of succession was invasion by *Dryas drummondi* and alder (*Alnus sp.*). Both these species have the ability to fix nitrogen in the soil (they have a symbiotic actinomycete, *Frankia,* in root nodules) and thus they build up soil fertility. Initially, alder shades out *Dryas* and is in turn outcompeted by the Sitka spruce which forms dense stands. Thus facilitation only truly occurs (according to definitions) when alder puts nitrogen into the soil as fixed nitrate. At other times, plants are outcompeted. If this is repeated elsewhere (and there is no reason to think that it is not) it throws into question both the nature of succession and the biogeographical implications in terms of mapping, etc. Given that successional changes can create considerable environmental stresses (Stiling 1999, p. 489) it is worth considering where one draws the boundary for vegetation mapping, i.e. which part of the sequence is held to be the true ecosystem? It also creates other, technical, difficulties in the interpretation of ecological change. To avoid this, some ecologists use the idea of 'assembly rules' (see Box 12.1) which outline how species are distributed through all the stages of the succession. The aim is to create a set of ideas which help us manage successional areas.

BOX 12.1

Assembly rules, guild rules and successional change

The examples given in this chapter all assume that there is some definable process as species give way to other species. The classic work by Connell and Slatyer published in 1977 (ideas of facilitation, tolerance and inhibition) is one of the most tempting but it is not without its critics. For example Horn, writing around the same time, suggested that succession was a matter of probability: the more seedlings of a tree that were present, the more likely that species would succeed (called a Markov chain, one of the important uses of statistics in biogeography). Other ideas can also present us with challenges. ▶

For example, the idea of assembly rules is that succession takes place in a given sequence. Work in New Guinea found that succession appeared to follow a set of rules – given combinations of species were allowed or rejected according to the existing biota. This made succession a more dynamic interplay of existing and potential communities rather than simple replacement. Dating from the late 1960s, Root considered that there were sets of species that operated in similar niches – he referred to these as guilds. In terms of succession one would assume that the guild as a whole would operate to include/exclude new species which again is a different idea from the usual notion of the individual species. For biogeographers, this leaves the interpretation of change wide open: there are still many important questions awaiting definitive answers.

Zonation is the distribution of organisms in space. Rocky shores provide an excellent visual reference usually because of the striking colour changes between species (Little and Kitching 1996). The idea is that those species such as kelp that can withstand little change are found in the lowest zone. A 'middle' zone of barnacles and limpets is next because they can withstand some drying out but not too much. Those species which can withstand almost complete exposure to air such as *Littorina* sp. and lichens dominate the top zone. Although it might be argued that coastal zones will remain static they do show change. What makes this interesting are the timescales over which such change operates. Rising and falling sea levels will affect water and nutrient supply for plants and alter the areas colonised by various species. Clearly this operates over a long timescale in the order of thousands of years. At the other end of the scale, it might only take a year to make complete changes. Studies quoted in Little and Kitching (1996) show the variation that can be seen:

> Over a period of 10 years, [it was found that] numbers in the *Mytilus/Semibalanus/Patella* assemblage at the bottom of an exposed shore fluctuated violently. These fluctuations were due to such factors as the irregular phases of predation . . . the settlement of barnacle spat . . . Thus predation . . . was limited in upshore extent by the relationship between their desiccation tolerance and the temperature regime in a local year. (Little and Kitching 1996, pp. 27–9)

The lesson here is that any zonation can show change not necessarily associated with succession. As with succession, it follows that in mapping the distribution of plants and animals one must be careful of the choice of boundaries. In addition, few zones are so clearly marked as those on rocky shores, making specific zonal rules even more difficult to establish.

(c) Short-term changes

Whereas changes in the biogeography of evolution and succession produce fundamental alterations in distribution, few causes can be more spectacular than the short-term changes wrought by catastrophic events such as storms and volcanoes. The much-researched Mount St Helens volcanic eruption of March–June 1980 provides an excellent case study (USFS 1999, Lenahan u/d, del Moral 1999). Mount St Helens has had a history of eruptions in the recent past (5,000 years) and had, according to US Geological Survey volcanologists, a 70 per cent chance of erupting again before the end of October 2004. The 1980 eruption blew away a portion of the cone of the old volcano and directed a blast of heat, gases and debris which affected over 500 km^2 of forest. Prior to the eruption, the area was part of the Pacific Northwest forest area described by Barbour and Billings (1988). It is characterised by a dominance of large and long-lived conifers such as Douglas fir (*Pseudotsuga menzeii*), western hemlock (*Tsuga heterophylla*), western red cedar (*Thuja plicata*), Sitka spruce (*Picea sitchensis*) and coast redwood (*Sequoia sempivirens*). The forest thrives in a climate of wet, mild winters and dry summers, giving it one of the highest biomass productivity rates of any ecosystem. The eruption which destroyed so much of the local area demonstrated both an environmental gradient in terms of volcanic deposits and a matched response from the wildlife. The United States Forestry Service (1999) describes six zones (see Figure 12.4):

1. Pyroclastic flow zone – an area of lava and gas reaching temperatures of 600 °C. Nothing survived in this area;
2. Debris zone – part of the cone slid down the mountain rapidly filling a 60 km^2 area to a depth of 45 m. Only some vegetation carried by this avalanche sprouted;
3. Mudflow zone – a flood of material that swept downslope in ribbons. Only a few fragments of plants survived to sprout (although the area recovered very quickly due to the influx of seeds from undisturbed areas);
4. Blast zone – a 25 km radius area of forest to the north was knocked over or stripped away in winds up to 1,000 km per hour. Recovery was variable depending on the damage to the particular organism;
5. Scorch zone – beyond the blast zone but with wind still hot enough to kill (but leaving nearby trees unaffected);
6. Ash zone – anything from a light covering hundreds of kilometres away to a 1m blanket nearer the volcano. The actual damage ranged from light retardation of growth to death in extreme cases.

Animal species associated with the forest were also affected, with reductions in local populations of deer, elk, bears and numerous bird species. Remarkably, researchers found that the area still had life in all but the most severe areas. Often small pockets of snow or a slight slope had protected the organisms where others perished. Burrowing mammals had survived where their ground-dwelling cousins had been killed. Insects were badly affected by heat and dust (although ants survived to be one of the first recolonisers). In terms of the physical environment the eruption had a considerable impact on atmospheric particulate matter (with dust from the eruption plotted around the world). Rivers near to the blast site were filled with debris which might lead to loss of fish stocks. Lakes were also affected differently with some being destroyed, while others were created by damming local rivers. Generally, the clear, low-nutrient lakes were replaced by ones rich in nutrients from the blast. In most cases, however, recovery to the initial state took only a few years.

This loss of old-growth forest should be seen as an outdoor experiment in recolonisation. Studies have shown that the return of species is determined by the way they were affected and not by the usual dictates of succession. The main species were initially Pacific silver fir (*Abies amabilis*), mountain hemlock (*Tsuga mertensiana*) and huckleberries (*Vaccinium spp.*). Fireweed (*Epilobium angustifolium*) and pearly everlasting (*Anaphalis margaritacea*) have been the main ground cover where the dark understorey of the woodlands has been replaced by a far more open setting. In the long term, researchers expect weedy plants like fireweed to continue to spread with shrubs colonising soon after that. Eventually, the forest will return.

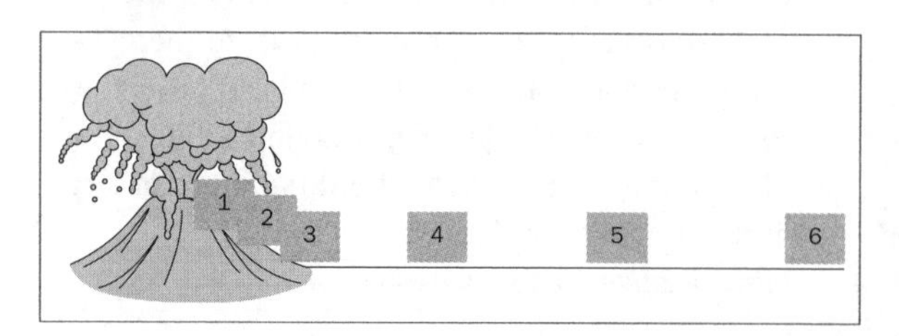

Figure 12.4 A schematic diagram of the Mount St Helens eruption and its impact on the local ecosystems. Numbers refer to the zones in the text. Note that the impact is reduced the further away one gets from the cone (referred to as the distance decay effect). Impact is further constrained by local weather conditions giving 'upwind' and 'downwind' patterns – effects are only regular in models!

Biogeographically, the eruption is an interesting case. The recolonisation of the area does not appear to have followed the conventional seral succession due, largely, to the differential effects of the various volcanic products. Studies carried out by del Moral (1999) over a (nearly) 20-year time span (see Box 12.2) have shown that recovery of both ground cover and species diversity is erratic and linked, again, to the volcanic materials in the given area. In the 1980s the main growth of vegetation was linked to the few survivors. In the 1990s this has changed, with an increasing number of species coming into the area and filling in spaces left.

BOX 12.2

Mount St Helens – models and species recovery

The case of Mount St Helens should continue to draw our attention. Time-series data like those gathered by del Moral (1999) shows how local events/features can alter the picture so that models might be useful to study but are no replacement for actual fieldwork. Using data from del Moral, one finds the graphs produced from three sites as shown in Figure 12.5. The three sites suffered different impacts from the eruption. The primary site had virtually no survival, the secondary site had significant understorey survival and the lahar (or mudflow) was covered by near intact vegetation. We get two main ideas from here: the gradual growth of plant cover and the differential survival/recovery of species. Data from del Moral (1999).

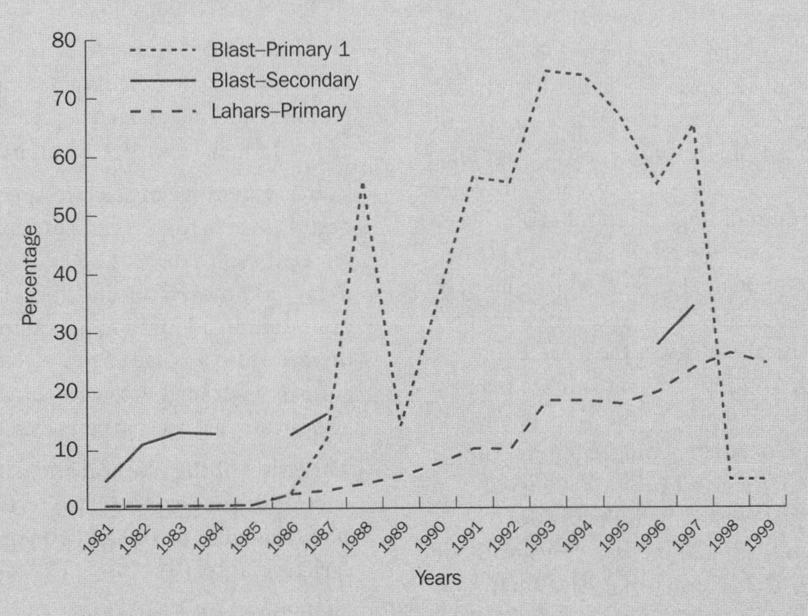

Figure 12.5 Percentage cover – three areas in Mount St Helens

12.2.2 Human threats

> The destruction of habitats is widely regarded to be the most severe threat to biological diversity, while fragmentation and degradation, which are often precursors of outright destruction, also present significant cause for concern. In many countries . . . where human population densities are high, most of the natural habitat has been destroyed to provide farmland, rangeland, and land for settlement and industry. (Middleton 1995, pp. 87–8)

While the eruption of Mount St Helens is a massive local problem, Middleton reminds us that it can pale into insignificance alongside the inexorable spread of human impact upon the environment. There are numerous examples at all scales illustrating anthropogenic impact, but they could all be considered to fall into one of three categories: ecosystem creation, modification or destruction. These can be seen as the human counterpart of the natural agencies discussed above. We hear most about destruction but, with the vast majority of the UK, for example, being modified through centuries of intensive land use, the other categories deserve attention also. One of the main problems for biogeographical mapping is the basis that is to be taken. For example, if you are mapping the South of England should you choose the actual vegetation (the result of human action, e.g. agriculture), potential vegetation (the natural environment if human action were removed) or the original natural ecology of the place (oak/ash/beech woodlands)? This is more than a technical argument: it goes to the heart of any mapping system of study of biogeography. The land use maps of the Land Utilisation Survey showed the actual vegetation at the time of recording but very little in the way of historical or ecological data. The Ministry of Agriculture, Fisheries and Food's agricultural land classification or land capability maps give a greater insight into current and potential land use but still fail to address the question of what is the actual biogeography of the area.

(a) Ecosystem creation

Agriculture is perhaps the greatest example of human ecosystem creation. As discussed above, its existence calls into question the way in which we ought to view the land. Since its spread over 8,000 years ago it has taken over an increasing proportion of our land surface, often in those most productive areas which could support alternative ecologies. In the last 300 years it has been estimated (Richards 1990) that forests and woodland have decreased globally by 18.7 per cent, with corresponding figures of –1.0 per cent and +466.4 per cent for grasslands/pasture and croplands respectively. This represents a considerable shift in species diversity and individual species' biogeography especially if we note that in the North America, croplands increased by 6666.7 per cent! Although the precise impact in any given area depends upon a number of variables such as agricultural practice, climate, economics and technology, it is possible to outline some of the common impacts:

- **Diminution of local species and species richness** – existing plants and animals will be removed and the preferred, agricultural species, allowed to take over the vacant habitats;
- **Increase in range of selected species** – as has been shown before, agriculture spreads from a few centres. It follows that those species used today were moved outside their ecological optimum. In addition to agricultural species there are also a range of other plants and animals associated with agriculture that have also spread, e.g. weeds, pests etc.
- **Alteration of the physical and chemical environments** (e.g. drainage, fertilisers) – to allow the more successful development of agricultural species;
- **Reduction in the gene pool** – apart from loss of local populations farming appears to be developing (at least in North America and Europe) a trend of replacing local agricultural species with far fewer approved varieties. In the

EU there has been much concern that 'old' varieties of, for example, potatoes and apples which have for centuries been locally derived, are being forced out of production in favour of a handful of EU-approved cultivars.

Lest it should be thought that habitat creation is a terrestrial occupation it should be noted that there are an increasing number of cases where old ships and even oil rigs are deliberately scuttled to create artificial reefs. These function in much the same way as their coral counterparts in that they provide shelter and a useful substrate to grow on. Given the increase in the popularity of diving and the concomitant pressure on natural reef areas, such artificial aids are becoming increasingly valuable as both an economic and ecological resource.

(b) Ecosystem modification

Here, an existing ecosystem is altered to allow some species to develop at the expense of others. Of the many examples that come to mind perhaps the two most common would be woodlands and grasslands. Both of these are natural ecosystems but they have been altered in the UK with over 1,000 years of management. Woodlands such as the New Forest in the South of England were originally wild areas just developing after the Ice Age. Slowly they became a hunting resource for the Mesolithic and Neolithic peoples. This situation hardly changed until the Norman French invasion of 1066 and the subsequent development of the New Forest out of the native region of Ytene. Initially, hunting was restricted and new species brought in. The development of forests into productive units started about this time. Trees were grown to specific heights/widths for purposes as diverse as broom handles and ships masts. By the fourteenth century the New Forest was a productive farmland and manufactory. The impact of this on the original flora was to alter the composition and density of species. For example, oak trees were grown to a standard size (12 per acre/30 per hectare) so that ships' timbers could be produced with minimum need for further shaping. Species not wanted were removed or, in the case of farm animals, fenced into a given area (the so-called wood pasture). This practice declined with the advent of the Agricultural and Industrial revolutions. Since 1950 there has been a revival of interest in the place as a tourist spot with another set of changes in species composition and distribution arising from the open spaces required by visitors.

The destruction of parts of the coastal mangrove communities in South-East Asia for fish farming lagoons is a worrying and increasing trend in environmental exploitation. Mangroves, such as the example in Plate 11, help stabilise coastal sediments and provide valuable littoral and estuarine habitats for plankton, crustaceans and fish. Their removal reduces the level of coastal protection against such potentially catastrophic events as tidal surges, typhoons and even *tsunami*.

Grasslands developed, at least in the South of England, from the enlarged clearings in the chalk downland. The specific grazing patterns of sheep removed vegetation differentially. Over generations this allowed some species, e.g. orchids, to thrive. This led to the development of a species-rich grass sward more diverse than the original woodland glade. However, it did become tied in with human action. This has been demonstrated several times when uniformed management has been employed. Often the first action has been to remove the sheep to protect orchids from being trampled. Unfortunately, this led to the restarting of the succession and the growth of scrub which effectively outcompeted the orchids. Today, most conservation organisations such as English Nature have a sheep programme to help manage grassland areas.

(c) Ecosystem destruction

This is by far the most serious threat to many species according to some researchers (e.g. Middleton 1995). Alongside the outright destruction there are also changes in the surrounding areas through fragmentation of ecosystems and degradation of habitats. Furthermore, these changes do not seem to be related to any given place or time although the rate of change often is. It is possible to study losses of habitat in preliterate societies and modern nations. The rate of change is usually thought to increase with time since modern techniques of land clearance are much faster than earlier methods.

12.3 Assessment of threat

Whatever the threat, the most important response today is to assess its impact and suggest ways in which this might be addressed. Over the past 30 years a considerable array of techniques have been used. Today, at all levels from global to local there is a reassessment of the techniques employed and a standardisation of methodology around the notion of environmental impact assessment. As threat becomes a legal issue so the definition of threat takes on a significance beyond the ecological. Some solutions to threat such as the Convention on the International Trade in Endangered Species (CITES) rely upon an organism being placed in a defined category for it to be classed as a protected species. In this context, Middleton (1995) outlines six different categories for threatened species:

- extinct – not seen in the wild
- endangered – unlikely to survive
- vulnerable – may well become endangered if conditions do not improve
- rare – small population not currently at risk
- indeterminate – under threat but not enough known to place it in a specific category
- insufficiently known – threat suspected but not enough information to be definite

The key implication of this is that it demands considerable knowledge of the species to appreciate which definition it goes under. It also takes biogeography out of the academic and into the legal arena which is where conservation issues are increasingly fought over.

We can see change but we cannot see threat. By this we mean that, for example, although we can see a tree being cut down we cannot assess the specific threat to the general environment as a result of that specific action. Numerous studies have been conducted to assess threat to organisms and environments. Middleton (1995) reports on examples as diverse as Australian and American mammals and pre-agricultural changes in Africa. All these studies share a common thread; they are based on individual methodologies which, while shedding light on a particular area, do little for the global picture. These could be referred to as *ad hoc* studies. In contrast, there is an increasing interest in impact assessment and the collection of baseline studies from which change can be more readily quantified (see e.g. Petts 1999, Barrow 1997, Treweek, 1999, Roberts and Roberts 1984 and Institute of Environmental Assessment 1995) and which could be considered systematic studies. Since the best ad hoc studies follow a common path (which has been built on for systematic studies such as ecological impact assessment – EcIA) it makes sense to provide a critique and overview of this concept.

One of the first attempts to measure threat was developed in the late 1940s by Leopold. Originally put forward as a way of highlighting interactions of development and environment (and thus being, strictly, a qualitative measure), it is based on a matrix pattern where axes for development and environment intersect and can be used to alert people to potential impacts (see Figure 12.6). It would be possible to rank the impact and so produce a rough quantitative guide. Its key problem is also its main advantage – simplicity. As a visual record, it draws people's attention to main areas of impact, but at the same time does not allow for accurate ranking of impacts nor does it allow for long-term impacts or impacts following on indirectly from development. Further development to allow it to be used in impact analysis (especially in the USA) has led to it being too large and cumbersome for rapid use. Other workers have developed similar systems but all suffer from the

	Modification of regime			
Proposed actions	Species introduction	Biological controls	Habitat modification	Alteration of ground cover
Mineral resources				
Construction material				
Land form				
Unique physical features				

Figure 12.6 Part of a Leopold matrix-type form. The idea is that each action (row) could generate a reaction (column). All that would be needed would be to assess the strength of that potential reaction to see where the greatest problems could be caused. In reality the form could generate over 450 possible impacts which provided coverage but lacked detail

trade-offs between usability and comprehensiveness, i.e. systems simple to use do not reflect the complexity of the environmental interactions.

Today, quantitative study has moved on to ecological impact assessment (EcIA). According to Treweek (1999, p. 6):

> The idealised EcIA process is flexible, iterative, proactive and based on accurate, consistent, transparent and defensible methods. This is only achievable if EcIA forms part of integrated systems of environmental regulation, based on strategic planning and regulation of environmental quality and operating standards in relation to quality objectives or standards.

Although we are still some way from Treweek's ideal view, the value of EcIA in our study of biogeography is that it allows a reasonable critique of the measurement of threat. From the quote above, three notions are put forward. By 'flexible' it assumes that there are a range of methodologies that can be used. Whereas this is useful in pinpointing threats in a specific environment it does less to provide the ecological baseline that should be needed. 'Iterative' suggests that the process is ongoing, a continuous cycle of monitoring. This might be the ideal, but increasing pressures of funding make this less likely. Finally, 'proactive' posits the view that this process should identify threats before they become too difficult to solve easily. As will be shown in Chapter 13 (fragile environments), this is not always the case.

The EcIA process consists of six main areas:

- **Scoping** – basically this involves delimiting the study – which area, what key species/ecosystems, etc. It is also possible to consider potential areas of threat and to suggest where any study be best focused. To this extent, scoping is much like the work of Leopold described above;
- **Focusing** – as the name suggests, this takes the preliminary study and identifies specific areas of interest, e.g. valuable habitats, rare ecosystems/species. It can also be used to refine the methodologies that can be used – specific ecosystems or genera will need specific sampling and monitoring procedures;
- **Impact assessment** – the first element here is the collection of baseline data, i.e. a control area where there is unlikely to be any threat and against which any changes can be checked. Given full site analysis it should be possible to predict likely changes to either habitats and/or species. It is in this area where EcIA differs from other ideas on the measurement of threat. Other studies note changes and suggest causes and may, as a by-product, highlight the importance of these changes and what can be done to ameliorate them, but EcIA demands that this be an integral component of the study. While this is its greatest strength it is also a potential weakness in that only known threats are likely to be checked for and that any reduction in scope due to commercial considerations would weaken the process;

- **Mitigation** – once the threat has been assessed it is possible to suggest ways in which it can be reduced or removed;
- **Evaluation** – the monitoring of the area to check upon the actual changes;
- **Feedback** – using the data from the evaluation to fine-tune the policies that should have been in place to mitigate the threat.

EcIA is obviously rigorous in its approach to measuring threat. This is both its great strength and main weakness. However desirable it might be to check entire regions, costs and practical considerations preclude such a vast scheme. If such schemes are only for specific developments how do we monitor threat on a more general level?

Fortunately, there are numerous examples of threat monitoring which, together, allow us a considerable insight into human impact. Whereas these do not add up to a comprehensive global monitoring system as yet, there is sufficient (with extrapolation and the use of key species) to present us with a guide. One of the best known of these is the Red Data Book series originated by the IUCN (Caughley and Gunn 1996).

Since 1994, when the IUCN adopted the Red List system there have been numerous problems mainly concerned with the volume of work, the comprehensive coverage of species and areas and the definitions of threat and the methodologies and criteria needed to study them. Despite these difficulties the Red Lists, as they are known, comprise assessments for about 5,000 animals and 30,000 plants (see Figure 12.7). Red Data Book concepts have been translated into numerous areas, scales, taxa and languages, leaving a considerable body of knowledge and ideas from which to assess threats and to guide the provision of solutions (Holdgate 1999).

12.4 Solutions to threat

Once it has been proven satisfactorily that threat exists to an area or species a series of solutions are

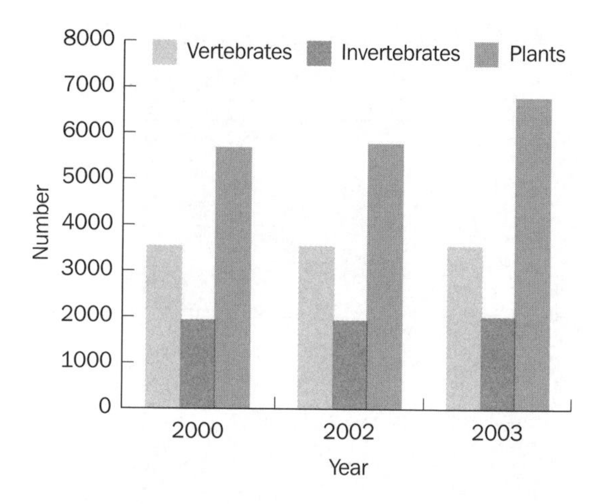

Figure 12.7 Number of threatened species by major groups of organisms, 2000–3. Although this number is small compared to the total number of species it represents a reliable measure of human impact. Taken from Red List data (http://www.redlist.org/info/tables/table1.html – accessed 10/4/04)

available. These can be divided into five categories – conservation, legal, educational, organisational and political (although many solutions will operate in more than one category). Each of these categories has the same ultimate aim – the reduction or removal of threat and the recovery of the species or area – but each goes about it in different ways.

12.4.1 Conservation

This acts directly on the species or area concerned. Because of this direct action it is more often seen in plant and animal recovery, with one of the best examples being the marine environment. For centuries the ocean has been treated as a free good and exploited without regard to impact. During the twentieth century it became increasingly obvious that fish and mammal stocks were declining, e.g. whales and tuna. The initial response was to either change species (in the case of whales) or increase the fishing effort (for tuna). Since the yield per unit effort decreased it is obvious that in order to meet demand, the overall effort would increase.

However, in doing this the total catch would eventually decline due to lack of restocking (which is exactly what has been seen in these cases and others). The solution has been to study the human impact, highlight the chief causes and to eliminate or reduce them (with stocks returning naturally if demand is removed). Whales became part of a global movement to ban all hunting while tuna are starting to recover with the application of fishing limits. Although the aim is to allow the species to recover it is entirely possible that some species will become extinct due to the very low numbers left.

12.4.2 Law

The law aims to reduce the threat by controlling human action. Although it might seem cynical to suggest that is only when the law is applied that any area or species receives protection it is undeniable that this method has had considerable success at all scales. It can be applied locally to protect small nature reserves; it can be used to secure the survival of nationally significant areas such as the UK National Parks and it is used to control global activity where, by the very nature of the legal situation, it provides support where no other category can. At the global level one of the best examples is CITES – the Convention of the Trade in Endangered Species – which came into force, when sufficient support had been gathered, on 1 July 1975 (Lyster 1985). Its aim is to regulate the trade in species. Endangered species can be placed in one of three categories (or appendices to the treaty). Appendix I species have an almost total ban in trade while Appendix II species can be traded within strict guidelines. Appendix III is used to bolster protection for locally endangered species not on Appendices I or II, i.e. it supports local legislation by stopping outside trading. Today a great number of states accept CITES – it can be regarded as one of the best examples of conservation law.

12.4.3 Education

In its widest sense education is becoming an increasingly important tool. The options in this category spread from a few local groups deciding to protect their area through to national school and university education schemes and global research programmes such as those supported by the OECD and UNESCO. In the last 50 years the growth in this field has been enormous, although it is fair to say that the returns have yet to be seen as correspondingly great. Its value can be seen as universal:

> There is no field of knowledge that is not being turned inside out by the environmental revolution. In nuclear physics, with its long-term pollution dangers; in biology, stressing man's place in the ecosystem; in the humanities, social sciences and religion, the inescapable human situation of exploding population and demand within the limited resources of a finite planet has become the reality to study. (Carson, 1978, p. vi)

The argument was that this reality could best be served through an environmental education programme:

> Environmental education is the process of recognizing values and clarifying concepts in order to develop skills and attitudes necessary to understand and appreciate the inter-relatedness among man, his culture and his biophysical surroundings. Environmental education also entails practice in decision-making and self-formulating of a code of behaviour about issues concerning environmental quality. (Carson, 1978, p. viii)

It is this second aspect that sets environmental education apart. It demands not only knowledge of subjects as diverse as ecology and physics but also that the student puts forward a plan of action based on the study of the environment. In its extreme form this would suggest living in a manner consistent with one's environmental education philosophy, e.g. in a self-sufficient, low-energy, low resource demand setting. Today, a large number of school and university courses have put some thought towards environmental education and most have environmental themes in their working practices (e.g. paper recycling). Increasingly, business is seeing the benefits of this way forward.

12.4.4 Organisations

These have been involved in various aspects of biogeography for over 150 years. Their activities cover all manner of issues and they can be seen at all spatial scales from local to global. Given the range of activities and their coverage it is difficult to produce a definitive typology, but it is possible to highlight the main actors involved in contributing to discussions on the biogeography of change:

- **International organisations** – primarily the United Nations and associated organisations. Their aim is to focus on issues of global importance of which the conference by UNCED in Rio de Janeiro in 1992 (Agenda 21) is just one of the latest and most important. Although it is possible to criticise them for lack of activity, working outside national legal frameworks does limit their choices. They do, however, have a chance to discuss issues and push national governments towards action;
- **Governments** – the primary legal body for the nation. Most nations have departments specifically looking at environmental issues (such as the UK Department of the Environment). They might also have specialist groups associated with them to provide detailed research, e.g. CEH (Centre for Ecology and Hydrology) in the UK or CSIRO in Australia;
- **Non-governmental organisations (NGOs)** – groups usually with an international focus but not allied to any given political force. The rise of NGOs has been best seen in the 1992 Rio conference where considerable prominence was given to them. The IUCN (International Union for the Conservation of Nature) is just one example which has worked since its inception in 1948 to counter threats to the environment;
- **Learned societies** – academic organisations which exist to disseminate research findings among like-minded professionals. Some have a global focus, e.g. British Ecological Society, Royal Geographical Society; others have started from amateur groups to become world respected, e.g. Royal Entomological Society;
- **Pressure groups** – difficult to define and covering a wide range of issues and scales, these are groups whose publicity might be said to outstrip their numbers. These are organisations trying to raise consciousness about environmental issues. Some groups are global in focus and finance such as Greenpeace and Friends of the Earth, while others exist just to protect and conserve the village pond or local wood. Given the rise of grassroots politics, pressure groups are becoming more important in local and national politics.

12.4.5 Politics

This is the final category. Conservationists might provide the skills, legislators the means, educators the knowledge and organisations the publicity, but it is in the political arena that the battles must be fought. Rarely can this be better seen than in the development of UK environmental policy in the 1940s (Cullingworth and Nadin 1994). During the Second World War there was a small group of individuals who, together in various committees and study groups, set the foundations for current conservation in the UK. After the war their plans were largely put in place including the British planning system, the Nature Conservancy and National Parks (not in Scotland). The National Parks and Access to the Countryside Act 1949 started to provide an effective solution to environments under threat which, despite setbacks in recent years, continues to provide a level of protection which has not been equalled in the UK.

12.5 Summary

Ecosystems are dynamic entities: change is the norm. However, we tend to distinguish between generational changes, i.e. the replacement of one population by its younger equivalent, and species changes where the characteristics of the ecosystem (or the entire ecosystem itself) might change. This provides the biogeographers with a series of insights and challenges. This chapter has assumed that anything potentially altering an ecosystem can be seen as a threat to its survival. This allows us to study natural as well as human actions to evaluate change. Natural systems operate over a range of spatial and temporal scales. Evolution, not initially seen as a biogeographical concept, is becoming increasingly important in looking at longer-term changes in our biota. The most common changes are those brought about by succession and zonation. These might be seen as simple regular changes but we have yet to capture a comprehensive model. Finally, there are the short-term, catastrophic changes brought about by volcanoes, earthquakes, landslides, etc.

Each of these changes operates on a different scale. In recent times we have added human action which can, in some circumstances, rival natural action for the changes it can bring. Humans can create, modify or destroy ecosystems with concomitant impacts on ecosystems both directly and indirectly involved (or, as in the case of Amazonia, some distance away).

Alongside increased alteration of ecosystems has been the development of systems which review and evaluate change. At the global level we have research in areas like global warming but below that, the use of types of environmental impact analysis seems to be the most promising. Such methods are now routinely used in development studies but they could also apply to our ideas in the next chapter – the problems of fragile environments.

APPLICATIONS – USING BIOGEOGRAPHY

Using ideas from this chapter in real-life situations

Given the current endangered status of a range of species it seems that an obvious application of biogeography is wildlife conservation. The approach could be seen as two-pronged. The theoretical side deals with species distribution and abundance. The practical side is focused towards maintaining biodiversity in specific situations. The aim of conservation is the maintenance of desired ecosystems and assemblages. Although this started in earnest in the nineteenth century in both Europe and North America the real growth occurred after the Second World War. Today, there are numerous organisations dedicated to conservation, ranging from the global giants such as Greenpeace and the World Wide Fund for Nature to more national ones like the Wilderness Society in Australia (www.wilderness.org.au). Their activities focus on a range of issues but one of the more persistent ones is the battle against the logging of old-growth forest in Tasmania. The argument is that this area is pristine temperate forest (and part of the South-west Tasmania World Heritage Area) and it should therefore be left untouched. Tasmania is also a place with a relatively low population and little in the way of large-scale employment other than the wood-chipping mills in the north of the island. It is not surprising that there is a good deal of local debate (sometimes heated!) between the two factions. For biogeography this means that there is more than one avenue to explore. For example, there is the work with the Wilderness Society or for its rival, the timber industry (both employ ecologists). The third way is to work for the local government conservation department to try to produce a plan for sustainable development.

Review questions

1. To what extent is any ecosystem static?
2. With reference to a named ecosystem describe the range of threats facing it.
3. Which are greater – natural or human threats?
4. Using a range of case studies show how changes can alter the ecosystems in an area.
5. Does the EU policy on older agricultural crops diminish our gene reserves?
6. Contrast changes in human impacts of lowland and upland ecosystems.
7. Evaluate the use of EIA at a range of scales.

Selected readings

Much of the work reported here is relatively basic ecology. One of the problems with biogeography is that workers tend to come from either a geographical or biological background. Modern research shows how an amalgamation of the two is needed to appreciate the development of ecological relationships through time and space, i.e. patterns and processes of vegetation. To do this from the references used, Stiling (1999 – in the third rather than fourth edition) is an excellent guide to ecology, with Forman (1995) as an authoritative (but not really introductory) guide to land mosaics. For the human dimension and integrated research, try Petts (1999).

References

Barbour MG and Billings WD. 1988. *North American Terrestrial Vegetation*. Cambridge University Press.

Barrow CJ. 1997. *Environmental and Social Impact Assessment*. Arnold.

Carson SMcB. (ed.) 1978. *Environmental Education: Principles and Practice*. London: Edward Arnold.

Caughley G and Gunn A. 1996. *Conservation Biology in Theory and Practice*. Blackwell Science.

Connell TH and Slatyer RO. 1977. Mechanisms of succession in natural communities and their role in community stability and organisation. *American Naturalist*, **III**: 119–44.

Cullingworth JB and Nadin V. 1994. *Town and Country Planning in Britain*. 11th edn. Routledge.

Del Moral R. 1999. Mount St. Helens permanent plots 1980 to 1999. http://www.biology.washington.edu/delmoral/ (accessed 8/4/04).

Forman RTT. 1995. *Land Mosaics*. Cambridge University Press.

Gilbert IR, Jarvis PG and Smith H. 2001. Proximity signal and shade avoidance differences between early and late successional trees. *Nature*, **411**: 792–95.

Holdgate M. 1999. *The Green Web*. Earthscan.

Institute of Environmental Assessment. 1995. *Guidelines for Baseline Ecological Assessment*. E & FN Spon.

Laurence WF *et al.* 2004. Pervasive alteration of tree communities in undisturbed Amazonian forests. *Nature*, **428**: 171–5.

Lenahan S. u/d. The eruption of Mount St. Helens: the impact on landscape, vegetation, mammals, and birds. http://www.personal.psu.edu/users/s/e/sel143/

Little C and Kitching JA. 1996. *The Biology of Rocky Shores*. Oxford University Press.

Lyster S. 1985. *International Wildlife Law*. Grotius.

Middleton N. 1995. *The Global Casino*. Edward Arnold.

Petts J. (ed.) 1999. *Handbook of Environmental Impact Assessment*. Blackwell Science.

Richards JF. 1990. Land transformation. In Turner BL *et al.* (eds) *The Earth as Transformed by Human Action*. Cambridge University Press.

Roberts RD and Roberts TM. (eds) 1945. *Planning and Ecology*. Chapman & Hall.

Stiling P. 1999. *Ecology – Theories and Applications*, 3rd edn. Prentice Hall.

Treweek J. 1999. *Ecological Impact Assessment*. Blackwell Science.

USFS. 1999. Biological responses to the 1980 eruptions of Mount St. Helens. http://www.fs.fed.us/gpnf/mshnvm/research/

Websites

This area is well supplied with examples. To follow one of the case studies mentioned here (Mt St Helens) consider the data sources on:
http://www.biology.washington.edu/delmoral/

And also:
http://www.fs.fed.us/gpnf/mshnvm/research/

To look at changes from a global and local perspective, start at either English Nature or the Red Lists of the IUCN:
http://www.english-nature.org.uk/
http://www.redlist.org/

CHAPTER 13

Fragile environments: the biogeography of existence

Key points

- Fragile environments are one of the commonest ideas in biogeography and yet the concept is not well defined in ecological terms;
- Fragility is better considered in terms of the notions of extinction (loss of species) and origination (gaining new species);
- Examination of the concept of species loss and gain shows it to be a complex system based upon the interplay of genetics, organisation and chaos dynamics;
- These factors can be placed in a model to explain more clearly the loss and gain of species;
- Examples of fragile ecosystems and threatened species need to be studied to better enable us to appreciate the factors which can cause catastrophic changes in ecosystems. It is clear that this will be one of the most urgent tasks facing biogeographers in the twenty-first century.

13.1 Introduction

Deep sea corals are under threat from bottom-trawling fishing methods. Unlike their tropical shallow-sea counterparts, deep-sea corals appear to have a greater diversity and take far longer to grow. In the cold deep waters it is possible to devastate an area such that regrowth is impossible (Amos 2004). It is not possible to state whether these species are endangered because little is known of the area. Ninety-nine per cent of species that have ever existed are already extinct (Jablonski 2004). Given this figure it seems difficult to argue in favour of conservation. Perhaps it might be better to be concerned about which species have become extinct recently and if this is different from the extinctions of the past. Alternatively, the issue might be not the percentage lost but the actual range of species under threat, i.e. not all species are of equal significance in terms of global biodiversity.

The following examples, each with their own perspective are yet both concerned about the survival of species. Many corals are endangered and their conservation can be justified on the grounds of the biodiversity of the area and its significance to other marine areas. Extinction is only a problem if we exploit certain taxa disproportionately (which is what appears to be happening at the moment). Another connection is that both corals and many species are under threat because they are found in fragile environments, i.e. those not robust enough to withstand change. Since it implies that ecosystems are not equal (i.e. in terms of species or ability to withstand change) it follows that we need to understand what happens in these areas. From there we can make better judgements about conservation and also extend our knowledge as to how the basic biogeographical building block operates. The whole notion of fragility is ill-defined and poorly 'parameterised'; it does not have the scientific rigour which is commonly found in other ecological concepts.

So, what is fragility? Initially, it could be argued that it is the unwanted loss of a species or ecosystem. This implies an anthropocentric perspective because it is people who define those losses. Alternatively, it could be seen as an ecological or ecocentric term. Here, fragility would be the outcompeting of a species or ecosystem as a result of changes in the physical or biological environments of that place. These alternative perspectives start to show how the debate can be broadened. Just because ecocentric and anthropocentric definitions share the common word of fragility does not mean that they are further associated. It could be argued that 'ecocentric' refers to an absolute loss of species and 'anthropocentric' to a relative loss. Either way, trends in biogeographical research demand that we explore this further (Figure 13.1).

'Fragile' is one of those terms which can be readily applied and almost self-evidently apparent in certain ecosystems. It is a common term used to describe an area that cannot hold up well under pressure. In this respect, coral reefs are described as fragile in that they become 'bleached' and die with even a slight rise in temperature (although current research is showing that this simplistic picture may not be entirely accurate). So common is the term that there is rarely an opportunity to consider what it actually means to the theory and practice of biogeography. As will be argued below, fragility is a complex term whose parameters require considerable thought. At one level, all ecosystems can be said to be fragile in that building a city over them will destroy them but where (and why) is the line between the usual dynamic equilibrium of one ecosystem and the complete removal of another?

Figure 13.1 The iconic Australian koala can be seen as a prime example of the problems both of conservation and fragility. Restricted to a narrow range of habitats due largely to a highly specific diet, the koala is finding far fewer areas to survive. Land clearance and attacks from introduced species have reduced numbers. Although the Eucalypt woodlands are not 'fragile' in the conventional sense there is a concern about the survival of this and other species. Australia has both a high level of endemicity and the greatest rate of species loss. These issues coupled with some very fragile environments (alpine areas are under extreme pressure) means that the issue of biodiversity has both academic and political dimensions

Many 'protected landscapes' are extremely fragile in the face of human activity and this is one of the reasons for their status. Plate 12 is a 'tongue-in-cheek' view of one of the major problems confronting the Burren National Park in Ireland. Others include agricultural subsidies for fertiliser and the removal of species-rich hazel (*Corylus*) scrub as part of agricultural improvement.

13.2 Thinking about fragility: concept and critique

The preparation of most writing requires some research and a reassessment of basic ideas so that they fit into current thinking. Occasionally, that research throws up some most unexpected aspects which require much more detailed analysis. The concept of fragility is one such case. For the student of biogeography there is often the thought that the textbook contains a solid core of truth, i.e. generally accepted aspects of the subject, and yet it is often the case that the cutting edge is only just beneath the surface. In the case of fragility this edge is the surface! Given this situation, it is worth following the development of this concept as a demonstration of the development of biogeographical ideas rather than a *fait accompli* (cf Rosenzweig 1995).

The fundamental question is about the nature of fragility. It is easy to find texts describing the loss of species and ecosystems and equally easy to find cause (usually some form of human action or inaction). It would be simple to list the main culprits: agriculture, urban expansion, pollution, ecosystem modification, etc. and yet this is a list of causes, not a theory explaining existence (or non-existence) in a given location. In other words there is an empirical idea of fragility but not one from ecological theory. If the researcher now turns to standard ecological texts (e.g. Begon *et al.* 1996, Colinvaux 1993 or Stiling 1999) there is no mention of fragility in the index. Thus this basic concept, vital to our understanding of human impact for example, has no common basis in ecological theory. The nearest we come to a model of fragility in early literature is the work of Connell and Slatyer whose ideas, modified by Holdgate (1979), have formed one of the more important pieces of work (see Figure 13.2).

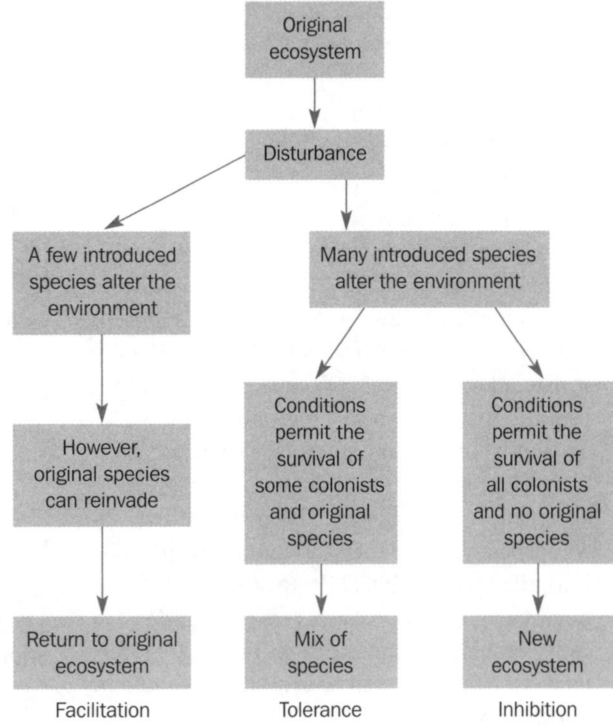

Figure 13.2 Pathways in response to invasion by new species. Depending on the conditions, the result can be either no change, complete change or a partial change. The notion of numbers of species should not be taken as precise: no exact figure was given originally.
(After Connell and Slatyer 1977, Holdgate 1979). See Box 13.1

If we cannot consider a species loss mechanism then why is it that some species continue to survive despite considerable odds? Organisms are seen in three contexts: species, community and ecosystem (Begon *et al.* 1996). Each species exists in versions of all three contexts – the problem here is the considerable variation in reaction between and within both species and contexts. This will need to be addressed, but first we need to see what key ecological elements are involved:

- At the *species level* we know that the continued survival of individual or population is the interaction between the external environment (climate, soils, etc.) and physiological response (which is in turn limited by the genetic structure of the individual). Acting through time this results in a species history (see Morin 1999). This produces a series of interacting species which together make up the community.

- The *community level* is subject to a series of forces which limit its development. Among the more important of these are population dynamics and successional changes which again interact within the same physical environment as the individuals and species. Yet the response is different. Species are limited by their ability to withstand the physical environment and yet the community can be more or less affected depending on the circumstances. There appears to be an additional dynamic when species aggregation is considered. The notion of the whole being greater than the sum of the parts is intriguing yet puzzling. Like the topic of fragility, this additional aspect is given little consideration. The implications of this are rarely pursued – this may well be crucial and should be borne in mind.
- Finally, these communities link up to the *ecosystem level* still affected by the same physical environment (or, more properly, environmental range since the area is greater). Additional population dynamics are seen governing the interspecific reactions. This model (albeit simplified) describes the basic ecological reactions between and within species and yet the linkages between the three scales are weakly developed and we are still no closer to finding the model which helps to explain why some species exist and others die out.

13.3 Towards a biogeography of existence

Perhaps the problem is not the search for the answer but the formulation of the question. It might be better to consider an ecology (or biogeography) of existence rather than loss through fragility? In other words why do organisms survive? The value of this approach is twofold: most texts focus on the growth of species in ecology and not their loss (giving a wider range of research material from which to draw) and it gives us an opportunity to see if existence and loss might be more closely connected (i.e is loss x balanced by gain y?). Research into the area of the loss and gain of species (extinction and origination respectively) appear to centre around three questions:

- What is the evidence for extinction and origination?
- What ecological strategies for extinction and origination can be found at the commonly studied scales of organism, community and habitat?
- What is the nature of the rate of change?

13.3.1 The evidence for extinction and origination

This has been a crucial topic for some years in palaeontology (Weill, pers. comm.). There is a large body of evidence which supports the idea of massive loss and origination events. One of the key elements is the work by Sepkoski (2000). In studying a range of fossil fauna, he noted that there were bursts of origination and some equally dramatic extinction phases. This shows an interesting picture – three events where each one has a growth and decline in rates of origination punctuated with massive losses in the late Ordovician, Permian/Triassic boundary and the much studied Cretaceous/Tertiary event. In each case there was an assemblage of fauna of an increasing number of species, i.e. increasing biodiversity. (Note that the division into faunal groups is at the genus or higher level – origination would be the introduction of a new group into an area but speciation would be the spread of species from the original area.) Sepkoski notes two phenomena attached to his calculation of diversity through time: rebound and origination rate. Rebound occurs when there is a loss of species – it suggests a renewed vigour following loss of diversity. The mechanism for this is not taken further by Sepkoski, but further work (e.g. Kirchner and Weill 2000, Irwin 2000) suggests that it may well be linked to niche differentiation. This in turn implies a variable origination rate depending upon the opportunities available. Further, Irwin presents evidence that extinction and speciation are not linked, i.e. loss can be immediate but it takes longer to recreate conditions for new species to flourish. Given the overwhelming evidence for a series of extinction and origination events it is possible to support the following conclusions:

- That extinction and origination are not linked by time, neither are the events symmetric;
- Evidence suggests that biodiversity is increasing but that its rate is declining. These conclusions would appear, initially, to be at variance and yet it is possible to suggest a link. If diversity is linked to niche production then as the number of niches increased so the rate would, logically, slow down (or else biodiversity would have peaked long ago). This in turn suggests that there is an upper level to biodiversity – that it is asymptotic. There is another argument that might also be accepted here: that as biodiversity increased alongside niche specialisation the whole system becomes less stable. Smaller niches suggest more specific conditions: minor changes could well remove the most specialised niches (again supporting the idea of an upper limit for niche production).

13.3.2 Ecological strategies for extinction and origination

The second area for study concerns the strategies employed for extinction and origination. The assumption here is that species do not just suddenly die out or appear without some facilitating mechanism. There is no need for this mechanism to be an internal factor of the organism or species. Work by Roberts (1981) on the Carboniferous brachiopod zones of eastern Australia puts forward external mechanisms, i.e. sea-floor lowering and climate as a reason for the loss of diversity. Presumably, those species that could not adapt fast enough died out. A highly detailed account of echinoid (sea urchin) loss sheds further light on this area (Smith and Jeffrey 1998). The approach adopted is similar to that posited here: in order to understand extinction mechanisms they studied survival. Their research shows that not all species survived the Cretaceous/Tertiary boundary extinction event: those that did had some inherent advantage. They found that survivorship was most closely correlated with the taxonomic structure (higher-order taxa surviving better than lower-order ones), feeding strategy and geographic region with no links to numbers of species, geographic distribution or feeding ranges and types. What is most significant is their main conclusion:

> It is possible that the final blow was dealt by asteroid impact, but there is direct evidence that conditions for plankton were becoming less favourable immediately before the K/T [Cretaceous/Tertiary] boundary. Climate was deteriorating rapidly and extinction of several major molluscan groups had already taken place. Furthermore, numerous lineages of echinoids independently switched to non-planktotrophic development in the Maastrictian, regardless of palaeolatitude or water depth, implying that survival for planktonic feeding larvae was becoming markedly less predictable. Furthermore, the fact that high levels of extinction continued into the Danian suggests a slow squeeze rather than an instantaneous catastrophe. (Smith and Jeffrey 1998, p. 71)

The implications of this for our study are considerable. If some molluscan (shellfish) species had already died out then it follows that extinction is not just a sudden, universal event (although in a severe catastrophe it might be). It also suggests that some species are inherently strong for the given conditions while others are weak. The reference to non-planktonic development is also crucial, for it argues that some species/genera can evolve rapidly to accommodate change. That conditions continued for some period of time might be used to infer that the rate of adaptation was not even for species as a whole.

At the species level it is crucial to consider that basic unit – what is a species? Although there are some arguments against a fixed categorising of organisms, most accept the principle and the fundamental upon which it must stand: reproductive isolation (Coyne and Orr 2000). The problem then turns to how this may be achieved. There are no easy answers here. A review by Harwood and Amos (2000) puts forward the idea that evolutionary change (i.e. speciation) is affected by population size. For our purposes their conclusion has most to offer:

> The relationship between population size and observed levels of neutral variability is not as straightforward as is often assumed. Some relatively abundant species have low genetic diversity while some endangered species have no more or less variability than their more abundant relatives. . . . Alternative explanations for apparently low levels of variability include the effects of inbreeding and inbreeding avoidance, spatial structure and the possibility that heterozygotes are more mutable than homozygotes. Some instances of extreme impoverishment in species whose close relatives have shown 'normal' levels of variability may be explained by natural selection resulting in a selective sweep rather than passive loss through drift. (Harwood and Amos 2000, p. 85)

This puts a different complexion on the matter, suggesting that speciation is due to a complex selection process at least part of which is genetic. Nor is this the only idea. Price (2000) in a study of bird populations has put forward the idea that sexual selection is important. There is also reference to population size and species abundance. In a related exercise Johnson (1998) examines the relationship between range size and abundance, concluding that there are positive relationships for recent species assemblages but negative ones for ancient assemblages. Such variation in result only serves to highlight the complexity of any answer.

Although species seem to get most of the attention, there are other scales to examine. It is difficult to characterise responses at the community level, given that this a very topical area, mainly because most workers tend to disagree on even basic theoretical points! (Weill, pers. comm.). Morin (1999) argues that there are two basic approaches to the subject of extinction and origination at the community level: equilibrium or non-equilibrium dynamics. Equilibrium dynamics is characterised by stable populations of more or less constant composition (although constancy might be more a question of perspective where the human lifespan is compared with that of tree species). Non-equilibrium situations have fluctuating compositions of a range of species. Explanations of speciation and extinction can take either of these perspectives (or even a mixture of the two). Whichever is chosen, there appear to be three main areas of study:

- **The nature of interactions**. Johst *et al.* (1999) argue that population dynamics are dependent upon demographics. By using a model which investigates phenotypes that respond differently to a common feature they come to the conclusion that:

  > Evolution can favour phenotypes that have the intrinsic potential for very complex dynamics provided that the environment is spatially structured and temporally variable. These phenotypes owe their evolutionary persistence to their large dispersal rates. They typically coexist with phenotypes that have low dispersal rates and that exhibit dynamic equilibrium when alone. This coexistence is brought about through the phenomenon of evolutionary branching. . . .

  This suggests that far from being a simple case of interacting species, the community is made up of species with different reaction responses and rates to different perturbations. Doebeli, in an earlier paper (1993), alludes to the idea of different rates of interactions existing in the same community. Judson (1994) asserts that local reactions between individuals are important in understanding community systems. Sole *et al.* (1996) echo this point but add that the reactions are not only between species but also seen in interactions with the physical environment. Are we seeing a series of small interactions that might have significance or could this be a unifying system? Odum *et al.* (1995) have argued for the latter by devising the pulsed ecosystem. Here, external pulses (such as storms) and internal pulses (such as predator–prey reactions – see also Pascual and Caswell 1997) form a reaction not dissimilar to wave interference. They argue that when external and internal pulses are coupled then species survival is enhanced. Although each paper quoted deals with a different aspect, the unifying concept seems to be chaos (see below) which takes interspecific competition out of conventional food web theory;

- **Feeding relationships**. Although chaos dynamics might be one of the more recent advances there is still work to be done on more conventional ecological topics. Ecosystems are bounded by predator–prey dynamics. The way in which food webs are constructed and the relative numbers of organisms and species is fundamental to our understanding. However, recent work has shown that patterns can be more subtle than previously thought. Sait *et al.* (2000) argue that in real systems containing more than two species there are several possible combinations depending upon initial conditions and the sequence in which they arrived. Thus the simple two-species system replicated at differing densities is replaced with a three-species system where species arrive in a different time order. It would appear that such temporal changes might be responsible for the relative numbers of organisms (see Hecnar and McCloskey 1996). McCann and Hastings (1997) change feeding type rather than time of introduction. Here, the addition of omnivory tends to produce a more stable ecosystem;
- **The nature of the dynamics**. Morin argues that there may well be more than one state possible in an ecosystem. This paves the way for the multi-phase ecosystem dynamic where different types and amounts of perturbations create different levels and rates of responses in the ecosystem. There is general agreement on the existence of chaos theory in community dynamics. There is less agreement on its actual role (see Allen *et al.* 1993, Sinha and Parthasarathy 1996, Huismann and Weissing 1999). There would appear to be two competing views. One holds that since responses are chaotic, extinction is caused when a small population is subject to large chaotic fluctuations. Alternatively chaos, by constantly changing the dynamics of reaction, is more likely to allow a small population to survive. Huismann and Weissing (1999), in addressing the paradox of the plankton, go so far as to argue that chaos keeps more species available because the changes of conditions are greater than the response times of the species, implying that interspecific dynamics is more akin to a juggling act. Whatever the final answer it is probable that the old stable-species model is no longer viable (see Box 13.1).

The change of scale to ecosystems does nothing to resolve the seemingly intractable and contradictory evidence we are presented with. There is some agreement that we can deal with ecosystem dynamics as examples of chaos theory (see e.g. Stone and Ezrati 1996) and that now the question becomes which of a range of competing theories can be best used (e.g. Vandermeer and Yodzis 1999). If it were just a question of reaction to perturbation then it would be easy enough to say that ecosystems that survived were those with sufficient resilience to withstand the perturbation and that those which became extinct lacked this

BOX 13.1

Resilience, elasticity, amplitude and perturbation

A number of terms have become popular in the literature about ecosystem dynamics particularly in relation to ecosystem changes such as were seen in Figure 13.2. As the diagram stands it could be interpreted as suggesting that any one of the three paths could be chosen. In reality it is going to be more complex. For example, feeding relationships are a question of stress between predator and prey. Success is not guaranteed. However, some forces may have greater impact at some times than others.

To understand this, it is possible to consider that an ecosystem, like an individual, has a range of tolerances within which it operates. Forces causing changes the ecosystem has to deal with result in *perturbation*, i.e. the system is operating outside its usual conditions. Mostly, the ecosystem responds and the situation is damped down. The extent to which this happens depends on the *resilience* of the species – the ability to withstand stress. Those with high resilience would show in an ecosystem with *elasticity* – the property of being able to recover from perturbations. However, if the *amplitude* of the stress is too great (through too great a force or a lesser force applied over a longer time span) then change will occur.

Although it is useful to consider these terms there is very little in the literature which directly measures them – these terms should be seen as a way of conceptualising changes rather than obtaining detailed figures.

resilience. This might be helpful, but it leaves unanswered the actual nature of the resilience and whether its presence or absence is uniform in all species. It also brings into question the nature of that most commonly discussed ecological principles, succession. Is succession a linear process as most texts would appear to describe it, or is it a more haphazard change based on the individualistic reactions of the populations present in a given time or space? Even this brief foray into the extinction or origination of ecosystems suggests that the subject centres around four key concepts:

- **The nature and role of the individual.** However we describe ecosystems we must not forget that they are composed of the same individuals that we have been discussing above. Some writers see the individual as vital to the success or otherwise of the ecosystem. Levin (1998) emphasises that ecosystems are complex and adaptive and that their patterns are derived not from that level but from the myriad interactions at lower (i.e. species) levels. Thus any ecosystem would be dependent upon the history of multiple outcomes from the species. Apart from putting the individual organism as a key player it also calls into question the dominant paradigm and philosophy through which current ecology is mediated (i.e. the control of a landscape by inherently underlying forces is a classic realist perspective which is contrary to the usual positivist science currently espoused!). Further, Levin argues that it is the aggregation of responses at individual (and higher) levels that determines the buffering properties against ecosystem change. Another viewpoint emphasising the importance of the individual is that of Barbault and Sastrapradja (1995). Here it is argued that response of the ecosystem is based not upon the individual but upon its specific location in the food web. Generally speaking, the less complex food webs would become extinct if plant species were lost, while more complex food webs would die out upon losing predator species.

- **The function and structure of the ecosystem.** Clearly not all ecosystems are fragile. Whereas some of this resilience is due to the species involved it must also have contributions from other, higher-order aspects which can be seen only by virtue of the organisation of species into communities. Holling *et al.* (1995) suggest that it is functional diversity that determines resilience. An ecosystem with a range of potential responses to perturbation would be more likely to survive. Part of this might well involve links with the individual level. Species redundancy is a controversial notion that argues that two or more species might coexist within the same range of environmental parameters and that the loss of one might be compensated for by the increase in the other. Holling adds weight to the idea of functional diversity by showing that spatial attributes are neither uniform nor scale invariant. This suggests that by virtue of its existence the ecosystem might have properties (of organisation?) needed to keep it functioning that help also to ensure its survival.
- **The role of community and population dynamics.** Interspecific dynamics is the 'glue' which holds the ecosystem together. Without it there is just a collection of plants and animals (although this perspective actually accurately reflects conditions in some marine ecosystems – Jackson 1994). It follows that this area is of importance in the extinction or origination of ecosystems. Harris (1994) suggests that at this scale many processes seem predictable because they are dealing with an aggregation, whereas the reality for the individual may be chaotic. A further problem would be to separate out the random noise – part of the natural fluctuations of population levels – from genuine chaotic responses. Work by Stone and Ezrati (1996) follows this idea but seems to add that not all situations are inherently chaotic. In their examples, non-linear (i.e. chaotic) situations occur when the rate of species increase is above a crucial point. If this could be repeated in the field then there would be a strong case for arguing that r-selection populations are chaotic while K-selection ones are more stable. Whether this could be taken further to community level, i.e. r- and K-selecting ecosystems, is another matter. Certainly Mooney *et al.* (1995) posit the view that because of human disturbance, ecosystems change from long-lived, large species to short-lived, small ones. The precise mechanism is not given, but presumably it would involve a change in both species composition and the dynamics of the ecosystem at times of change. A different perspective is taken by Golding (1994). The basic idea put forward is that evolution can provide a range of possibilities. Some of these are not used but survive (neutral evolution); some survive and are used (non-neutral evolution). Such thoughts return us to ideas of redundancy in species but in an evolutionary context (see also Holmes 1998). If this were taken further then a fragile ecosystem would be one with insufficient back-up species!
- **The role of space.** Ecosystems exist in a spatial sense even if, to some workers, they are no more than connections of patches or sub-ecosystems bound together (Forman 1995). The location of the ecosystem can be a key factor in its survival. It may be no coincidence that some of the most fragile ecosystems are in the most extreme environmental conditions, e.g. montane and arctic ecosystems. Of course, this does not help to explain the fragility of the coral reef but it does provide one avenue. Holling *et al.* (1995) go so far as to suggest that ecosystem stress is a greater determinant of survival than ecosystem process. Although these arguments concern entire ecosystems the nature of different scales can be seen here also. Laurence *et al.* (2000) report that tree survival in rainforests depends on the location of the tree relative to the exposed margin. Exposed trees are less likely to survive. A similar situation is seen in the UK where conifer

plantations are often worst affected at the edge following any storm (the damage caused by direct wind speed – windthrow – and indirect tree vibration – waggle).

13.3.3 The nature of the rate of change

Finally, we turn to the nature of the rate of change. It must be clear from what has been written above that there are a number of perspectives that can be used here. Conventional notions of succession and zonation paint a picture of orderly change, whereas this cannot be the case if we are dealing with the extinction and origination of individuals, communities or ecosystems. There are two lines of argument to follow. The first, as exemplified by Morin (1999) and Levin (1998), suggests that history is a crucial component. Levin refers to it as historical dependency – that whatever the rate and nature of change it is a function of the time frame of that particular ecosystem/community, etc. Alternatively, we have workers like Stone and Ezrati (1996) who consider that change is neither regular nor predictable because it follows the ideas of chaos theory. To them chaotic change can occur if:

1. Populations are density-dependent;
2. There is sensitivity to initial conditions;
3. Parameters will fluctuate between fixed boundaries although within that, the precise nature of change may be difficult to determine.

There is the evidence. What can we make of it? In summary, research seems to be suggesting:

- There is no precise ecological link between extinction and origination/speciation. This appears to be fundamental to the work described here. Evidence seems to show that the process leading to extinction is not the same as that leading to origination, neither are there linkages in time or space between the two. Thus the question of species loss should be considered quite separately from that of the creation of new species. However, this is not to say, paradoxically, that the two processes may not share many of the same process types but in different combinations, rates, etc.
- The processes of extinction and origination operate at three contexts with three key sets of processes: individual/species, community and ecosystem contexts and, respectively, genetics, organisational responses and dynamics. We can see many of the key processes at any given level but the linkages between contexts are more implicit-given than explicit. These linkages must be present for at each context the same individual organism is seen – it exists in all three states at once;
- The processes are highly variable. The initial conditions seem to set the scene for the development of the pattern. Given this sensitivity it is reasonable to suggest that the processes exhibit chaotic dynamics. One of the consequences of this is that it is not possible to predict, with precise accuracy, the way in which a situation will develop. In addition this makes the actual rate of loss and gain very difficult to calculate. Both processes may take such a long time that they are not easily perceived in terms of a human lifespan.

13.3.4 Putting the evidence together

How can this be aggregated into a model of the biogeography of fragility or existence? Figure 13.3 shows the arrangement of processes. It can be best seen as a tri-level system with each level affecting the level above (and/or below).

Firstly, we have the species. The most important aspect is sensitivity to perturbation. If the individual cannot survive then the whole system will collapse here. Individual sensitivity is a function of taxonomy which in turn relies on genetic (species) and phenotypic (individual) abilities. These attributes will determine the types and rates of adaptation possible, feeding strategy, population size (reproductive strategy) and behaviour. The net response of this level is to

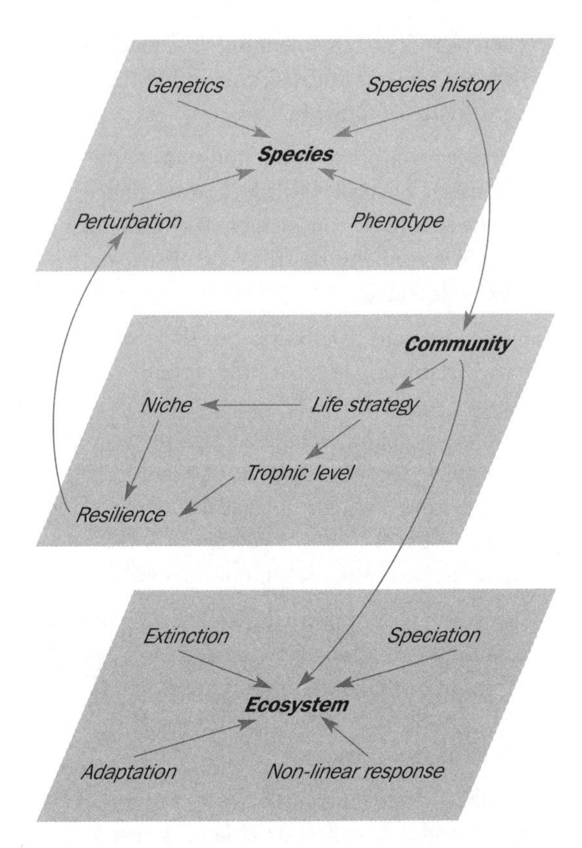

Figure 13.3 The relationship between species, community and ecosystem. For detailed explanation, see text

produce a set of species (a species history) which is acted upon by the next level. This is not a one-way, single response: individuals are born and die and the reactions at the community and ecosystem levels are fed back to this level which informs the parameters for the next generation (if there is to be one!). This is a level of pulsed reactions to external forces mediated through genetics.

The next level contains the community organisation – the way in which individuals and species interact. The community is handed a set of historically determined organisms upon which to apply community rules. At this level we could argue for a systems-type diagram where the input is the individuals received, the order in which they arrived and their life strategies of r- or K-response. This is then organised into a trophic level and niche. The reactions created here provide the community with its functional diversity, one crucial element of which is buffering – an anti-response which may well stop the community operating outside its bounds and thus losing (or gaining?) species. As with the individual/species level, the responses of the ecosystem level feed back into the community. This level is not infinite in its responses. For example, it is clear that there is an upper limit to biodiversity.

These communities interact within a given space – the ecosystem. This level contains the dynamics of the system. It is characterised by chaotic complexity where the outcome of the interplay of the major forces is far from certain and only marginally predictable. It is likely that chaotic reactions determine the rates of change of conditions which in turn determines the rate of extinction. Susceptibility to extinction and the ability to adapt (and the rate of that adaptation) is part of the individual/species reaction so the connections between these two levels are clear. In most cases the ecosystem operates within certain limits (unless the change is so catastrophic as to remove most/all species, e.g. meteorite impact, volcano), but these limits might actually contain two or more viable states. The nature of this chaotic reaction would be determined by the lower levels thus making it unique in each situation. The historic situation portrayed in Plate 13 is an example where a vulnerable ecosystem – in this case coastal grassland on a crumbling Jurassic cliff, heavily influenced by wind and salt-spray – was further compromised by the construction of anti-invasion fortifications in 1943–44 and the chaos produced by severe aerial bombardment in 1944–45. In the post-war period, the amount of 'chaos' has decreased but the climatic and recreational pressures on the site remain, maintaining the immaturity of the ecosystem.

How would this work for the biogeography of existence? Firstly, given the problems of actually viewing speciation it would be better to focus on extinction. Thus a series of species, each with their own genetic/phenotypic limits would, as they arrive in the area, form a community. The community would form an organisation of organisms based on feeding relationships. This would, in turn, form with other similar populations to form an ecosystem. The system would operate within certain limits. However, if the rates or nature of changes moved towards the edge of the chaotic system then the whole would become unstable. Depending on the nature of the changes and the role of the organisms affected then it is possible that loss of species would occur. If the position of the organism was central enough, e.g. keystone species, then the ecosystem would be replaced by another. (Of course, it is entirely possible that this mechanism would work for ordinary successional sequences as well!)

From both previous work and ideas put forward here it is evident that there are certain areas of this topic where a critical response is needed. Overall, criticisms of the evidence and interpretations can be summarised thus:

- This is a very new science as the references given here will attest. The amount and type of new information needed is, at present, greater than our ability to produce them. Thus all aspects mentioned here are, to an extent, speculative;
- Are species actually vulnerable to change or is it a function of human perception? We might see a small mammal becoming extinct from its final habitat because it happens within our lifespan but how do we see extinction of trees? The science has not been in action for long enough to enable us to get all the answers we need;
- Is fragility a purely anthropocentric term? Are we only concerned about those species we seem to want? Early conservation activities on English chalk downlands did as much to remove species as retain them! Alternatively, fragility could be seen as a more adaptable community – one ready to lose species to let others use the niches made available;
- There is lack of agreement over the parameters and mechanisms of extinction and of the significance of individual aspects;
- There is no clearly defined notion as to the percentage of species which need to change for the area to be considerably fragile, neither is there agreement over the type of species most vulnerable;
- Our evidence is patchy. Most of the geological record is missing because only certain characteristics can be fossilised. These must then be discovered and researched making it a very incomplete record. To this must be added the interest (and ability from available evidence) for the newest aspects of the geological record – often referred to as the 'pull of the new';
- Speciation is even more difficult than extinction because the divisions between species are often the result of human opinion rather than fact and we might be making two species where only one exists (or vice versa);
- Finally, it could all come down to experimental design and data collection (Magnuson *et al.* 1994).

13.4 Case studies in fragility

Irrespective of the debate about the nature of extinction and origination it is clear that we are losing species due mainly to anthropogenic action. Further, it is usually accepted that these losses are unacceptable and that it is important to conserve these species and ecosystems. Numerous reasons can be put forward to support this action (see e.g. Given 1994). What concerns us here is neither the cause of extinction nor our reaction to it but the ecological processes and biogeographical implications of such loss. To illustrate our case we take four examples: coral reef and alpine grasslands for ecosystems and tigers and manatees for species.

13.4.1 Coral reefs

> Coral reefs, which are among the most biologically diverse ecosystems on the planet, are also among the most ancient. They first appeared in the Mesozoic era some 225 million years ago and some living corals may be as much as 2.5 million years old. . . . Yet, in just a few decades, human activities have devastated many of these biologically rich, ancient ecosystems. In the next two or three decades, more are destined for destruction. (Bryant *et al.* 1998, p. 5)

We could go back even further and say that ancestors of today's reefs were among the first rock-forming organisms. The statistics of richness and threat are well catalogued – 1 per cent of the marine environment is coral and yet it has more than 25 per cent of the fish. At the same time (and usually because of this) 58 per cent of reefs worldwide are under threat (and these are rarely given a comprehensive conservation programme – see Figure 13.4). However, to some extent this is speculative because reef ecology and functioning are not well known. What is known suggests a highly complex habitat of considerable diversity and fine environmental gradients supporting numerous niche-specific organisms. The actual coral itself is made up of millions of coral polyps similar to miniature sea anemones (see e.g. Hughes *et al.* 1999). These polyps are linked symbiotically to algae called zooxanthellae. This interaction is vital, for the polyp cannot function without the algae's photosynthetic products neither can the algae exist without coral metabolites. It is also this interaction which provides the limestone secretion which forms the coral's rock structure. Further, the interaction limits coral production to a specific set of (quite limited) environmental parameters. The practical upshot of this is that the vast majority of reefs are in warm, shallow tropical water (see Figure 13.5) although the details can hide a multitude of differences, e.g. Florida's reefs are limited by winter temperature, while reefs off Brazil are influenced by silt and fresh water from the Amazon. One of the implications of this quite restricted range is that any minor alteration can cause changes in the reef structure. For example, the Red Sea reefs can show considerable physical variation even within one reef type. There is a tendency to try to group reefs according to Charles Darwin's approach, i.e. fringing, barrier and atoll, but this is more a question of location in relation to other features than a true ecological differentiation.

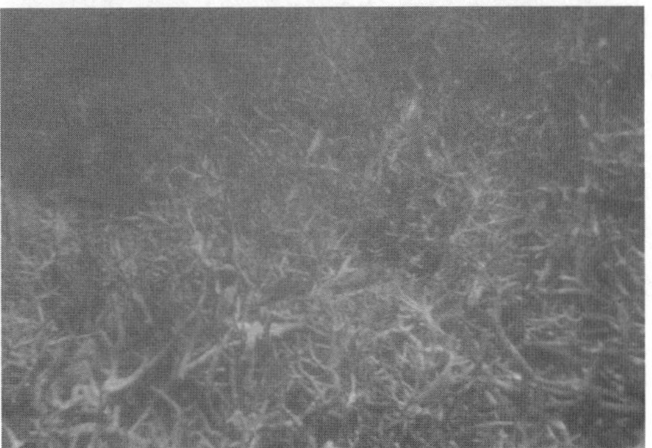

Figure 13.4 Two aspects of the Great Barrier Reef, Australia. Increase in interest in the reef has led to tourist pressure. Top, an area of reef still reasonably intact, while below, the area is a mass of dead coral. Environmental events such as coral bleaching and global warming threaten to exacerbate the situation

Figure 13.5 The Great Barrier Reef, Australia. In reality this reef is a series of over 2,600 reefs stretching 2,000 km. Like most other reefs it occupies warm shallow-water areas (other key areas include the Caribbean, SE Asia and the islands of the South Pacific)

Some important areas, e.g. Australia's Great Barrier Reef, do not fit into this scheme. Here, the reef stretches for over 2,000 km with over 2,600 individual reefs. Alternatively it might be more productive to study the variations in ecology between the seaward reefs and those on the leeward side. This suggests that physical factors are crucial to coral growth. Light, for example, has been shown to limit total species in the Indian Ocean, while tidal action and storm damage can produce significant disturbance even deeper in the water. Despite these variations (or perhaps because of them) reef productivity and diversity remain extremely high. Productivity may be as high as two orders of magnitude greater than the surrounding water. Diversity has been linked at least in part to extreme niche specialisation. It might also be linked to the position of the reef within the global system. The Great Barrier Reef is a vital component not only because of its size but by virtue of its location within the Indo-Pacific centre of diversity (Vernon 1999).

One of the greatest threats to the continuing existence of reefs is human activity. Coastal development, tourism, overexploitation of resources, pollution, erosion and eutrophication all play their part in degrading reef structures. Recent research by the World Resources Institute (Bryant *et al.* 1998) produced a threat indicator system based on coastal development, exploitation and pollution which showed that a majority of reefs were at risk. This is particularly serious because of related work that shows the reefs under greatest threat are least protected, and that many of these reefs are biodiversity hotspots which means that these areas are even more crucial by virtue of their extremely high number of species. However, we should resist the too easy equation: people = risk (Kasperson *et al.* 1995). The actual picture is likely to be far more complex. For example, Roberts (1997) reports that because of the high connectivity of ocean areas it is possible that some reefs are supplied with more larvae than other areas – these areas being more resilient (see also Hughes *et al.* 1999, Barnes and Hughes 1999). Sale (1999) notes that recruitment rates on the Great Barrier Reef do not always bear direct relationship to coral density (suggesting more factors at work). There is always the possibility that we are not seeing a structure/response at all but that it is merely a problem of poor sampling technique (Mumby and Harborne 1999).

What can coral reefs add to the model described above? Here is an area of seeming paradox – high diversity suggesting a built-in redundancy factor coupled with high fragility risk. At first it could be argued that human action alone has removed coral and yet there are areas dying out where human impact is not as great. It would appear that high niche specialisation coupled with the coral's keystone position is a key. With greater specialisation comes the need to more closely differentiate oneself from one's neighbours. This can only be done in finite space by partitioning up the environmental gradient into smaller spaces. Thus any change, however small, is likely to affect some species. If the coral is dominant then anything that disturbs its environment is likely to lead to disproportionate damage. Slight changes in temperature have led to coral bleaching, while the crown of thorns starfish has predated on large areas with human agencies virtually powerless to prevent it.

13.4.2 Australian Alps

In contrast to the global distribution of coral reefs, the complexes that make up the Australian alpine ecosystem are restricted to a very small area in the south-east of Australia and Tasmania (approximately 11,500 km^2 or 0.15 per cent – see Figure 13.6). Typically, this means areas above 1,400 m on the mainland and 900 m in Tasmania. Abiotically it is characterised by low temperatures, frequent frosts and snow for at least one month per year (Williams and Costin 1994). The deeply dissected valleys might also be subject to unfavourable conditions which would further restrict the distribution of vegetation, especially trees.

Although the central area of Tasmania, verging on the South-west Tasmania World Heritage Area, is critical for that island, the mainland area focusing on the Snowy Mountains is perhaps the more instructive, being both highly fragile as an ecosystem and subject to heavy tourist pressure (mainly skiing). Geologically, the area is a mixture of igneous and sedimentary rocks which has sustained several periods of Tertiary uplift. Geomorphologically, the area is a dissected plateau which was subjected to glacial and periglacial activity in the Pleistocene (at its greatest *c.*20,000 years ago). As one would expect, the current climatic conditions centre around heavy seasonal snowfall (the main source of water) and persistent low temperatures as a result of both altitude and trend of the major valleys. Unlike the alpine areas of Europe, the soils are both varied and relatively deep. Unlike much of Australia, fires are infrequent.

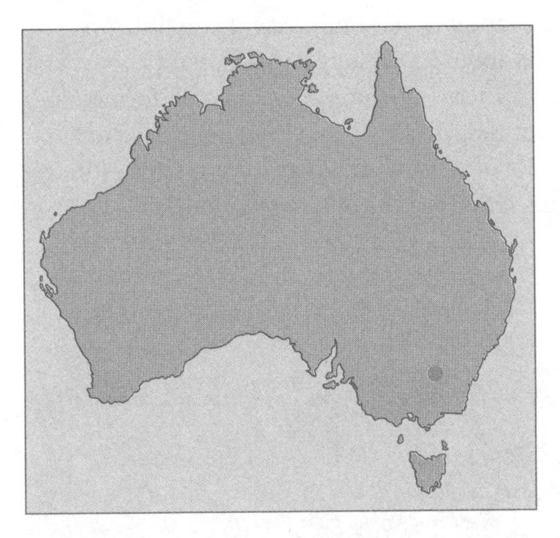

Figure 13.6 The Australian Alps. Given the rarity and fragility of the ecosystem this area is less than suggested by the map, being restricted to the highest peaks around Australia's highest mountain, Mount Kosciuszko. Semi-alpine areas stretch into Victoria to the south

Biogeographically, the area can be characterised as a continuum from open woodland to herbfield and feldmark with some areas supporting bogs. Community locations are determined by abiotic factors (especially soils) tempered by microclimate. Community dynamics centre on the interactions between climatic fluctuations, the life history of the vegetation, site factors and disturbances. Human disturbance from the Aboriginal communities has been restricted to the last few thousand years. Food seems to be the main use with the bogong moth (*Agrotis infusa*) as a major source. European settlement introduced grazing (altering species composition and clearing fragile vegetation), water usage (from the Snowy Mountains scheme) and, latterly, tourism (principally skiing). All of these activities have had an increasing impact upon the area even with current trends towards ecotourism. It is probable that this extra pressure is sufficient to alter species composition (see Figure 13.7).

This pattern of ecosystem distribution and disturbance also appears to fit into the model described above. Both Williams and Costin (1994) and Specht and Specht (1999) agree that the community dynamics play a considerable role. Habitat diversity would be increased with the range of micro-scale features on the dissected plateau. This would be moderated further by the impact of elevation. Given that the plants are adapted to the

Figure 13.7 The Australian Alps, view towards Carruther's Peak. This small area of alpine vegetation contains numerous endemic plants and unique ecosystems. The environmental conditions are extreme making this a very fragile area easily damaged

Figure 13.8 The tiger is one of the more threatened species today. Usually human population pressure and habitat destruction are blamed but this can also be seen as a question of fragility where a large species cannot adapt to new situations

conditions, it follows that any loss must be due to an increase in the already marginal dynamics of the community. Certainly there is evidence of the effects of grazing and trampling at major tourist/farming sites in the area.

13.4.3 The tiger (*Panthera tigris*)

> At the turn of this century, by general estimate, 100,000 tigers shambled over a vast range of Asia extending from Southern India to the Siberian taiga and from the equatorial tropics of Java and Bali to the Transcaucasus and eastern Turkey. By 1950 one authority was already predicting that the species is now on its way to extinction; three years from now, when the present century ends, perhaps no more than 2,500 scattered individuals of *Panthera tigris* will still wander a few isolated regions of their former range. (Matthiessen 1997, p. 54)

Although more recent estimates put the final figure at between 5,000 and 7,000, this sad summary is probably a reasonable statement of the fate of one of the most potent symbols for conservation. The tiger has held the human imagination in a way few other species have (Figure 13.8). It has been part of the folklore, medicine and lives of countless generations of Asian peoples and yet today it is facing extinction. What is it about the tiger that gives rise to such a prediction?

One of the main problems facing tiger conservation is that until recently, so little was known about its ecology and what was known was patchy. There was some agreement on former and present ranges (see Figure 13.9) and based on a number of research findings a reasonable estimate of the various tiger populations (see Table 13.1).

Table 13.1 Tiger (*Panthera tigris*) populations by subspecies. Mid 1998 estimate.

Tiger subspecies	Approximate number
Bengal (*P. t. tigris*)	3,900
Amur (*P .t. altaica*)	380
S. China (*P. t.amoyensis*)	25
Indo-Chinese (*P. t.corbetti*)	1,500
Sumatran (*P. t. sumatrae*)	450

NB Caspian, Javan and Bali subspecies are extinct.
From Seidensticker *et al.* (1999)

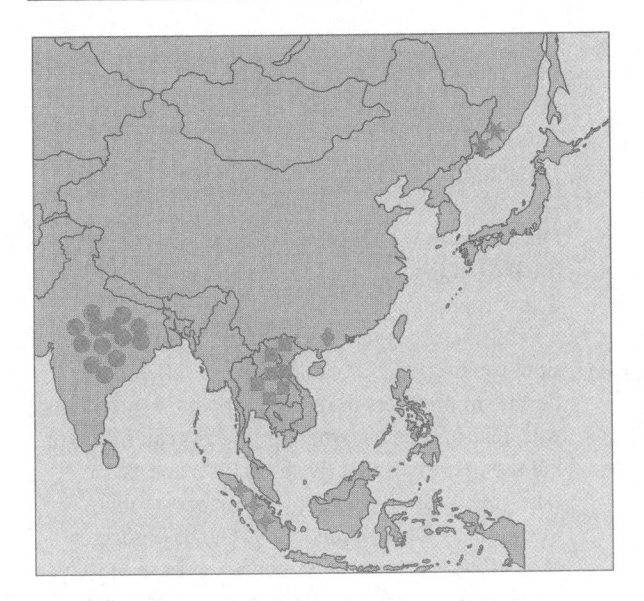

Figure 13.9 Approximate range of extant tiger subspecies. Circles represent the Bengal tiger, squares the Indo-Chinese, the diamond the S. China, 4-point stars the Sumatran and 6-point stars the Amur. Although current ranges are restricted, the original range is thought to be the majority of the SE Asian nations.
After Seidensticker *et al.* (1999)

Current research focusing on the Tiger 2000 project has enabled us to get a greater understanding of the detailed ecology upon which the existence of the species depends (Seidensticker *et al.* 1999). A certain amount of information is already well known: it lives near water in a range of habitats and climates but appears in greatest density where forest and grassland mosaics provide the greatest diversity. Its main food source is ungulates (ox, deer, etc.). This is a crucial factor: tigers do not survive readily without access to such animals. It is probable that the tiger evolved in the light of the Pleistocene radiation of its prey. This Ice Age also meant a considerable increase in speciation and extinction as climates and sea levels changed relatively rapidly. Of course, the changes in sea levels also opened up passages for tigers and their prey to migrate – one of the factors behind the very wide distribution of tigers today. However, it requires more than food supply to keep the tiger – reproductive capacity is vital. Given their relatively short gestation period (103 days), short inter-birth interval and a litter size varying between two and five, it is obvious that tiger populations can recover rapidly from any setback. Another key factor for tiger success is its ability to disperse. Clearly there is historical evidence given the range in which tigers have been found but here the data are sparse. In the few studies that exist, dispersal distances range from 30 to 60 km – very modest for such a large predator. Further, such data are gender biased; males travel far further than females (meaning that death rates are also biased). This is one aspect of behaviour which focuses on tiger numbers; another is predatory capabilities. Here it would appear that tigers can be adaptable, changing tactics to suit conditions and prey source. There is more to tiger ecology than just predator–prey relations. One important element is phenotypic variation. While most populations with such a wide distribution would have a high variability, the tiger's genetic stock is very low (possibly linked to a bottleneck, i.e. restriction of population in the Pleistocene). It would appear that the connectivity between tracts of tiger country was sufficient to allow survival of all the tiger populations.

All these data suggest a powerful predator which can adapt to a range of circumstances and which has the ability to reproduce its way out of trouble. So why have three tiger subpopulations (Caspian, Balinese and Javan) died out already (Sunquist *et al.* 1999)? The Caspian tiger was only ever associated with river courses, making a sparse population in an arid environment. Human colonisation and usage of the area fragmented the landscape. The preferred prey species, wild boar, while being often plentiful, could show considerable swings in numbers. In Java we see another marginal habitat for tigers. This, coupled with exploitation of the forest and human hunting of prey species, meant that the tiger could no longer survive. Does this mean that other populations are in danger?

Certainly there is concern about the fate of all tigers (which was the driving force behind conservation movements in the 1970s and 1980s). However, current research suggests that it is prey availability rather than habitat per se that is the key determinant. Tigers survive in tiger reserves not because they have a sufficient range (although that is important in the increasingly fragmented vegetation types of the tiger's current range) nor because of low genetic variation (although some might argue this case, e.g. Woodroffe and Ginsburg 1998) but because habitat conservation takes the pressure off prey numbers. In terms of the model proposed above, the tiger is controlled less by its species determinants and more by the community dynamics. Thus by limiting prey, human action limits tigers.

13.4.4 The dugong (*Dugong dugon*)

> The Dugong was widely distributed in the tropical and sub-tropical protected coastal areas of the Indian Ocean and the south west Pacific. . . . It has been exterminated or is now extremely rare, largely because of overhunting, in much of this former range. Dugong inhabit protected, mainly shallow waters and feed selectively on marine plants, mainly sea-grasses . . . but this distribution is discontinuous. Some migration probably occurs between adjacent populations. . . . Dugong populations in some parts of Australia are clearly in decline and these declines appear unsustainable. In other significant areas, however, populations are not currently threatened. (Australian Minister for the Environment 1997)

The dugong is one of four members of the order Sirenia, the other three being species of manatee. A fifth species – Steller's sea-cow – was hunted to extinction within a few decades of its discovery in the eighteenth century. The quotation above encapsulates our problems with these rare marine mammals: there is little information about the biology and ecology and insufficient data on their population demography. What little is known suggests a long-lived (about 70 years) animal with a 13–14 month gestation period. Their low numbers (Australia is estimated to have 90 per cent of the world's dugong population) and their slow recruitment rate makes the dugong a prime candidate for extinction. Difficulties in measuring population size and tracking individuals make any estimate of demography tentative at best (Marsh *et al.* 1999).

This picture is repeated with the manatee. The slow-moving, migratory species has suffered a decline in numbers in recent years due to accidents with nets and boats (where propellers can injure or kill the mammals that live in shallow coastal or estuarine waters). Gene flow is not substantial. Research on DNA sequences by Garcia-Rodriguez *et al.* (1998) suggests that open water poses a considerable barrier for the movement of species. While this affects populations on a global scale there is also concern about adult survival rates (Langtimm *et al.* 1998). However, the greatest concern (according to even a minor literature search) is the complete lack of data about manatee ecology.

Given the limited responses noted above, where does this place the Sirenia in our model of extinction? Its greatest value lies in the repeated concern for paucity of data. We do not even know how many species are on this planet and we cannot say with any great degree of certainty that we know about the ecology of more than a minute fraction of these. This point must be stressed because it is fundamental to all research (and of course it points the way forward for biogeography!). In terms of testing the model it appears from the data that the greatest concern is at the individual/species level, with population demographics and genetics at the forefront. The habitat is there (although it is under threat) but the variability of subpopulation numbers suggests an inherent problem in keeping the species viable.

Although few in number these case studies, chosen for the range of conditions they represent, do allow for the proposed model to be given some small degree of testing. As stated at the beginning of this chapter, this has been a journey through the available literature. This is not a conclusion but a way-marker on the journey.

13.5 Summary

One of today's most crucial questions is that of the survival of species. Only a few years ago this would be seen as a question of conservation (we ought to keep endangered species alive) or even of ethics. New evidence shows that the loss of species is both accelerating and is restricted to a specific range. The implications are that we might lose some crucial species upon which our current ecosystem services depend. The focus in this chapter is that species loss or survival is a question of fragility – the ability to withstand disturbance. This of course begs the question as to the nature of the disturbance, its duration and magnitude, along with the range of responses that the species has in order to counter this.

Initially, fragility was seen as a series of responses to outside pressure (*inside* pressure from the ecosystem being, presumably, within ecological tolerances), but it was also shown that it depended on the scale at which we were looking. This moved on to a detailed examination of the nature of speciation (or origination) and extinction. There have been several major events during geological time: research on these has shown considerable losses over large periods of time. However, in keeping with other aspects of research in this area it must be borne in mind that experimental error and faulty data collection can create false results. Even allowing for this there is now general agreement that current species loss is not sustainable.

How does this relate to current situations? The case studies shown here cover a range of examples and areas and yet there is a common theme. A range of external influences has created a situation where the desired species has a reduced chance of survival. What must also be noted is that these situations do have some solutions and it is one of the roles of biogeography to investigate them.

APPLICATIONS – USING BIOGEOGRAPHY

Using ideas from this chapter in real-life situations

Human action is creating a number of problems especially in areas that have too little resistance to stress. In such areas (usually referred to as fragile environments) people can find that getting a living is difficult. The aim is to break the cycle which almost demands that people further degrade their environment in order to grow or find food. One such scheme is conservation agriculture as demonstrated by the UN Food and Agriculture Organisation (FAO – www.fao.org). The aim was to allow farming in marginal areas using some of the basic principles derived from studies in similar locations: minimal soil disturbance, direct planting and the maintenance of vegetation cover. This is a complete contrast to conventional agriculture where the land is cleared, prepared and then planted, often leaving it exposed to soil erosion and moisture loss. In comparison, conservation agriculture minimises disturbance: plant residues must remain on/in the ground and there should be every attempt to conserve moisture and nutrients. The scheme has been tried in the drier Mediterranean areas with success, but the test will come when it is applied to the driest areas of North Africa. The interest for biogeographers is that it moves the subject out of its more traditional role in 'natural' ecosystems to look more closely at agro-ecosystems. The benefit is not just that people can be fed but that it reduces the pressure on other areas (see http://www.fao.org/english/newsroom/news/2002/10502-en.html – accessed 17/9/04).

Review questions

1. Distinguish between anthropocentric and ecocentric perspectives on species loss.

2. Outline some of the ecological problems that can be associated with the concept of fragility.

3. To what extent is the Connell and Slatyer model realistic?

4. What is the role of community in resilience?

5. Describe the role played by palaeontology in our understanding of origination and extinction.

6. With reference to a named ecosystem, outline the factors influencing its resilience.

7. Do non-linear responses (chaotic reactions) limit our ultimate knowledge of ecosystem function?

8. Describe in detail a fragile ecosystem. Evaluate the effectiveness of conservation measures used to protect it.

Selected readings

This is one of the faster-changing aspects of biogeography. Much of the key work is to be found in the papers used here. However, two excellent overviews are Magurran and May (1999) and Rosenzweig (1995).

References

Allen JC, Schaffer WM and Rosko D. 1993. Chaos reduces species extinction by amplifying local population noise. *Nature*, **364**(6434): 229–32.

Amos J. 2004. Deep sea corals protection call. BBC News. Accessed online at: http://news.bbc.co.uk/1/hi/sci/tech/3491501.stm on 17/2/04.

Australian Minister for the Environment. 1997. Advice to the Minister for the Environment from the Endangered Species Scientific Subcommittee (ESSS) on a proposal to add a species to Schedule 1 of the Endangered Species Protection Act 1992 (the Act). http://www.environment.gov.au/bg/threaten/lists/advices/fauna/dugong.htm.

Barbault R and Sastrapradja SD. 1995. Generation, maintenance and loss of biodiversity. In Heywood VH and Watson RT. (eds) *Global Biodiversity Assessment*. Cambridge University Press.

Barnes RSK and Hughes RN. 1999. *An Introduction to Marine Ecology*, 3rd edn. Blackwell Science.

Begon M, Harper JL and Townsend CR. 1996. *Ecology*, 3rd edn. Blackwell Science.

Bryant D, Burke L, McManus J and Spalding M. 1998. *Reefs at Risk*. Earthscan.

Colinvaux P. 1993. *Ecology*, 2nd edn. Wiley.

Connell TH and Slatyer RO. 1977. Mechanisms of succession in natural communities and their role in community stability and organisation. *American Naturalist*, **111**: 1199–44.

Coyne JA and Orr HA. 2000. The evolutionary genetics of speciation. In Maguran AE and May RM. (eds) *Evolution of Biological Diversity*. Oxford University Press.

Doebeli M. 1993. The evolutionary advantage of controlled chaos. *Proc. Royal Soc. Lond. B: Biological Sciences*, **254**(1341): 281–85.

Forman RTT. 1995. *Land Mosaics*. Cambridge University Press.

Garcia-Rodriguez AI *et al.* 1998. Phylogeography of the West Indian manatee *Trichechus manatus*: how many populations and how many taxa? *Molecular Ecology*, **7**(9): 1137–49.

Given DR. 1994. *Principles and Practice of Plant Conservation*. Chapman & Hall.

Golding B. (ed.) 1994. *Non-neutral Evolution*. Chapman & Hall.

Harris GP. 1994. Pattern, process and prediction in aquatic ecology: a limnological view of some general ecological problems. *Freshwater Biology*, **32**(1): 143–60.

Harwood J and Amos W. 2000. Genetic diversity in natural populations. In Maguran AE and May RM. (eds) *Evolution of Biological Diversity*. Oxford University Press.

Hecnar SJ and McCloskey RT. 1996. Regional dynamics and the status of amphibians. *Ecology*, **77**(7): 2091.

Holdgate M. 1979. *A Perspective of Environmental Pollution*. Cambridge University Press.

Holling CS *et al.* 1995. Biodiversity in the functioning of ecosystems: an ecological synthesis. In Perrings C *et al.* 1995. *Biodiversity Loss – Economic and Ecological Issues*. Cambridge University Press.

Holmes B. 1998. Life support. *New Scientist*. 15 August. (www.newscientist.com/ns/980815/lifesupport.html).

Hughes TP *et al.* 1999. Patterns of recruitment and abundance of corals along the Great Barrier Reef. *Nature*, **397**(6714): 59–63.

Huismann J and Weissing FJ. 1999. Biodiversity of plankton by species oscillations and chaos. *Nature*, **402**(6760): 407–10.

Irwin D. 2000. Life's downs and ups. *Nature*, **404**: 129–30.

Jablonski D. 2004. Extinction: past and present. *Nature*, **427**: 589.

Jackson JBC. 1994. Community unity? *Science*, **264**(5164): 1412–13.

Johnson CN. 1998. Species extinction and the relationship between distribution and abundance. *Nature*, **394**(6690): 272–4.

Johst K, Doebeli M and Brandl R. 1999. Evolution of complex dynamics in spatially structured populations. *Proc. Royal Soc. Lond. B: Biological Sciences*, **266**(1424): 1147–54.

Judson OP. 1994. The rise of the individual-based model in ecology. *Trends in Ecology and Evolution*, **9**(1): 9–14.

Kasperson JX, Kasperson RE and Turner BL. (eds) 1995. *Regions at Risk: Comparisons of Threatened Environments*. United Nations University Press.

Kirchner JW and Weill A. 2000. Delayed biological recovery from extinctions throughout the fossil record. *Nature*, **404**: 177–80.

Langtimm CA *et al.* 1998. Estimates of annual survival probabilities for adult Florida manatees *Trichechus manatus latirostris*. *Ecology*, **79**(3): 981–97.

Laurence WF *et al.* 2000. Rainforest fragmentation kills big trees. *Nature*, **404**: 836.

Levin SA. 1998. Ecosystems and the biosphere as complex adaptive systems. *Ecosystems*, **1**(5): 431–36.

McCann K and Hastings A. 1997. Re-evaluating the omnivory–stability relationship in food webs. *Proc. Royal Soc. Lond. B: Biological Sciences*, **264**(1385): 1249–54.

Magnuson JJ, Benson BJ and McLain AS. 1994. Insights on species richness and turnover from long-term ecological research: fishes in North Temperate Lakes. *American Zoologist*, **34**(3): 463.

Margurran AE and May RM. 1999. *Evolution of Biological Diversity*. Oxford University Press.

Marsh H, Corkeron P, Limpus CJ, Shaughnessy PD and Ward TM. 1999. The reptiles and mammals in Australian seas: their status and management. http://www.environment.gov.au/marine/publications/somer/somer_annex1/som_ann14.html.

Matthiessen P. 1997. The last wild tigers. *Audubon*, **99**(2): 54.

Mooney HA, Lunbchenco J, Dirzo R and Sala OE. 1995. Biodiversity and ecosystem functioning. In Heywood VH and Watson RJ. (eds) *Global Biodiversity Assessment*. Cambridge University Press.

Morin PJ. 1999. *Community Ecology*. Blackwell Science.

Morris SC. 2000. The evolution of diversity in ancient ecosystems: a review. In Maguran AE and May RM. (eds) *Evolution of Biological Diversity*. Oxford University Press.

Mumby PJ and Harborne AR. 1999. Development of a systematic classification scheme of marine habitats to facilitate regional management and mapping of Caribbean coral reefs. *Biological Conservation*, **88**(2): 155–63.

Odum WE, Odum EP and Odum HT. 1995. Nature's pulsing paradigm. Pulsed ecosystems: a new paradigm? *Proceedings of a Symposium Held in Hilton Head, November 1993*: 547–55.

Pascual M and Caswell H. 1997. Environmental heterogeneity and biological pattern in a chaotic predator–prey system. J. *Theoretical Biology*, **185**(1): 1–13.

Price T. 2000. Genetic diversity in natural populations. In Maguran AE and May RM. (eds) *Evolution of Biological Diversity*. Oxford University Press.

Roberts CM. 1997. Connectivity and management of Caribbean coral reefs. *Science*, **278**(5342): 1454–7.
Roberts J. 1981. Control mechanisms of Carboniferous brachiopod zones in eastern Australia. *Lethaia*, **14**: 123–34.
Rosenzweig M. 1995. *Species Diversity in Space and Time*. Cambridge University Press.
Sait SM *et al.* 2000. Invasion sequence affects predator–prey dynamics in a multi-species interaction. *Nature*, **405**(6785): 448–50.
Sale PF. 1999. Cord reefs: recruitment in time and space. *Nature*, **397**(6714): 25–7.
Seidensticker J, Christie S and Jackson P. (eds) 1999. *Riding the Tiger: Tiger Conservation in Human-dominated Landscapes*. Cambridge University Press.
Sepkoski JJ Jr. 2000. Rates of speciation in the fossil record. In Maguran AE and May RM. (eds) *Evolution of Biological Diversity*. Oxford University Press.
Sinha S and Parthasarathy S. 1996. Unusual dynamics of extinction in a simple ecological model. *Proc. Nat. Acad. Sci. USA*, **93**(4): 1504–08.
Smith AB and Jeffrey CH. 1998. Selectivity of extinction among sea urchins at the end of the Cretaceous. *Nature*, **392**: 69–71.
Sole RV, Bascompte J and Manrubia SC. 1996. Extinction: bad genes or weak chaos? *Proc. Royal Soc. Lond. B: Biological Sciences*, **263**(1375): 1407–13.
Specht RL and Specht A. 1999. *Australian Plant Communities*. Oxford University Press.
Stiling P. 1999. *Ecology*, 3rd edn. Prentice Hall.
Stone L and Ezrati S. 1996. Chaos, cycles and spatiotemporal dynamics in plant ecology. *J. of Ecology*, **84**: 279–91.
Sunquist M *et al.* 1999. Ecology, behaviour and resilience of the tiger. In Seidensticker J, Christie S and Jackson P. (eds) *Riding the Tiger: Tiger Conservation in Human-dominated Landscapes*. Cambridge University Press.
Vandermeer J and Yodzis P. 1999. Basin boundary collision as a model of discontinuous change in ecosystems. *Ecology*, **80**(6): 1817–27.
Vernon JEN. 1999. Technical Annex 1: Coral reefs – an overview. State of the Marine Environment Report for Australia. www.environment.gov.au/marine/ publications/somer/somer_annex1/som_ann 12.html.
Williams RJ and Costin AB. 1994. Alpine and sub-alpine vegetation. In Groves RH. (ed.) *Australian Vegetation*, 2nd edn. Cambridge University Press.
Woodroffe R and Ginsburg JR. 1998. Edge effects and the extinction of populations inside protected areas. *Science*, **280**(5372): 2126–8.

Websites

Much of the work in this chapter was focused on the ongoing debate in key areas of this topic. A new website which seeks to continue work on the science/ecology boundary including biogeography is the Science and Development Network (http://www.scidev.net/index.cfm).

CHAPTER 14

Looking at the past – human impact on biogeography

Key points

- There are probably very few areas left in the world where there has been no impact from human activity. Modification is greatest where human occupation has been longest;
- The lack of a baseline record and the amount of modification makes it difficult to define precisely human-induced biogeographical change. Our knowledge of species is highly variable, with lower taxa being the least well known or appreciated;
- Human impact can be seen to have three aspects: spatial (the area covered by change), ecological (species composition and genetic characteristics) and temporal (the rate of different changes through time);
- Human impact has been subject (at least in early times) to the same forces that affect other mammals, e.g. tectonic and oceanic barriers, land bridges and climatic cycles (Ice Ages);
- The type of change at any one place depends on the culture/society of the people and the available biophysical resources;
- Research suggests that all societies through time have been able to have considerable impact on species distribution – it is not a modern phenomenon;
- Today, human impact is so great as to be considered a global-scale force.

14.1 Introduction

> Annual wild rice is a marsh plant which responds to the monsoon climate by a period of rapid growth and seeding in the wet season and dormancy during the dry. It also derives nutrients from water rather than soil. Unlike seed agriculture in the Near East or Europe, therefore, soil exhaustion and field rotation are not issues where rice is grown in naturally marshy habitats. This is exactly the situation preferred by the earliest known agricultural communities in SE Asia. Wherever possible they occupied slightly elevated situations near low-lying land where flooding was predictable and within which rice would flourish. The stability of settlements over several centuries was, it is argued, underwritten by this regime, which was integrated with the maintenance of domestic pigs and cattle, much fishing and collecting, as well as trapping and hunting. (Higham 1996, pp. 322–3)

This quotation encapsulates almost every aspect of the interaction between people and biogeography. It shows a group of people using the natural resources of the area to their advantage, the unconscious result of which is the complete alteration of the flora and fauna at every level of interaction from population to individual. But what elements can we identify here? Higham starts by commenting on the ecology and physiology of the rice plant. Its easy-growing habit and ubiquity in the area make it an ideal candidate for domestication. However, any conscious usage of the plant, e.g. cultivation, may alter the genetic composition of the wild stock. Further, he talks about pigs and cattle – suggesting domestication and another alteration of wild genes through selective breeding. Thus human impact has an ecological perspective (see Figure 14.1). The note about village situation relative to the rice fields gives human impact a spatial aspect. Finally, since this took place over hundreds of years there must be a temporal or time dimension to the impact. Given that we do not know the precise genetic make-up of the first wild stock they used, it follows that we will have difficulty measuring the extent of that impact because there is no reliable baseline from which to take it. We know from other sources that rice was traded and grown over a very wide area (far wider than its original habitat), thus altering the biogeography not just of a region but internationally.

This chapter starts with an examination of the three impacts – spatial, ecological and temporal – and the problems of recording them. Such matters are critical in our understanding of modern biogeographical patterns. There has been much controversy about the impact of people at various stages in history. The old idea that early groups such as the Mesolithic (and by extension, modern indigenous cultures) had minimal impact is not borne out by current trends in research (Flannery 1996). Neither can we support the notion that all modern activity is creating enormous change (although that is certainly the case in some areas). One recent 'discovery' highlights this point and also shows us how rapidly biogeographical research is moving. Global warming has been assumed to be enhanced by modern industrial processes in the last 200 years. Now there is evidence that human impact

Figure 14.1 Human impact on biogeography. The Old Whaling Station at Tathra, NSW Australia. Originally part of the Southern Ocean's exploitation of a wild population, this area has now become a tourist area where fishing (foreground) and coastal development (background) have continued to alter the natural environment. There is some argument supporting the idea that such biogeographical change has been happening for the past 50,000 years since the arrival of the Aboriginal populations

may have started 8,000 years earlier, giving Flannery's comments (originally aimed at Australian research) a new global currency (Ruddiman 2003).

We are at the stage now in biogeographical research that we need a more structured overview of human impact. Although a comprehensive overview is beyond the scope of this chapter it can at least put forward the framework under which the analysis should proceed.

14.2 The spatial extent of human impact

There can be few areas of the world where there has been no human impact upon the plant and animal life of the region. However, some areas have seen waves of modification while some have seen few, if any. In nations as densely populated as the UK there are no areas without human-induced modification. The aim here is to show the growth of human-induced biogeographical change and the implications flowing from this. There is an obvious linkage to time in this (see below for temporal changes) but for the purposes of this work, spatial changes will relate to the areas used and temporal aspects to the rate of change. There are two elements in the issue of spatial change in biogeography: area of land and the scale involved. Put simply, we are having an impact over a growing portion of the earth's surface and this impact, although not evenly distributed, is increasing (as is the rate at which this impact is increasing). It is possible to distinguish between global, national, regional and local scales of change, each with their own implications.

Changes in species distribution and abundance follow human expansion around the globe. When this started is a matter for conjecture. It depends upon which ancestor is taken to be not only the forerunner of modern humans but also present in sufficient numbers to cause a change which would be discernible from normal animal food acquisition. Although this is an interesting point it is, for our purposes, largely academic – the spread of *Homo sapiens* up to 300,000 years ago from a supposed single centre in the middle of Africa (the Out-of-Africa hypothesis – see Box 14.1) correlates sufficiently with the loss of large mammals (the so-called megafauna) for the correlation to be noted (Brown and Lomolino 1998, Atkins *et al.* 1998).

Figure 14.2 shows the major changes in the percentage of megafauna through time. The sudden drop is being linked to the advent of humans, although with Africa it could be either co-evolution or a change more ancient than that noted by the data. However, changes in species should not automatically be linked to human impact. Much of the change in North America was due to the Ice Ages. Minnesota, for example, being linked to the Great Lakes drainage system, was considerably affected by glacial erosion and deposition to the extent that subsequent human occupation would be limited by it (Tester 1995). Neither should the expansion of the human population be linked, automatically, to species loss (see below).

BOX 14.1

The Out-of-Africa hypothesis

Where did we come from? That depends on how far back you wish to take the origin. Although the origins of the first human were noted above to be largely academic there is a lesson here for biogeography. Two hypotheses have been put forward to explain the spread of the human population. The first, and at this time the most supported, is the Out-of-Africa hypothesis, which postulates that our ancestors all came from one area in Africa. From here, humans spread out and changed physical characteristics (i.e. evolved) to take account of their new environments. In the multiple-centre hypothesis humans started in a number of places which later interbreeding made into one human species. The advent of genetic testing with DNA means that this problem is well on the way to being solved but it does raise an issue that should be of interest to us.

We do know that however humans developed they had an increasing impact on their surroundings. If we accept the Africa hypothesis then it follows that we can chart the loss of species and, if this fits neatly enough into the evidence of human spread, accept it. Currently, there is an argument in Australia about the impact of the first Aboriginal peoples on the local megafauna. Because the timing of the loss of megafauna is so close to that of human occupation, many see it as directly linked. An alternative idea being considered more recently in that the change in species was due to climate changes which may have had nothing to do with humans. The lesson is that we need to be very careful in ascribing every change to human action.

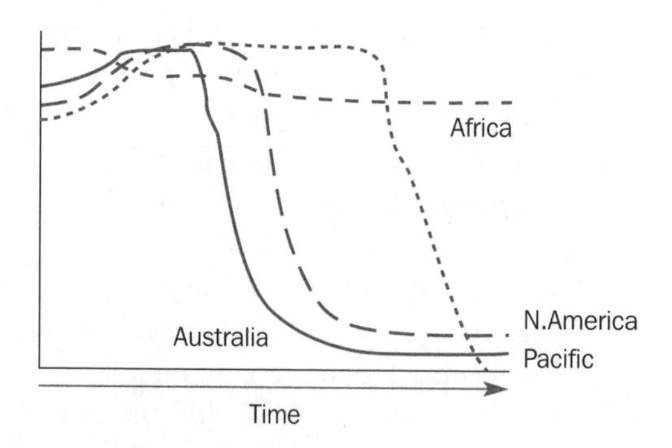

Figure 14.2 Human impact and the loss of megafauna. The four lines represent the loss of megafauna through time. With the exception of Africa (see text) there are sharp losses associated with human migration across the continents
After Atkins *et al.* (1998)

At the global level, human population growth and movement were the primary cause of biogeographic impact – it led to a greater area being affected by human activity. By changing scale the detail becomes both clearer and more complex. For example, at the global scale, the UK was populated from about 450,000 BP. Once this is translated to the national scale, variations can be seen. For example, the earliest people, the Palaeolithic (up to 12,000 BC) lived mainly in the Thames valley until glacial conditions improved and allowed them to move north. The Mesolithic (*c.*12,000 to 3,000 BC) preferred the heaths and moorlands of southern and eastern England while the Neolithic (from 3,000 BC) preferred the lighter chalk soils for their early forms of agriculture.

At the regional scale this broad sweep of people becomes more closely defined with the landscape. For example, the spread of human population from the Mesolithic period onwards started with the more accessible places, usually along riverbanks such as the Test and Itchen in Hampshire. The increasing sophistication of Mesolithic and Neolithic cultures led to the opening of new lands. The regional concentration of this is shown strikingly in the Sussex Weald where there is a considerable correlation between Mesolithic sites and sandy soil (Evans 1975). Further west and in a later time medieval settlement patterns differ (as would the sites of human impact), depending on whether the underlying lithology is chalk or clay. Similar regional spatial variations are seen in most parts of the world. One example is the hypothesis of westward extinction put forward by Lomolino and Channel (1995, quoted in Brown and Lomolino 1998). Studying mammal range 'collapses' (i.e. reductions in area) they noticed that a higher number than expected were taking place westward as in this example of Australian mammals. The explanation is that colonisation took place from the east (as it did also in Canada and the USA).

These examples demonstrate three aspects of the spatial extent of human impact: its complexity, pervasiveness and focus (on favoured areas). The final example takes the local scale. The UK is fortunate in having a written record of changes for almost 1,000 years. Maps show how local areas can be influenced by human action. One classic case is the work carried out by Oliver Rackham (Rackham 1975) in Hayley Wood, Cambridgeshire. In the early part of the twentieth century ecologists considered that there were many parts of the UK that were natural. Studies in East Anglia disabused them of this notion and indeed, the complexity of human action at the local scale was revealed:

> . . . there was soon added the need to recognize biotic effects such as the rabbit grazing . . . in Breckland, and various forms of human interference such as the felling of woodland, pasturing and arable cultivation, with the attendant rapid vegetational change following natural recolonization. . . . [T]he great heathlands of the East Anglian Breck had originated in Neolithic forest clearance; that technique has subsequently disclosed the outlines of progressive woodland destruction and alteration in this country throughout at least five millennia. (Rackham, 1975, p. xiv)

The changes of size and surroundings since AD 1251 demonstrate the changes that have taken place. Of more interest to the biogeographer are the changes that such alterations make on the local ecology. For example, dog's mercury (*Mercurialis perennis*) is found in only certain woodland conditions. These might indicate woodland disturbance (Rackham 1981) or, if found outside wooded areas, that the area was originally woodland and that the plant is part of a relict vegetation type.

14.3 The ecological extent of human impact

Discussions about the spatial changes in woodlands inevitably bring to the fore questions of ecology. Perhaps the most useful way to proceed here is to take a classification of successional mechanisms as a starting point (see Chapter 13) and demonstrate firstly how it is adapted to reflect human impact. The models assume the creation of an open space (e.g. forest clearance). From this three situations can develop. The 'inhibition' model for human impact suggests that once the area is cleared, preferred species are put in place and controlled so that there is no chance of native vegetation re-establishing itself. The 'tolerance' model would allow for some species to coexist with either those existing on the site or those that colonise accidentally or are deliberately brought in later. Essentially, there would be no loss of species due to this invasion. The facilitation model suggests that there is action which allows for the partial alteration of species at one or more of the genetic, species, population or community levels. Whichever route is taken, the key point to note is that there is an alteration of the existing species composition. This means that the area now contains a different biogeography than that originally seen. It might well be that such a change becomes so established as to seem natural, but this does not alter the fact that human interference with the natural biota is global-scale. Three examples illustrate these models: agriculture (inhibition), woodland management (tolerance) and restoration ecology (facilitation).

Despite the vast range of species which are more or less edible, humans tend to use very few. Wheat, maize, yams and rice are major plant species with sheep, cattle, goats, pigs and chickens for the animal kingdom (see Figure 14.3). Even allowing for a few others, staple food sources represent less than 30 species. Given that the number of species is quoted as being between 3.5 and 115 million then the paucity of our current usage can be seen. However, these species (and their numerous selectively bred descendants) do represent a spectacularly successful exploitation which now supports a human population of over 6 billion (see below). For this to be successful, however, the area must be cleared of competing species (or, for modern agribusiness, all species) to allow for the successful development of the desired crop or stock as a monoculture. This reduces the biodiversity of the farming area dramatically but does allow for maximum biomass yield on the site. The inhibition of competing species has grown today to be a multi-billion dollar market for many nations (for more details, see below). One of the more interesting debates on this topic is the concept of conservation. For years farming has reduced the species diversity of an area. However, the problem of transporting produce meant that local varieties were developed. Under EU rules, it is now illegal to sell seed if it comes from a non-approved variety. The result seems to be a massive reduction in

Figure 14.3 Farming in west Australia. Although much of the native vegetation has been suppressed, some native species (such as the emus here) are seen increasingly as farming diversifies, adding yet another biogeographical dimension to the area

cultivars (approved apple varieties have decreased from over 100 to less than a dozen in the UK alone). Taken over the whole of the EU, the resultant loss in genetic diversity of staple food crops is nothing less than scandalous and a typical result of an over-regulated community!

Woodland management illustrates the tolerance model (see Figure 14.4). Areas such as Hatfield Forest have been managed since Saxon times (with evidence of human occupation going back to 4,500 BC (Rackham 1989)), with new species added as needs dictated. For Hatfield this meant pheasant, rabbit and deer (brought by the Norman French land owners for hunting and food supply) along with sweet chestnut and sycamore. Later still, the Victorians added black pine, horse chestnut and London plane. Such additions were part of the extensive use of woodlands and their products to which all ancient forest in Britain were subject. The diversity of use can be seen in Hatfield as five principal activities – woodland (managed for timber), wood pasture (areas used for grazing and timber production), hedgerow trees and plantations (deliberately planted areas contrasting with the natural growth of the woodland) and exotics (specimen trees as part of a landscaping process). Although some of the trees would be at the limits of their ecological ranges (e.g. London plane) and

Figure 14.4 Wealden woodland near Capel, West Sussex. Although this looks a 'natural' scene, the tree management system known as coppice shows this to be little more than a medieval timber factory!

need renewing rather than naturally regenerating, most of the plants would be able to coexist. So successful is this that most visitors to places like Hatfield or the New Forest today would consider the landscape to be completely natural rather than the medieval industrial area it was in reality. Lest this should be seen as a purely western European tradition, Peterken (1996, p. 230) makes the same comments about the woodland of the USA:

> In fact, very little of this woodland [of the North-Eastern seaboard] is totally natural. It has been inhabited by Indians for thousands of years and, from the sixteenth century onwards, most has been logged and cleared. Virtually all the present-day eastern woods are either second-growth stands healing the scars of nineteenth- and twentieth-century logging or secondary woods growing on farmland which was abandoned during the last century or so.

Like Hatfield, we see a range of activities which alter the original ecology of the woodland with the addition of some species and the loss of others. Despite all these changes the essential forest nature remains along with the ecology and structure.

One of the more recent branches of ecology is that which deals with the restoration of land damaged by previous human activity such as mining and quarrying (see Figure 14.5). It provides us with our last example of human interference. In this case (the 'facilitation' model) human activity alters the composition of the vegetation. Such alteration works at a number of levels: genetic (by the selection of specific tolerances possessed by only a few individuals), specific (selecting key species for the task) and community (by the artificial selection of the species diversity of the area). The basis of this effect is easy to understand. Every population has a range of genotypes each of which is best adapted for certain specific conditions (see, for example, the effect of growing rice in different areas – as quoted in Bradshaw and McNeilly 1981). Under most circumstances, small variations will be tolerated within the phenotypic range of the species. However, if the change is too

Figure 14.5 Minnesota, USA. Restoration ecology is an increasingly important area of applied ecology but its impact on biogeography at species and genetic level can be considerable

great then some of the least suitable will die out and the genotypic mean (i.e. the most typical condition) will shift towards the new equilibrium. The effect on the individual is to select for specific characteristics; for the population it changes the mean and skews the distribution of genotypes. So far, this discussion has been linked to existence rather than characteristics as if to suggest that plant cover is the major feature. Studies on the effects of copper pollution on plants has shown that it can affect plant growth. Given sufficient time it is possible for plant populations to respond by becoming more tolerant. The biogeographical implication of this is that plant distributions can be considerably altered as they respond to human-induced changes in the physical environment. Even on a small scale, the effect of human activities such as mining can persist for many years. Plate 14 shows the environmental damage caused to its surrounding by one small Welsh copper mine.

14.4 The temporal extent of human impact

If spatial impact deals with the area covered by change then temporal impact is the rate at which such a change has been accomplished. While it is clear that there is a considerable interrelationship between the two (e.g. as above, we could say that the amount of land affected and the rate of change are both increasing), it is useful to separate them because, in the same area, we can have vast periods of time where little seems to change and we can have considerable change in only a few years. Such perception results in commonly held but increasingly contentious views that, for example, for the first 50,000 years in Australia the Aboriginal people had less impact than the Europeans post-1788 (see Flannery 1996). Should it be conclusively proven that, as some argue, Australia was more of a tropical woodland than the dry sclerophyll forest seen today, when the first Aboriginal settlers arrived, then one has to revisit the old low-impact perspective of indigenous groups.

However, until such matters are resolved there are plenty of cases which illustrate changes through time and also at different scales of activity. At the global level, bird extinctions provide us with another example of the rates of change (Peters and Lovejoy 1990, Birdlife International 2004). Like the São Paulo example noted below, the rate of change is increasing through time – up to one-eighth of bird species are currently thought to be under threat of extinction. Although obviously a result of human action, the precise cause is less certain but considered to be due largely to deforestation. On a national scale, studies of the Brazilian state of São Paulo (Peters and Lovejoy 1990) show that the rate of change has varied considerably during the deforestation of the state. Here, a main problem has been the fragmentation of the area as much as the amount of land lost. However, it is still possible to see the basic ground clearance. There was only the most minor clearance between 1500 and 1845, nearly 350 years. By 1907 there was a considerable loss in the north-east of the state, much of it due to not only clearance but also the fragmentation of the remainder. This last point is crucial. It is not just sheer rate of loss but also the way in which the loss has occurred (smaller areas are more prone to alteration than bigger ones). By 1952 the loss is considerable with only coastal and inland strips left in any proportion. It was estimated that by 2000 there would be only a strip left along the coast (that this has not happened shows the weakness inherent in any predictions but does not necessarily lessen the threat on the area). At the regional level, a study of the Caucasus area demonstrates the impact of changing human pressure through time (Badenov *et al.* 1990, p. 522):

> The main sources of anthropogenic impact on the Caucasian soil cover . . . were and still are animal-grazing, forestry and farming. These impacts have dramatically accelerated since the middle of the nineteenth century. The forested area has decreased by a factor of two, plowed area has increased nearly five times, and the population of domesticated animals has increased significantly.

Not only has the rate of change altered but so have the causes of that change and the impact on various parts of the biosphere, making the biogeography of the landscape a highly complex picture.

At the local level, one area that has undergone significant change has been the chalk downland in Dorset. Originally lowland woodland, it was one of the first to be cleared in the Neolithic era (*c.*3000 BC) and farmed. By medieval times it had become a sheep grazing area. This state of affairs was still recognisable in 1793 when Claridge wrote (quoted in Jones 1973, p. 1):

> The most striking feature of the county is the open and unenclosed parts, covered by numerous flocks of sheep, scattered over the Downs, which are in general of a delightful verdure and smoothness affording a scene beautifully picturesque.

This scene was not to last much beyond that time. Increased demands for food and the mechanisation of agriculture were to put many areas of Dorset downland under the plough. The final result has been similar to the São Paulo case – absolute loss of area and an increased fragmentation. Jones's study highlights this loss but also puts forward methodological and definitional difficulties in arriving at an accurate picture of change. For example, data are not always recorded either accurately or using the same units. This is compounded by changes in the definition of downland where some surveys record it as rough grazing and others use permanent pasture or open country. Although the rate of loss as calculated between the surveys used in the study can still be seen, it does highlight the problems we have in attempting to catalogue any change in biogeography.

14.5 Case studies in human impact

The cases mentioned above serve to illustrate the three main strands of human impact. This section takes a more complex perspective by adding rates of change in different times with the effect of changing the scale of activity:

- At the *global scale*, agriculture (from its earliest to medieval times) proves to be the greatest force in altering the distribution of plants and animals (both adding and losing species). Prior to agriculture, the hunting of selected groups of animals in Europe altered significantly the megafauna seen today in this region. Long gone are the beaver and bear from Britain and even the wolf was extinct in Ireland by 1770.
- Conservation alters biogeography at the *national scale* in post-modern Britain. Calls to reintroduce species today show another change in our approach to species distribution.
- Finally, the *local scale* change is demonstrated in seventeenth- and eighteenth-century garden design. Although small in area, the changes wrought by the introduction (and subsequent escape) of foreign species has become a major issue in both ecology and biogeography (where, for example, so-called exotic plants and non-indigenous feral animals such as cats have devastated the bushland around most Australian cities).

Agriculture remains one of the greatest examples of human impact upon biogeography at the global scale. It has translated species from their original areas to all parts of the world and in so doing has altered the ecology in the new areas – adding and reducing relevant species and altering the ethology and gene flows of others. So successful has this been that any map of natural vegetation is almost certainly incorrect in that it rarely gives due credit to heavily agricultural areas, preferring to subsume them under the label given to the native flora and fauna. In considering the impact of agriculture it is necessary to ask three questions: how did it start, how did it spread and what implications are raised by the first two questions?

Firstly, how did agriculture start? The initial view that it was started primarily in the Near East has been altered. We now think that there were seven regions of the world where agriculture started independently (Wenke 1999, Smith 1995 – see Figure 14.6). The first area still appears to be the Near East, the so-called Fertile Crescent, approximately 10,000 years ago. Initial evidence suggests two sites in China forming the basis of an agricultural society around 8,500 to 7,800 BP, with the remaining areas of sub-Saharan Africa, North and South America starting to convert to the agricultural lifestyle between 4,800 and 4,000 BP. Of course these dates are subject to considerable doubt. The start of agriculture in a region may be no more than a slight shift in a pollen diagram suggesting a move from general to more selective gathering. This in turn demands knowledge of the early plants and animals selected for initial agricultural purposes. Put simply, why are so few plants and animal species used in agriculture? The answer to this question helps explain the spread of species and their resistance to the local ecology. To be a good agricultural plant (Hawkes 1969) it would be necessary to:

- be weedy, i.e. have an ability to colonise disturbed areas quickly and outcompete the opposition;
- contain suitable food sources, e.g. tubers, edible seeds;
- contain food sources that are able to be stored without significant deterioration;
- have a wide ecological tolerance especially to ranges of climate;
- grow near a settlement in high-nutrient areas.

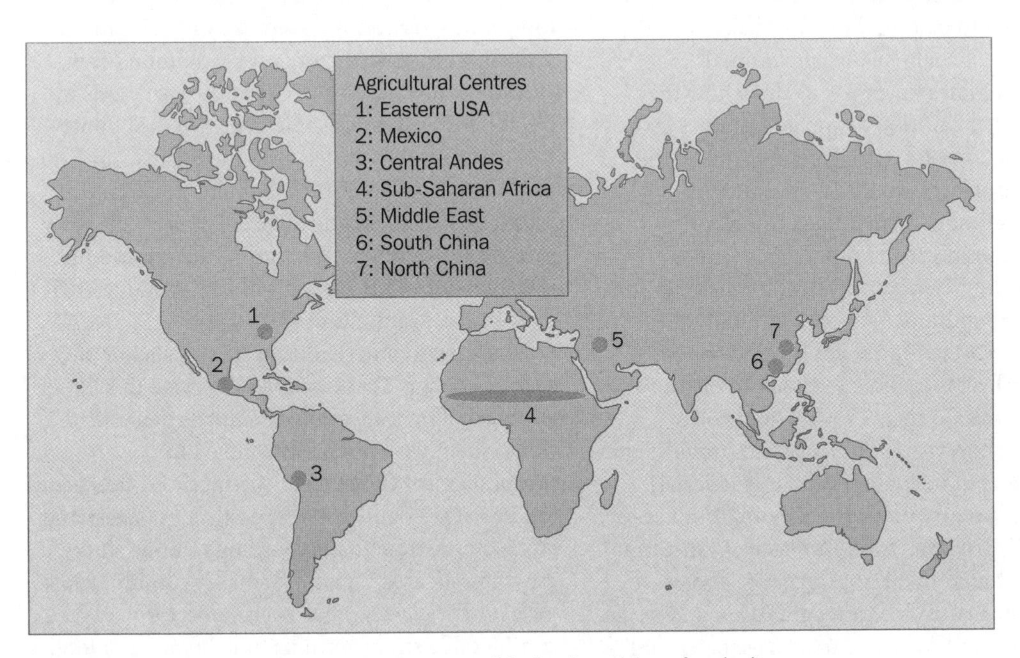

Figure 14.6 The seven areas said to be responsible for the origins of agriculture

Sometimes referred to as the rubbish heap hypothesis, it is an attractive idea because it supposes that any useful plant growing around the camp would be gathered preferentially to one growing many kilometres away.

From this list, plants such as corn/maize (*Zea mays*) and wheat (*Triticum* sp.) are ideally suited. Once these wild plants were established then evidence suggests that they would be altered morphologically as a response to the new environment. For example, in the new agricultural setting, plants that kept their seeds would be preferentially planted in next season's fields, thus altering the genetic material used. In addition, it seems that such plants would also change their seed locations from a more open pattern to terminal clusters (Smith 1995). Similar requirements would exist for animals, i.e. be hardy, adaptable, be able to be tamed, breed freely, be useful (i.e. have useful products such as meat and hide) (Clutton-Brock 1987). This would again limit the number of species (sheep, cattle, pigs and goats) and would also affect the preferred morphology (size, fat/lean meat ratio, length of wool etc.).

The origins of an agricultural biogeography would appear to be found in a mutualistic relationship. Once the plants and animals are selected it becomes necessary to show how they spread. Here, the notion of spread becomes less distinctive. It appears that, depending upon the species, some plants would be cultivated *in situ* (spread genetically) while others would be physically translocated to other areas (spread geographically). Two plants illustrate these notions: yams and groundnuts. Yams (strictly, only the genus *Dioscorea*) comprise about 600 species within which all the edible yams are to be found (Alexander and Coursey 1969). They mostly develop an underground tuber which provides the food supply (but in some species this is aerial). Although the centre of origin is thought to have been South-East Asia, by the Cretaceous ancestral forms were found in all tropical areas. Today, we find three main areas of yam cultivation and usage – Asia, Africa and America. Yams have been, and are, both gathered wild and cultivated, thus giving the biogeography of the yam several difficult areas to study. In all areas the wild yams are used as a food source but the cultivated yam is only a staple in South-East Asia, being of lesser importance in South America and Africa. Cultivated forms appear to have arisen through hybridisation from wild species. For example, the cultivar *Dioscorea alata* is thought to have originated from hybridisation of *D. hamiltonii* and *D. persimilis* in South-East Asia. From its origin in the northern Vietnam area it spread both eastwards to the Pacific islands and westwards to India and possibly West Africa by the sixteenth century. Thus in the yam we have an example of both wild and cultivated use with a complex interaction of the two being spread throughout the tropical regions.

In contrast, the groundnut or peanut (*Arachis hypogaea*) is restricted, in the wild form, to South America (Krapovickas 1969) with a probable centre of origin in the Andean foothills of Bolivia or Brazil. After the flower is pollinated the ovary elongates and buries itself in the soil, where the 'nut' develops. Studies have shown that there were five early genocentres (specific areas of distinctive genotypes) closely linked to subsistence use by indigenous groups. It stayed like this until the early sixteenth century whence, post-Columbus, it was spread to the Old World. Given the variations in other nations there is some evidence to support two methods of transfer – via the Spanish and via the Portuguese. By 1580 it was almost certainly found in Malaysia and reached China by the early seventeenth century. Apparently, the groundnut was introduced into the USA in 1871 and to Africa early in the twentieth-century.

The spread of agricultural species also implies a genetic change. Two examples illustrate this approach. Domestication of animals demanded traits which were apparently linked to morphological elements in the species concerned. Studies of pre- and post-domestication sheep and goat demonstrate that, according to bone studies, the domesticated forms are smaller (Smith 1995). Wheat (*Triticum sp.*) has been subject to considerable study given its importance as a food source (Langer and Hill 1991). It was one of the first plants to be cultivated. However, the early forms of wheat had a very effective dispersal mechanism – seed spikes broke off at maturity. This would be good for a wild plant but far less useful for a cultivar. One can imagine early farmers collecting those individuals less likely to drop spikes and thus effectively selecting for this trait. Soon after the domestication of this (diploid) einkorn wheat (*T. monococcum*), a tetraploid wheat was developed (*T. timopheevii*) which developed into the cultivated emmer wheat. The next stage was the hexaploid bread wheat (*T. aestivum*) from which most modern forms are derived.

Wheat, sheep and goats demonstrate just some of the changes that have been seen in agricultural biogeography. The initial collection and eventual domestication of a limited number of species have considerable implications for the natural biogeography of the planet of which the more crucial are:

- There has been a *considerable loss of species* following the spread of agriculture. Vast tracts of land have been replaced by a reduced number of species. In addition, the control of these alien species has led to the need for the control of native species through the use of pesticides and fertilisers;
- There has been a *change in the distribution of species*. Although many species have been displaced or removed by agricultural practices, it should also be borne in mind that the agricultural species themselves have undergone massive increase in their ecological ranges (and beyond if we count intensive agriculture as a use of species);
- The agricultural species themselves have seen a great *increase in subspecies and varieties*. At one time the UK had regional breeds of most of the main species – sheep, pigs, cattle, etc. Current commercial pressures within the EU, together with production subsidies may well lead to the abandonment of the less productive breeds. This is probably the main reason for calls for conservation of old breeds ('rare breeds trusts').

Hunting was one of the first large-scale activities carried out by people. It was the dominant form of human life support by 15,000 BC. It fell dramatically with the advent of agriculture from 3,000 BC onwards but even in AD 1500 there were still areas where it survived (Atkins *et al.* 1988). There is sufficient evidence to support the notion that the advent of hunting was the start of extinction of the megafauna. This is compounded by the range of foodstuffs consumed and the areas used to procure them. It is possible, following patterns seen today, that the Mesolithic hunters had several areas for food acquisition, including summer and winter areas and possible coastline for molluscs and fish. Atkins *et al.* suggest that inland areas were more greatly affected because of the use of fire to clear areas (an explanation also used in a US study – Dorney 1981). This activity, while small scale in any one area, might have significant impact if done over a large enough area (leading to a more humanised landscape). As tempting as this picture might be, for Europe at least there are more factors at work than the purely anthropogenic.

Around 10,000 BC, Europe was emerging from the last glacial period. Within as little as 50 years the temperature rose an average of 9°C (Yalden 1999). Sites such as Star Carr in Yorkshire provide evidence of the Mesolithic hunter's prey – deer, elk, aurochs and wild boar. Elsewhere there is evidence for a wide range of large mammals including tarpan (wild horses) and lynx. The extinctions in Mesolithic times are enigmatic. Certainly there are records of loss for lynx, reindeer, tarpan and elk but the extent of human action is uncertain. Three reasons for this loss suggest themselves:

- Firstly, it could be due to the rapid changes in climate. Vegetation would have started to change in character and composition and there would be fewer resources for the remaining animals.
- Secondly, changes induced by human occupation might have reduced the area for these animals to mere islands upon which the pressure became so intense that the species died out.
- Thirdly, there is the possibility of human action, such as hunting, wiping out the final individuals. Of course it is possible, given the paucity of data, that all three aspects were

involved. For the rest of Europe the changes following the last glacial period were also marked. For the UK the rise of sea level meant that any new terrestrial and freshwater species were stopped at the new Channel (ca. 8000 BP) (although this did mean the spread of marine fauna!). Many mammal species which had spread from the south or east were stopped by this barrier, thus reducing the species diversity of the UK relative to the rest of Europe at this time.

If the impact on the megafauna of Europe by the Mesolithic hunter is fraught with difficulties of data and interpretation there is less uncertainty about our third case. Conservation is often regarded as a help for the countryside, a chance to keep certain areas for their scientific or economic benefits or just for their aesthetic value. However, it must still be regarded as human impact in time (albeit a post-modern perspective of Britain). The pattern of conservation as we know it today is only about 50 years old. Prior to that time there were conservation schemes and organisations interested in nature dating back to the early nineteenth century. Most of the examples in this chapter have been concerned with the change of species (implying loss), whereas conservation deals with the maintenance of species (and even the possible reintroduction of creatures from butterflies to lynx – Yalden 1999). Perhaps the best example of this is the chalk downland. Originally a medieval (11th century onwards) industrial landscape dedicated to sheep production in southern England, it has become a major conservation area. Initial attempts to conserve it led to the removal of the very agent (sheep) that caused it! Chalk downland is a plagioclimax brought about by the clearance of woodland on light and well-drained chalk soils and the subsequent soil loss leading to the establishment of grassland on the remaining thin soil cover. Conservation seeks to maintain that which nature would wish to remove as part of the succession of the area. Such effects of nature are easy to see when a downland area is devoid of sheep. Very soon hawthorn and blackthorn take over and the area reverts to scrub. If this continued then after a few decades woodland would start to develop. However, nature conservation seeks to stop the natural succession and maintain the desired (artificial) one. Lest this should be seen as a minor impact, National Parks and Areas of Outstanding Natural Beauty made up 15 per cent of England and Wales in 1997 (DETR 1998) and from 1982 to 1997 the areas of Sites of Special Scientific Interest rose from 3,000 to 331,000 ha! Many areas of the UK are subject to some sort of conservation control making this a widespread human impact.

Although some form of artificial environment in the shape of the garden seems to have been present in most societies since the start of agriculture (Huxley 1978) it is in the designs of seventeenth- and eighteenth-century Europe that we find them at their height (see Figure 14.7). Their origins were in the practical gardens of the monastery which served as a medicinal supply for the hospital. Gradually, the gardens became larger and less designed for supplying herbs and food (although these remained important, such activities were carried out in separate areas from the main garden). These early gardens sought to grow common materials close to where they were

Figure 14.7 A formal garden at Hampton Court Palace, near London, UK. The introduction of formal gardens created an artificial nature 'tamed' by people but it also had the implications of suggesting a mindset towards wilderness as being fit for nothing but being tamed while allowing imported species to spread into otherwise original ecosystems

needed (although the Egyptians may have imported some fruit trees). In the biogeographical sense they hardly altered the situation. Even the formal gardens of China and Persia were only extensions of the areas' ecology.

This situation changed, for England at least, with the work of John Evelyn (Fleming and Gore 1979). His early attempts at garden design were widely copied as were his ideas about combating early deforestation by iron works. It was in 1664 with the publication of his highly important work *Sylva, a Discourse on Forest Trees* that he was taken seriously by the government and landowners. He introduced a few species from France to help in his garden designs. His importance to us is not so much the plants he introduced but the mindset that such a work created. Here was Nature to be tamed for people's benefit. The Court was instrumental in spreading the new interest in gardening through such schemes as the garden at Hampton Court (Figure 14.7). The highly formalised style seen here was part of the fashion for an organisation of plants and animals. However, by the early part of the eighteenth century this had started to change. Defoe (quoted in Fleming and Gore 1979, pp. 87–8) writes of his view of the moorlands and the new garden of Chatsworth in Derbyshire:

> Upon the top of that mountain begins a vast extended Moor or Waste, which, for fifteen or sixteen miles together due North, presents you with neither Hedge, House or Tree, but a waste and houling wilderness . . . on a sudden the Guide brings him to this Precipice, where he looks down from a frightful height, and a comfortless, barren, and, as he thought endless Moor, into the delightful valley, with the most pleasant Garden. . . .

The attitude is obvious. The implication was of the uselessness of nature and the advantages of a well-tended garden. In addition to the native trees and a few flowers there would be an increasing range of exotic imports, e.g. pineapple, orange, artichoke, guava, hibiscus. However, this system was evolving. Alongside the rural garden boundary the agricultural scene was shifting as well. The medieval open fields were being replaced with more compact (but enclosed) field systems. The highly formal garden of Bridgeman gave way to the more natural designs of people like Kent. The impact he made can be seen in these plans of Stowe House. Gone was the formal layout to be replaced by a far more open design. By the 1750s a new force was to be seen in gardening – Capability Brown. He continued with the ideas of Kent but enlarged upon them. He often dammed streams to make lakes and would be known to have over 100,000 trees (including the importation of larch and cedar of Lebanon) planted at one place to improve it. So great was the concern for the landscape that even villages were demolished and rebuilt elsewhere if they spoilt the view! From this time, gardens developed in four ways: the increasingly picturesque garden of successors like Repton; the increased interest in the import of foreign plants (particularly Chinese and Himalayan plants like rhododendron) and the collection of rare species (e.g. trees in an arboretum); the development of the public park especially in Victorian towns and the rise of the household garden; and, fourthly, the development of new and imported plant species and varieties. Thus the small plot to aid the sick became a clearinghouse of hybrids and exotics.

What of the implications for biogeography? Although individually small in size, the actual area of gardens can be considerable. If we assume that about one-third of a plot is devoted to garden, then it suggests that 3 per cent of the UK surface is given over to the activity. What happens there is of significance to both local and regional biogeography:

- It suggests a perspective about biogeography where nature is subordinate. Such a worldview, common to the Judaeo-Christian perspective, makes nature merely a part of human activity. It could be argued that with such a perspective we run the risk of failing to maintain the biodiversity of the area;
- Many garden plants will thrive excessively outside the garden setting. The best plants are

often those from foreign parts which have little local herbivory. This can lead to invasive plants dominating the local ecology (e.g. *Lantana camara* – Cronk and Fuller 1995);

- By encouraging certain species and discouraging others – flowers versus weeds – we are altering the ecology and species richness of an area;
- By breeding new species and varieties we alter the genetic composition of the remaining stock. If this is then subject to legislation or food industry pressures (as is currently happening with agricultural and garden crop species in the EU), then it could well reduce the gene pool to a critically low level of diversity.

14.6 A critique of human impact

The study of human impact is not without controversy. Theoretical and practical problems mean that even the best data provide only a tentative understanding of human impact and that when time is taken into account this becomes even more difficult. In addition to data collection and analysis there is the whole notion of human impact itself to take into account. Some would argue that early hunter-gatherer impact was limited (with implications for such indigenous groups today, suggesting a cultural/political interest), while others see a considerable change even in the Mesolithic. There are studies pointing to incorrect analysis of data (Marean *et al.* 1998, Brady and Scott 1997), the need for better techniques (Grayson 1981, Leach 1986, Moore 1998) and the need for greater theoretical understanding of the subject (Meltzer *et al.* 1992). It suggests that the greatest care be taken in the interpretation of human action on biogeography.

14.7 Summary

Although it is common to talk about biogeography as being concerned with natural distributions of plants and animals, it becomes increasingly untenable as we move into the twenty-first century. The impact of people on species distribution is considerable. It is reasonable to consider that as human populations spread and developed so did the impact on the biosphere. This is not to say that early people had no impact (in many areas such as the UK it was considerable) but that the impact was more gradual and, to a certain extent, natural systems could adapt. The impact was more than just spatial, i.e. the spread of species associated with human activity. There is also an ecological dimension in that only certain species were preferred which led inevitably to the change of biodiversity. Time and rate of change are also key factors. The history of a species distribution can tell us much about the ecosystem especially in community terms (assuming guild rules – see Chapter 13). Given the range of possible factors, detailed case study analysis is perhaps the key way forward in understanding the complex relationships between humans and other species.

APPLICATIONS – USING BIOGEOGRAPHY

Using ideas from this chapter in real-life situations

In the early 1970s the field of archaeology was opened up considerably with the addition of new areas of study, principally scientific. Although some work in this area had been going on before that, the 1970s saw a revolution in what evidence could be gathered from sites. As the fields grew so the subdisciplines started to create their own areas of expertise and university units were set up. Today, there are many specialist units catering for all manner of archaeological evidence from bones to beetles. Biogeographers have a strong presence through bioarchaeology. A brief overview of two units will put this in context. The George Pitt-Rivers Laboratory for Bioarchaeology is part of the University of Cambridge (http://www.arch.cam.ac.uk/pittrivers/index.html). Its researchers have projects spread around the world with a main focus of providing environmental evidence on how the people lived in the area and interacted with it. Similar themes are explored by the Bradford University Bioarchaeology Unit (http://www.bradford.ac.uk/acad/archsci/depart/resgrp/bioarch/) although their current research interests include biological anthropology. In addition to this line of work there are also groups involved in the study of palaeoecological remains not necessarily associated with human activity. For example, Wilf *et al.* (2003) have analysed plant remains dating back into the Eocene, and there are numerous reports of groups involved in climate change studies making increasingly detailed comments about biological material to aid their work.

Review questions

1. To what extent is the use of the term 'untouched wilderness' justifiable given current rates of human impact?

2. Discuss the limitations on our analysis caused by incomplete taxon data.

3. Show how human impact has changed with regards to area, ecology and time.

4. With reference to a named case study, show how changes have been made to the biogeography of the area through time due to human activity.

5. Is biogeographical change culturally biased? Describe using specific cases.

6. What is more pervasive: long-term change by early human populations or rapid change by modern ones?

7. Are gardens a positive or negative force in biogeography?

8. Within a given area analyse the degree to which human-induced change matches natural changes.

Selected readings

Direct human action is not often seen as a part of biogeography. Although there is a great deal written about change there is less on the theory and mechanisms behind it. Thus often we need to look outside the traditional biogeography areas. Although quite old, a classic text in terms of breadth and depth of coverage is Ucko and Dimbleby (1969) – few texts today match it. A very good general introduction is Atkins et al. (1998). Woodland is well served with some excellent researchers: Rackham (especially 1975) and Peterken (1996) are the most useful here.

References

Alexander J and Coursey DG. 1969. The origins of yam cultivation. In Ucko PJ and Dimbleby GW. (eds) *The Domestication and Exploitation of Plants and Animals*. Duckworth.

Atkins P, Simmons I and Roberts B. 1998. *People, Land and Time*. Arnold.

Badenov YP *et al.* 1990. Caucasia. In Turner BL III *et al.* 1990. *The Earth as Transformed by Human Action*. Cambridge University Press.

Birdlife International. 2004. State of the world's birds 2004. Birdlife International, Cambridge, UK. Accessed online: http://www.birdlife.org/action/science/species/ sowb/index.html 8/3/04.

Bradshaw AD and McNeilly T. 1981. *Evolution and Pollution*. Edward Arnold.

Brady JE and Scott A. 1997. Excavations in buried caves deposits: implications for interpretation. *Journal of Caves and Karst Studies*, **59**(1): 15–21.

Brown JH and Lomolino MV. 1998. *Biogeography*. Sinaeur.

Clutton-Brock J. 1987. *A Natural History of Domesticated Mammals*. British Museum of Natural History.

Cronk QCB and Fuller JL. 1995. *Plant Invaders*. Chapman & Hall.

DETR. 1998. *Digest of Environmental Statistics*, no. 20. TSO.

Dorney JR. 1981. The impact of native Americans on presettlement vegetation in southeastern Wisconsin, USA. *Transactions of the Wisconsin Academy of Sciences Arts and Letters*, **69**: 26–36.

Evans JG. 1975. *The Environment of Early Man in the British Isles*. Paul Elek.

Flannery T. 1996. *Future Eaters*. Reed Books.

Fleming L and Gore A. 1979. *The English Garden*. Michael Joseph.

Ganderton PS. 2000. *Mastering Geography*. Macmillan.

Grayson DK. 1981. Effects of sample size on some derived measures in vertebrate faunal analysis. *Journal of Archaeological Science*, **8**(1): 77–88.

Hawkes JG. 1969. The ecological background of plant domestication. In Ucko PJ and Dimbleby GW. (eds) *The Domestication and Exploitation of Plants and Animals*. Duckworth.

Higham C. 1996. *The Bronze Age in SE Asia*. Cambridge University Press.

Huxley A. 1978. *An Illustrated History of Gardening*. Macmillan.

Jones C. 1973. *The Conservation of Chalk Downland in Dorset*. Dorset County Council.

Krapovickas A. 1969. The origin, variability and spread of the groundnut (*Arachis hypogaea*). In Ucko PJ and Dimbleby GW. (eds) *The Domestication and Exploitation of Plants and Animals*. Duckworth.

Langer RHM and Hill GD. 1991. *Agricultural Plants*, 2nd edn. Cambridge University Press.

Leach F. 1986. A method for the analysis of Pacific island fish bone assemblages and an associated database management system. *Journal of Archaeological Science*, **13**(2): 147–60.

Marean CW *et al.* 1998. Mousterian large-mammal remains from Kobeh cave. *Current Anthropology*, **39**(3): p. S79(35).

Meltzer DJ, Leonard RD and Stratton SK. 1992. The relationship between sample size and diversity in archaeological assemblages. *Journal of Archaeological Science*, **19**(4): 375–87.

Moore PD. 1998. Plant domestication: getting to the roots of tubers. *Nature*, **395**(6700): 330–1.

Peterken GF. 1996. *Natural Woodland: Ecology and Conservation in Northern Temperate Regions*. Cambridge University Press.

Peters RL and Lovejoy TE. 1990. Terrestrial fauna. In Turner BL III *et al.* (eds) *The Earth as Transformed by Human Action*. Cambridge University Press.

Rackham O. 1975. *Hayley Wood: Its History and Ecology*. Cambridgeshire and Isle of Ely Naturalists Trust.

Rackham O. 1981. *Trees and Woodland in the British Landscape*. JM Dent.

Rackham O. 1989. *The Last Forest*. JM Dent.

Ruddiman WF. 2003. The anthropogenic greenhouse era began thousands of years ago. *Climatic Change*, **61**: 261–93.

Smith BD. 1995. *The Emergence of Agriculture*. Scientific American.

Tester JR. 1995. *Minnesota's Natural Heritage: An Ecological Perspective*. University of Minnesota Press.

Ucko PJ and Dimbleby GW. (eds) *The Domestication of Plants and Animals*. Duckworth.

Wenke RJ. 1999. *Patterns in Prehistory*. Oxford University Press.

Wilf P *et al.* 2003. High plant diversity in Eocene South America: evidence from Patagonia. *Science*, **300**: 122–5.

Yalden D. 1999. *The History of British Mammals*. T & AD Poyser.

Websites

The Biodiversity Web has a huge range of resources but one area worth starting from looks at human impact in detail:
http://www.biodiversity.nl/encyclopedia.htm

As noted above, case studies are crucial. One very new version is the State of the World's Birds 2004 from Birdlife International:
http://www.birdlife.org/action/science/species/sowb/index.html

CHAPTER 15

Environmental change and conservation

Key points

- In terms of biogeography, conservation can be seen as one more element in the dynamic interactions between species and their natural environment;
- As interest in conservation has developed over the last 100 years, so has its impact upon ecosystems;
- Conservation seeks to maintain a desired range of species and habitats often against prevailing ecological and environmental forces;
- Current concern with changing environmental parameters, especially global climate change, will mean that, if it is to remain true to its origins, conservation must rethink some of its strategies (which is not without some difficulty).

15.1 Introduction

This chapter is not about conservation or conservation biology: there are a range of excellent texts already dealing with these topics (e.g. Caughley and Gunn 1996, Given 1994, Pullin 2002, Hunter 1996, Bolen and Robinson 1999, Meffe and Carroll 1997, Goldsmith and Warren 1973). Rather, it is a chapter which seeks to place conservation as just one more impact upon the distribution and abundance of species. It also aims to examine the way in which conservation reacts to ecological and biogeographical dynamics especially in the context of changing environmental parameters.

Since conservation is a human activity, it seems right to start with a consideration of human impact on the environment:

> We outnumber every wild mammal found in Britain, with the possible exception of the field vole. We outnumber the commonest wild bird by about five to one. If we all had a decent-sized garden, there would be no countryside. (Marren 2002)

This quotation actually refers to the way in which the population dynamics of the UK has changed since the 1960s, but it also encapsulates the issues facing conservation today and the problems and prospects we face in the future. It is obviously anthropocentric in perspective: people are measured against other species. This is important because conservation is essentially a human characteristic: wild species tend not to worry about the population dynamic of their next meal! This human concern for nature also initiated the conservation movement, started in the 1880s, which has had some great successes in keeping habitats and species that would otherwise be lost. Whether that is a good thing in terms of ecology or human action is, and rightly should be, open for debate. What is less open for debate is the impact of human action upon 'natural' environments. Leaving aside changes of species distribution and abundance for human survival (e.g. agriculture), conservation is one of the few activities, along with landscape gardening, that seeks to determine the location of species. This is done for a range of reasons but the key point as far as biogeography is concerned is that we are determining which species shall be found where. Such activity can have considerable impact (Low 2002). As this chapter is being written, large bushfires have already damaged sections of Canberra, Australia. One aspect of the debate is the conservation issue where one group of people wants regular burning to maintain low ground biomass (and thus reduce fuel for the fires) whereas another section of the community wants less burning. What might seem interesting to the outsider is that the anti-burning section is often comprised of numbers of conservationists (or members of the public who support conservation – which may be an entirely different issue!) and yet the Australian eucalypt forest (i.e. bush) is a fire-formed landscape. Even more worrying is that there appears to be little agreement about what natural fire levels are anyway (but see, for example, Setterfield 2003). Thus the anthropocentric side of conservation is based on limited science although in the last 10 years this situation has changed dramatically.

Marren notes that populations of many wild species in the UK have declined. The inference is that we have lost species that we should keep for the health of the ecosystem. But, as Low (2002) has noted, many species have increased. The rarer ones favoured by conservation programmes tend to be less able to withstand human and other pressures whereas pest and common species (i.e. opportunistic species doing well in whatever circumstances are available) have been able to increase. If we take the grey and red squirrels as one case we find the introduced grey has gained at the expense of the native red. It is possible that this has caused changes in other parts of the ecosystem as well.

Finally, there is the idea of space. The quotation talks about gardens but any human-produced space would do as well. Conservation seeks to keep specific habitats in specific conditions. The argument is that this is the best way of conserving species. Certainly, habitat conservation (i.e. maintaining ecosystems) is a broader and often more cost-effective approach than species conservation. However, the argument is not about relative costs but about the use of space. If the conserved area is big enough then it is possible to have a biogeographic impact. The spatial element has another aspect as well. Changes take place in time. There is no doubt that, given time, species would adapt to human-made environments as many do now in urban settings. However, it is also possible that changes might be too rapid and over too great a scale to allow species to respond. Here, we are faced with the possibility of losing ranges of species to human-induced change (e.g. global warming). This puts a different perspective on the matter. If we lose one or two species then the overall impact is probably limited. However, if we change environmental conditions too quickly then species cannot adapt appropriately and widespread problems could arise. Is this important? Recent work on species response to global warming gives us this conclusion:

> Projected future rapid climate change could soon become a more looming concern. During rapid climatic changes in the past, species showed differential movements rather than shifting together – it behoves us to increase our understanding about the responses of plants and animals to a changing climate. (Root *et al.* 2003)

15.2 Conservation as an anthropocentric dynamic

The notion of conservation has changed through time and there are many definitions in use. In the first of a series of four texts, Warren and

Goldsmith (1974) stated that conservation is 'a responsible attitude to natural resources' (p. 1). In 1983 they noted (p. 7) that conservation was the planned use of functioning natural systems. By 1993 there was no such introduction and it is left to Harrison to offer conservation as stewardship of nature (Goldsmith and Warren 1993, p. 39). The most recent text (actually edited by Warren and French 2001) argues that conservation is also aimed at the physical environment. These four ideas are given not because they encompass all the possible views (the authors are adamant that they do not) but that they show the changes in perspective (and therefore of impact) upon ecosystems that have happened in just 20 years. The first two definitions point towards the idea of conservation as being the active management of resources. In this context it differs markedly from preservation which is passive management (i.e. in conservation you make sure the species survive, in preservation you make sure species do not change). Since ecosystems are dynamic it follows that preservation would eventually kill the system (as limited experiments have shown). By 1993 we see conservation linked to the idea of sustainability – that ecosystems and species need looking after for future generations. The final definition is perhaps the odd one out. Note that from 1973 to 1993 the emphasis is on *nature* conservation. It is only recently that other aspects of the natural environment have been given the same degree of coverage (and yet the early works on conservation saw it as management of natural resources, i.e. minerals as well as species – see, for example, the discussion in Warren and Goldsmith (1974)).

We could extend this initial discussion into a fuller exploration of conservation philosophy but that would defeat the point: conservation is a human activity whose focus changes through time and for which there are no firmly established set of universally agreed rules. Nature conservation (the main branch, currently) seeks to maintain a specific set of species and/or habitats for some purpose which, as we have seen above, clearly changes through time (and would also vary between different societies). In keeping a given set, it works against other dynamics – to this extent, nature conservation is anti-nature!

To further add confusion, conservation does not seek to change the dynamics of all ecosystems, just those selected. Although the rationale behind conservation can seem inclusive, in reality it is highly selective. Species are conserved (adapted from Given 1994, p. 3) because of a mixture of their:

- economic value
- environmental role
- scientific/research value
- maintenance of future stocks of species
- moral and aesthetic value for people.

A look at some aspects of the early days of British nature conservation will highlight this point. Sheail, in his excellent history (1998), notes (p. 2) that one of the first conservation Acts was passed on humanitarian grounds and to protect those bird species of value to farmers and fishermen. There is no mention of habitat here and the species chosen are highly selective. Later, in 1912, the Society for the Promotion of Nature Reserves had, as one of its objectives (ibid., p. 5) 'to collect and collate information as to the areas which retain primitive conditions and contain rare and local species liable to extinction.' What is defined as primitive is open to question, but the idea of rare and local species (presumably small population and endemic species respectively) still has favour today. If we need further proof we only have to look at the distribution of National Parks in the UK to see that the vast majority are to the west and north and that nearly all are mountainous. Although it could be argued that this has occurred because these areas are the least populated, it is in part a mindset that saw such areas as having natural beauty. The subsequent visitor pressures on these areas have created numerous problems, leading to far more control than was envisaged!

Yet another aspect of conservation in England was that of woodland resource management from the eleventh until the early nineteenth century when wood was a strategic resource for building houses and ships as well as providing raw material for charcoal and the peasants' fuel. Plate 15 shows contemporary management of a coppice woodland with over 800 years of documented history.

So much for the case for conservation being ecologically selective. What about the ecological concepts used in conservation biology? Using the contents pages of some of the conservation texts mentioned above, a basic topic analysis was carried out. We can see that the ecological dimensions are limited and often social and political elements are mentioned. This is not to decry the situation because in a human-made space such as the UK social and political elements are often crucial, but it does highlight the relative reduction in theory compared with ecology *sensu stricto*.

Table 15.1 lists some of the basic areas covered by texts' authors. Leaving aside the obvious academic caveats such as restricted scope of the categories and the fact that no two authors saw the topic the same way, this basic analysis still manages to create some instructive ideas. The only two areas where all agreed were the basic rationale behind conservation and the basic tenets, i.e. what the subject was dealing with – the limits of the topic.

Table 15.1 Some basic areas covered by authors

Concept/author	Caughley	Hunter	Pullin	Bolen	Given	Meffe
Basic concepts, etc.						
Rationale	X	X	X	X	X	X
Basic tenets	X	X	X	X	X	X
Ecological concepts						
Population dynamics	X			X		X
Extinction/speciation	X	X			X	X
Ecosystems		X	X	X		X
Genetics		X	X			X
Biodiversity		X				
Behaviour				X		
Abiotic environment				X		
Disease				X		
Human-oriented concepts						
Risk	X					
Economics	X	X		X	X	
Trade	X					
Legislation	X				X	
Human impact		X	X	X		
Politics		X				X
Society		X		X		
Sustainabillity			X			
Management concepts						
Management theory		X	X	X	X	X
Education					X	
Reserve design	X	X	X		X	
Habitat loss		X	X			X

NB Authors: Caughley (Caughley and Gunn 1996); Hunter (Hunter 1996); Pullin (Pullin 2002); Bolen (Bolen and Robinson 1999); Given (Given 1994); Meffe (Meffe and Carroll 1997).

After that, the table is most obvious by a lack of agreement. Even though the books covered different continents and slightly differing foci of topic, you would still expect the books to cluster around a few basic principles and yet this is not the case. It could be argued that population dynamics would be a vital topic in terms of population viability and yet this is only chosen as a key topic by 50 per cent. Given that conservation tries to keep species, the study of extinction and speciation by two-thirds of texts is surprising. Many other topics reach the two-thirds level including ecosystems, economics and management.

Although it is possible that these differences were due more to the analytical method than any real distinctions it is also possible to put an alternative perspective. Much practical conservation work is done without full knowledge of the complex dynamics of the species/ecosystem under review. There have been numerous successes but equally many failures. Conservation science is a new topic: it would be possible to characterise it as a topic still trying to find its paradigm. In these circumstances it is likely that there would be disagreement about key topics. It is also likely that there are large areas of ecology that have received insufficient attention from conservation scientists, e.g. soil ecology. Whatever the reasons, the outcomes for biogeography are a series of ecosystems where human conservation activity has created a situation other than that which would occur naturally.

15.3 Elements of environmental change

Conservation seeks to address aspects of change in the natural environment so that a range of desired species/habitats might be kept. With ideas of environmental change within the range of a human lifespan (e.g. global warming) it seems reasonable to ask how conservation might deal with a dynamic picture rather than the static (preservationist?) one. Firstly we need to see what changes are possible and what their impact might be. For ease of discussion it is possible to divide changes into two categories: those caused by natural phenomena and those as the result of human action.

Despite the ecological concept of dynamic equilibrium suggesting that there is slight change around a mean value, it is unlikely that there have been many times when there were not conditions that could force the adaptation or extinction of species. Further, we can identify different forces operating at different scales which, together, would suggest a multiplicity of phenomena affecting plant and animal distributions. In terms of natural phenomena we can see the impact of forces at a global (e.g. geological change), regional (e.g. Ice Age) and local scale (e.g. biological and physicochemical changes) (see Park 1997). Anthropogenic change has tended to operate on a local scale (e.g. pollution and urban ecosystems) although we are now seeing evidence to suggest a regional (e.g. acid rain pollution) or even global scale (e.g. climate change). There is evidence to suggest that it is the movement of human activity into the global scale that is causing some of the more dramatic global changes we are now seeing (e.g. see Levy 2002).

Figure 15.1 suggests that there are a series of forces operating at different scales (both spatially and temporally). The major forces are geological and tectonic. Two examples commonly noted (e.g. in Ernst 2000) highlight this. Firstly, the first species were likely to be unicellular and anaerobes – a situation that did not change radically for the first 2 billion years. These were gradually replaced by aerobes, but it is not until the Cambrian that we see larger remains which make up our ideas of fossils. At this scale it is reasonable to note that we have lost entire sets of lifeforms during geological time rather than just a few species. The second example shows the development of the modern horse from ancestral forms. Although this is often presented as an evolutionary picture it must also be linked to the environments in which the animal found itself. Thus the development of the modern horse must also be linked with both the development of its modern environment and its

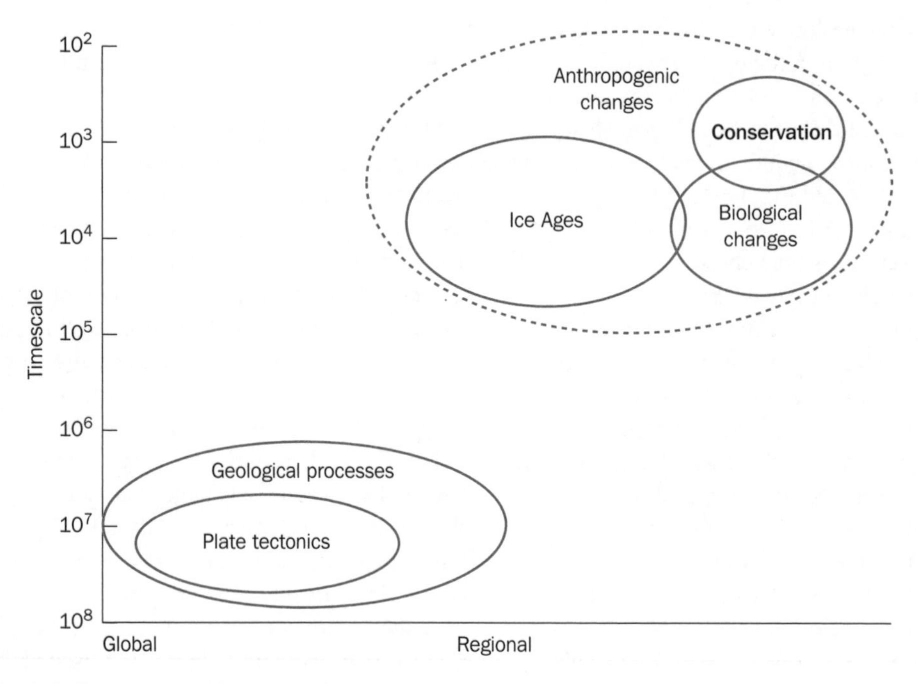

Figure 15.1 Environmental forces and their timescales.

relative position within the food web. Once this is understood then it is easier to appreciate palaeoecology.

The development of species is only one part of the geological story. Recent geological time (i.e. the past 400 million years) has also seen significant movement of the surface due to the actions of plate tectonic movement. It is probable that there were changes prior to that time but our focus has been on the more recent past partly because of the geological evidence available and partly because of the impact this has had on the distribution and subsequent development of flora and fauna. The process has been described elsewhere in this book (see Chapter 3): what is instructive here is to highlight the changes to biogeography that have been brought about by plate tectonics:

- changes in the distribution of land masses
- changes in altitudes/latitudes/depths
- changes in isolation/opportunities/joining
- changes in land mass areas
- changes in palaeocurrents both atmospheric and oceanic
- changes in meteorological phenomena
- changes in continentality/aridity

These changes have timescales of tens of millions of years. On a smaller scale we have the Ice Ages. Whereas these might be seen as less important geologically they have a significant impact ecologically. Taking the start of the most recent Ice Ages at 2.6 My BP we see a series of changes (probably more than 50 cold–temperate cycles – see Lowe and Walker 1996). The impact of these coupled with the advance of the ice into Europe had the effect of moving the biogeographic zones southward. These were not isolated events: there is considerable evidence of climatic changes in the periods before the Pleistocene (referred to as the Neogene 14–25 My BP – see Agustí *et al.* 1999). As with the larger geological factors there are problems of interpretation of evidence, not least the selection of the correct taxonomic level to analyse, the potential rates of change of physiological parameters for species and the assumption that the populations had reached some sort of equilibrium (Lowe and Walker 1996). The effect for biogeography is a lack of confidence in more geologically recent material. The effect for conservation is that it makes decisions on alpine relict species more difficult to assess accurately.

At a smaller scale are the changes brought about by ecological change, e.g. succession, zonation, trophic interrelationships, behaviour (see e.g. Begon *et al.* 1996). This area is critical to biogeography because it impinges on the actual distribution and abundance of species.

Currently, at the smallest scale we have conservation. As stated above, it has the aim to maintain a specific range of species in a given area. Given that the subject is relatively new, it follows that it is at present untested against the larger-scale forces. Its main work is aimed towards a relatively few species and carefully defined areas. The focus on conserving *past* ecosystems (if they had not been under threat of change we would not have considered conserving them!) means that we can exert an influence on local areas.

Into this situation we can bring anthropogenic changes. The dashed oval in Figure 15.1 indicates that the scale and thus impact can (and probably will) change. Although most changes are localised at the moment, anthropogenically enhanced global warming has the potential to spread human impact to a wider scale. The start of human impact cannot be determined accurately but we can say with some certainty that it paralleled human development (Mannion 1997). As long as the population was small this impact was both localised and relatively insignificant, but as society developed so did the range and nature of change. By medieval times, much of the UK was deforested. Since that time the area coming under impact has grown globally.

However, in the last 30 years we can see something of a radical shift in human impact. Until this time the impact stayed approximately in the area it was created. This is not to say that the impact was insignificant but that the reach was local-regional rather than global. Since the 1970s we have seen the rise of a number of instances which show the global reach of human action. Part of this rise must be due, in part, to improved science. The so-called ozone hole was probably in existence (Andersen and Sarma 2002) before it was discovered (Roan 1990). Global warming or, more correctly, human-enhanced global warming has now reached such a pitch that the considerable works of the International Panel on Climate Change are sufficiently convinced that such a phenomenon does exist (Watson *et al.* 2001) and that it does create a noticeable shift (e.g. Parmesan and Yohe 2003). Acid rain may not be the global problem it was once thought to be but we are finding that human impact on biodiversity has reached such a level that its effects can be considered at the global scale (Heywood and Watson 1995).

Human impact has not just affected species and atmosphere. Urban areas are ecosystems in their own right. As such, they have unique physical conditions, plant and animal species. Cities are often warmer than their surroundings and the resulting heat island effect can create differences of up to 3 °C. Artificial surfaces have different radiant properties than natural ones and the distribution of buildings (especially taller buildings) can create unusual wind patterns (Bonan 2002). These factors contribute to the changing hydrological and biogeochemical cycles in urban areas, and whereas urban sprawl was once a major concern in terms of loss of land (Nebel and Wright 2000) we now see changing household demographics (nuclear families splitting into two) as a new factor which could contribute to the enlargement of urban areas (Liu *et al.* 2003).

It is these challenges that conservation needs to meet once we have assessed the current paradigm.

15.4 Conservation and biogeography

Currently, conservation is carried out by defining specific politico-ecological areas (i.e. defined by political process within specific boundaries and ecological constraints) which fit the criteria of the agency involved. Internationally, the United Nations

Environment Programme (UNEP) and the World Conservation Union (IUCN) are the key bodies, although others such as UNESCO are involved with World Heritage Sites. Those sites recognised as being of international importance are placed in one of eight categories (Lucas 1992 and Gubbay 1995 for the marine equivalent). The definitions provided can be informative in terms of the perspectives which drive their designation. For example, category I (scientific reserve) sites are set up to:

> protect nature and maintain natural processes in an undisturbed state in order to have ecologically representative examples of the natural environment available for scientific study, environmental monitoring, education, and for the maintenance of genetic resources in a dynamic and evolutionary state. (Lucas 1992, p. 164)

Category II sites (National Parks) have fewer constraints. They are set up to:

> protect natural and scenic areas of national and international significance for scientific, educational and recreational uses. (*Ibid.* p. 164)

Further designations permit the increased presence of human impact. Although it is clear that the perspectives chosen depend on an anthropocentric viewpoint there are sufficient sites to make at least a local impact on biogeography and biodiversity. Globally, these sites have been growing: in 1970 there were 3,392 sites covering 2.78 million km^2, whereas by 2000 this had reached 11,496 sites and 12.18 million km^2 (UNEP 2002). Analysis of the UNEP GEO3 data (UNEP 2002) shows a more complex situation. The area has increased for each of the four recording times with Asia-Pacific and Central America (including the Caribbean) showing the greatest percentage growth. However, when focusing on the number of sites we find that Europe has by far the most growth. This suggests that the area per site is relatively small in Europe but could be far larger elsewhere (e.g. Africa). The implication is that in Europe we are concerned with conserving those few remnants left, while we can tackle a larger picture in less developed areas. It could also mean that site selection is more a question of human interest than biological diversity or imperative (see Figure 15.2).

The situation in the UK is broadly similar. The amount of land given over to some form of conservation has increased (although the increase between 1990 and 2002 is limited). Conversely, the number of sites has grown steadily, suggesting a reduction in mean site size. Again, this could be explained by a need to protect even small remnants but it could also mean a fragmentation of effort. Either way, it is far from clear that purely scientific concerns for biodiversity are paramount.

Unlike many international cases, in the UK much of the fight for National Parks was started as a question of access (which eventually became embodied in the 1949 National Parks and Access to the Countryside Act). Through a series of committees during the Second World War the UK government started to plan for these beautiful and relatively wild areas. It is no surprise that the first few National Parks were western uplands and mountain areas. Even today, the National Park concept in the UK still retains some of this initial interest. The Environment Act 1995 notes that National Parks should be involved in:

> Conserving and enhancing the natural beauty, wildlife and cultural heritage of the areas promoting opportunities for the understanding and enjoyment of the special qualities of those areas by the public. (Quoted in Cullingworth and Nadin 2002)

Today, the UK has a hierarchical network of sites covering a variety of ecological, biogeographical and cultural perspectives. The existence of these sites owes much to the work of those early Parks committees and their subsequent developments. While they are undoubtedly of benefit to nature conservation in the UK, their contribution to biogeography and especially biogeography in a time of rapid change is open to question. Sites are subject to organisation-specific criteria which may (or may not) relate to potential change. The challenge facing us is to see how and if they can respond to different criteria.

Yakushima is clearly an exceptional place with all the characteristics of a world class site for

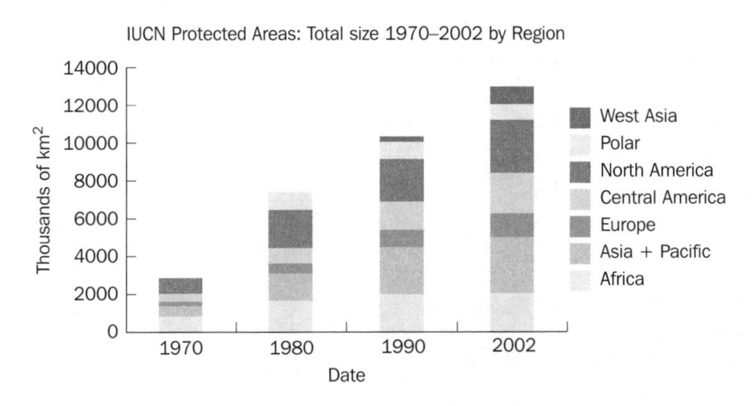

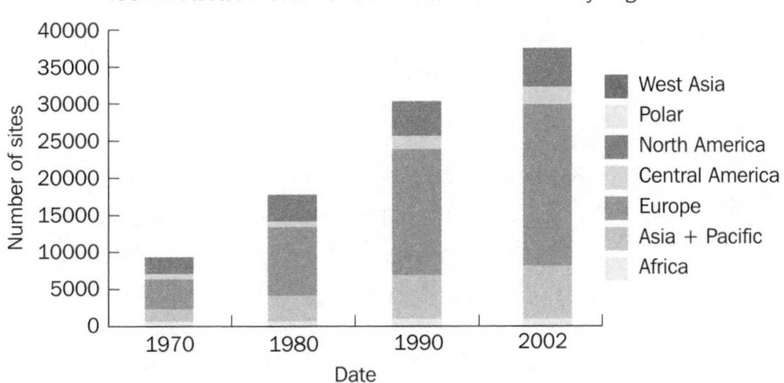

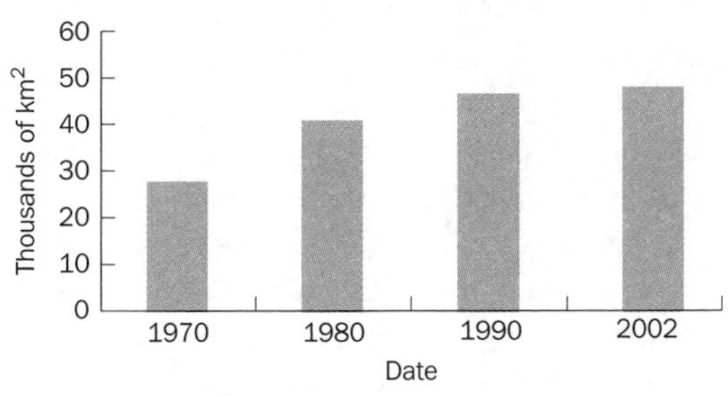

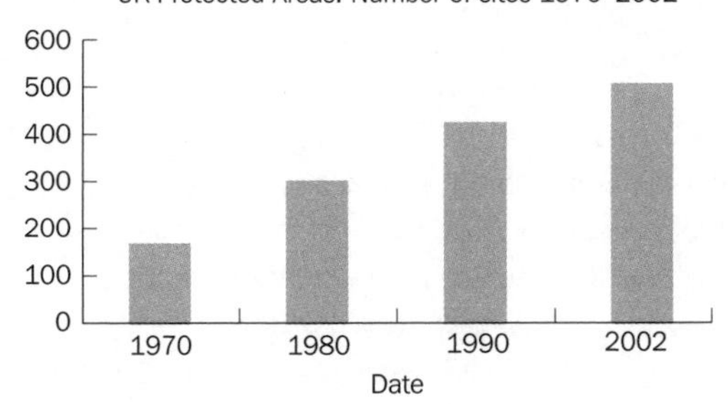

Figure 15.2 Protected Areas by size and number 1970–2002: global and UK data

Data from UNEP (2002 – *GEO3 Data Compendium – Protected Areas*)

conservation (see Box 15.1). The high level of endemism (around 5 per cent of the vascular flora) illustrates precisely the special importance of isolation in the development of genetic diversity. Isolation has resulted in the development of several subspecies of animal that occur on the mainland. There could well be more, since relatively little scientific and taxonomic work has been done on the invertebrates and lower plants.

Human exploitation of the coastal strip and lower slopes has meant that much of the warm temperate broadleaf forest has been replaced by banana, soft fruit (such as citrus fruits) and sweet potato cultivation. The high value of cedar timber

BOX 15.1

Yakushima (Yaku Island), Japan

Please note that this material has been taken from the published Protected Areas Programme with appropriate updating and amendment.

This island lies approximately 40 km south of the southernmost point of Kyushu, Japan. There are about 14,000 inhabitants (1993), living almost exclusively in the coastal areas, and the protected area has a complex, sinuous boundary, which is less than 1 km wide in some places. It lies in the centre of the island, with arms stretching west, south and east. The western arm extends down to the coast at 30° 20'N, 130° 30'E. The major vegetation type is Japanese evergreen forest. The highest point of the island is some 2,000 m although the highest point of the protected area is just over 900 m. The resident population is declining but visitor numbers are increasing annually because of the scenery and great natural history interest of the island.

The following data were extracted from the formal description of the area.

IUCN management category

Kirishima-Yaku National Park, II (National Park)
Yaku Island Forest Ecosystem Reserve, Ia (Strict Nature Reserve)
Yakushima Wilderness Area, Ia (Strict Nature Reserve)
Biosphere Reserve
Natural World Heritage Site – Criteria ii, iii

Biogeographical province

2.02.02 (Japanese evergreen forest)

Date and history of establishment

Kirishima-Yaku National Park was gazetted in 1964 under the Japanese National Parks Law, comprising land on Yaku Island and the Kirishima National Park on Kyushu mainland. A wilderness area designated under the Nature Conservation Law in 1975, forms a small part of the central World Heritage Site. Under the Law of Protection of Cultural Properties, 4,300 ha were established as a special natural monument area and lies entirely within the World Heritage property. A forest ecosystem reserve was established in 1991, and comprises the World Heritage area and various adjacent blocks of land. The centre of Yakushima island, and parts of the island's southern and western coastal lowlands, were internationally recognised as a Biosphere Reserve under UNESCO's Man and the Biosphere Programme in 1980. Yakushima was inscribed on the World Heritage List in 1993.

Area

Kirishima-Yaku National Park: 54,833 ha
Yakushima World Heritage Site: 10,747 ha
Yakushima Island Biosphere Reserve: 19,000 ha
Yakushima Wilderness Area: 1,219 ha
Yaku Island Forest Ecosystem Reserve: 14,600 ha

Land tenure

96 per cent government owned; 4 per cent private ownership (anon 1992).

Altitude

Sea level to 935 m

Physical features

Yakushima island is almost 2,000 m high and is the highest mountain in southern Japan. Several peaks are over 1,800 m with mountain ridges over 1,000 m surrounding these central high peaks. Topography from coastline to the mountainous summits is extremely steep. The predominant bedrock is granite, but small areas of sandstone and shale occur at the foot of the central mountain (Anon 1992; Numata 1986).

Climate

Varies with altitude from subtropical, warm temperate to cool temperate, tending to subalpine. Mean annual temperature is 20 °C in coastal areas, decreasing to 15 °C inland. Air temperature can fall below zero in the mountain summit area and snowfall is common in winter. Annual precipitation is very high, varying with altitude and aspect, from 4,000 mm along the coast to 10,000 mm inland. Humidity is also very high, averaging 73–75 per cent, and in the rainy season, June, exceeding 80 per cent.

Vegetation

Vegetation is significantly different from the mainland. Vertical vegetation distribution is distinct, with subtropical vegetation near the coastline, and warm temperate, temperate, cool temperate and subalpine species further inland as altitude increases. Cool temperate zone coniferous forest, characterised by *Abies firma, Tsuga sieboldii* and *Cryptomeria japonica* occurs, rather than the cool temperate beech forests (*Fagus crenata*) typical of the mainland. Warm temperate broadleaved forest previously covered extensive areas of south Japan. This has largely been removed, due to high human population pressure, and the warm temperate forest trees in Yakushima are thus some of the few remaining in Japan.

Of great significance to the area is the presence of indigenous Japanese cedar *Cryptomeria japonica*, known colloquially as *sugi* (Figure 15.3). *Sugi* can reach more than a thousand years old on stable sites under the climate of the island; specimens younger than 1,000 years are known as *Kosugi*; older specimens, which may reach 3,000 years, are known as *Yakusugi*, and are found between 600 and 1800 m (*Numata* 1986). *Sugi* wood is very resistant to decay and the only current

Figure 15.3 Japanese cedars (*Sugi*) grow to a great size on Yakushima and the timber (highly resistant to decay) was regularly felled and the largest trees were usually selected. In some cases, old tree stumps can be seen surrounded by newer growth derived from basal buds which started into growth once the main trunk had been removed

Figure 15.4 *Yakusugi* are probably one of the oldest remaining cedar trees on the island and are likely to be more than a thousand years old, although it is highly unlikely to be as old as some local legends say. They are specially protected trees and some, such as the huge *Jomon-sugi*, are venerated by many of the visitors who make the trek to see it. This tree has a circumference of 16.4 m and is over 25 m high

source of *Yakusugi* timber is from buried logs or old stumps (Figure 15.4).

The flora is very diverse for such a small island, comprising more than 1,900 species and subspecies. Of these, 94 are endemic, mostly concentrated in the central high mountains. More than 200 species are at the southern limit of their natural distribution and a number are at their northern limit. A distinctive characteristic of the vegetation is the exuberance of epiphytes, particularly mosses and liverworts, at higher elevations. The island has a small population of the mangrove species *Kandelia candel* which is the most northerly natural site for mangroves.

Fauna

The fauna of the island is diverse, with 16 mammal species. Four mammal subspecies, including Japanese macaque *Macaca fuscata yakui* and sika deer *Cervus nippon yakushimae*, are endemic to the island. Population size of both species is approximately 3,000. A further four sub-species are endemic to both Yaku island and the neighbouring island of Tanegashima, including a wood mouse, *Apodemus speciosus dorsalis*. Among the 150 bird species present, 4, including the Ryukyu robin *Erithacus komadori komadori* and Japanese wood pigeon *Columba janthina janthina*, have been designated as Natural Monuments. There are also 15 species of reptile, 8 species of amphibians and about 1,900 species of insects. The Yakushima macaque differs significantly from the mainland subspecies, being darker haired, smaller and stockier in build than mainland forms. One beach is particularly important for loggerhead and green turtles which come ashore at Nagata Inaka-hama to lay eggs in June–July.

Cultural heritage

Traditionally, the island mountains have been considered to have a spiritual value and the *Yakusugi* were revered as sacred trees.

Local human population

None inside the area. The population of Yaku island is 14,000.

Visitors and visitor facilities

Within the nominated area, a limited number of hikers' paths and two huts are maintained, but no other man-made constructions exist. There is no visitor centre, although a new 'Yakusugi Museum' has been built on one of the main access roads to the buffer of the nominated area. The island is accessible by air or by a four-hour ferry trip from Kagushima (Sutherland and Britton 1980). Each year 13,000 visitors walk the trails inside the area.

Scientific research and facilities

The Yakushima forests have been the subject of detailed ecological studies since the area was selected as a Biosphere Reserve (Tagawa and Yoda 1985). An ecological study of Japanese macaque in the nominated site has been carried out since 1975 by members of Kyoto University.

Conservation value

Yakushima occupies a strategic situation on the boundary between Holoarctic and Palaeotropic biogeographical regions. Much of its conservation value is reflected in the 200 plant species which have the southern limit of their natural distribution on the island. The altitudinal continuum of the forests across nearly 2,000 m is considered to be not only the best in the Japanese archipelago, but the best remaining in East Asia.

Ancient *Yakusugi Trees* are of prime conservation value to the island. Individual trees are known by name, and details of height and age are given in Anon (1992) and Chyo (1989).

Conservation management

A managment plan for the World Heritage Area was prepared in 1995. Managment of the site is the responsibility of the Environment and Forestry Agency, the Agency for Cultural Affairs and Kagoshima Prefecture within whose jurisdiction it lies. To provide more effective cooperation and collaboration between these agencies, a World Heritage Area Liaison Committee is being established. It is also intended that local people become involved in the implementation of management objectives.

Management priorities are to strictly conserve the area so as to prevent the loss of its World Heritage values. Activities which may threaten the integrity of the site, such as tree felling, construction of buildings, collection of soil, stones and rocks are strictly regulated (Anon 1995). A total ban on cutting of *Yakusugi* now exists. The only utilisation of Yakusugi permitted today is by unearthing of buried timber and stumps (Anon, 1992).

Management constraints

High precipitation and susceptibility of sandy soils to water erosion, places constraints on trail construction and maintenance. The main threat to this area is the proposal to widen the 'Seibu-Rindoh' road, which would damage the surrounding forest and could also cause landslides.

Staff

Three permanent rangers. Forestry and environment agency staff regularly visit the area on foot.

Budget

No formal budget allocated to the site.

Date

February 1993, revised June 1993.

http://www.wcmc.org.uk/protected_areas/data/wh/yaku.html

for building temples and official structures in other parts of southern Japan meant that for at least 400 years, timber was extracted from the cedar forests on a significant scale.

Although access to significant parts of the 'Protected Area' is not permitted, and visitors are kept to marked paths and trails, there are signs that increased tourism (and ecotourism) will have deleterious effects on the area and not only through pressures on accommodation and waste disposal facilities. Although litter and deliberate damage are very uncommon on Yakushima, the sheer numbers passing along trails and trampling vegetation will reduce the ecological value of some parts of the island. Allied to this is the understandable but unwise habit that tourists have of giving food to, or leaving the remains of their lunches for, mammals and birds. The macaque is quite likely to come into close contact with humans by stealing food, particularly from children, and raiding tourists' belongings for snack foods. This loss of the instinct to keep away from humans is an increasing problem throughout Japan where tourists come to watch groups of these highly intelligent primates. Similar problems exist with other animal groups, particularly in zoos, safari parks or similar situations elsewhere in the world. The instinct of human visitors to feed the animals can have serious implications for the social structure of the animal group, and can trigger inappropriate and potentially dangerous responses, particularly from larger animals. The predilection of bears for easily available human foodstuffs, whether as part of camping supplies or of domestic wastes, is a problem in areas such as Alaska, where polar bears or grizzly bears come into close contact with human populations.

Should conservation be equated with the complete removal of human influences from nature reserves?

15.5 The impact of rapid environmental change – can conservation respond?

Conservation has faced a series of natural and anthropogenic factors in establishing the ecological agenda it sees as necessary for the protection of preferred species. What we need to see now is if it can adapt to meet the challenges of an increasingly changeable situation brought about by anthropogenic influences. To this extent the argument is about the practical application of theoretical biogeography. It is also about the evidence we can gather to assess this perspective: the likelihood and extent of change, our ability to accurately assess change and the way in which we respond.

There are now very few voices who do not argue for anthropogenic change in the environment. Common among the arguments are those of Vitousek (1994) who considers that carbon dioxide, nitrogen biogeochemistry and changes in land use are the greatest threats we face. Carbon dioxide is part of the global warming problem; nitrogen (as in nitrogen-based fertiliser) is being added at such a rate that it threatens to overtake natural production, while land-use change disrupts biodiversity and distribution patterns. After years of gathering data the Intergovernmental Panel on Climate Change have also become more certain that change is both noticeable and anthropogenic (IPCC 2001). Thus in terms of our first question we can answer that there is a great likelihood of change. The second part of the question relates to the extent of the change and here we have less agreement (as one would expect from a range of models). The IPCC (2001) note that the response to change is not uniform in either species, time or space – we are dealing with a set of ecosystems reacting at different rates. Along similar lines, to quote Root *et al.* (2003, p. 57):

> Consequently, the balance of evidence from these studies strongly suggests that a significant impact of global warming is already discernible in animal and plant populations. The synergism of rapid temperature rise and other stresses, in particular, habitat destruction, could easily disrupt the connectedness among species and lead to a reformulation of species communities, reflecting differential changes in species, and to numerous extirpations and possibly extinctions.

Despite the agreement about overall changes there is less agreement about the extent of the change. This may lie partly with the lack of accurate data at the correct scale needed to understand the changes. It might also be due to the variable environmental response of species noted by both the IPCC (2001) and Parmesan and Yohe (2003). The latter research is particularly interesting because it argues not only for the differential response between species (just as the IPCC does) but also for a differential response between different academic interests. The significance of this to our argument is profound. Given that many groups such as the IPCC have workers from a variety of backgrounds it is crucial that they speak the same language, i.e. interpret the significance of data in the same way; any intradisciplinary problems will affect the outcome.

From this brief review we can conclude that the majority of evidence points to certain changes in global terms even if the regional and local detail may be missing. This brings us to our next point: if we are going to deal with change we need to be able to detect it. There has been a considerable volume of work on potential bioindicators of environmental change. What is instructive is that there is no one solution and that those noted here are subject to a wide range of caveats (see Table 15.2).

From the table we can see that there have been a wide range of studies and analyses. The relatively recent date of these is partly due to the recent interest in this area but might also be due to the search strategy used (implying that this area has yet to have a coherent topic base and so relevant papers might be stored under different keywords). The wide range is also interesting from the point of view of organisms used. Diatoms have long been used in palaeoenvironmental analysis, but there are other perspectives commonly seen in environmental archaeology that might be adaptable (see e.g. Dincauze 2000). The potential for genetic changes opens up possibilities especially as DNA can be viable after 30,000–40,000 years (Bowler *et al.* 2003). The geographical areas involved are widespread, suggesting that some form of analysis can be used in most locations. Summarising, we can

Table 15.2 Potential bioindicators

Author/date	Bioindicator	Comment
Flowers (2001)	Wetland ecosystem	Interdisciplinary approach for the study of N. African lakes. Biological groups reacted differently
Chamberlain and Fuller (2001)	Farmland birds	Might be useful but differences in farm ecosystems could also affect results
Tegler *et al.* (2001)	National monitoring	A trans-Canadian scheme to monitor change through the use of simple but widespread tests of 'core monitoring variables'
Slate and Stevenson (2000)	Siliceous microfossil analysis	Relative abundance of key microfossils, e.g. diatoms and sponges mirror key environmental changes
Eley *et al.* (1999)	Forest sub-ecosystems	Analysis of forest sub-ecosystems using climatic data and statistical tests shows forest types sensitive to change in South African climate
Bennion and Appleby (1999)	Diatom analysis	Study of a Welsh lake shows that diatom assemblages change with changing environmental conditions, most notably the increase in sewage
Hogg *et al.* (1998)	Population genetic structure	The absence of genetic analysis might hide potential examples of environmental change
Bright and Morris (1996)	Dormouse populations	Dormice and some other animals can act as 'mine canaries' because they are so sensitive to environmental change
Kushlan (1993)	Waterbirds	The use of waterbird ecology and biology can produce accurate results which may demonstrate their sensitivity to environmental change

say with some degree of certainty that detection is possible to the accuracy we need.

What ecological responses are possible in the face of this change? Fundamental to the work of palaeontology and palaeoenvironmental/palaeoecological reconstructions is the tenet that species do not change. This *principle of uniformitarianism*, put forward in the nineteenth century has been a cornerstone. By asserting that the present is the key to the past we assume that there are no changes in species responses to environmental conditions. Paradoxically, we are also aware of Darwin's central message of 'adapt or die' in terms of natural selection. What we are now finding is a far more complex picture – one which could force us to re-evaluate our knowledge of past environments and our perception of future ecosystems. Work by Jablonski and Sepkoski (1996, but see also Sepkoski 1999) shows that throughout the fossil record there have been cases on assemblages forming, breaking down and re-forming with different species. This suggests that community structure is more variable than we might formerly have believed. Recent work by Realé *et al.* (2003) shows that we get not just changes in groups of species but that there is evidence of microevolutionary change (phenotypic plasticity) within the scale of a few decades, with a strong suggestion that it is linked to anthropogenic climate change. If this is to be replicated elsewhere then it means we need to re-evaluate our understanding of community stability and environmental reconstruction.

We have gathered the evidence of environmental change, its nature, rate, our detection of it and some ideas of ecological response. The final question is whether conservation can also adapt. The initial picture is far from healthy. Standard texts such as the *Global Environmental Outlook 3* (UNEP 2002) are concerned about the loss of biodiversity but say very little about any potential changes. The idea that there is not enough done to consider change factors in conservation is a small but recurrent theme (see for example Hannah *et al.* 2002). Part of the problem might be the nature of that change. We focus on climate change and yet that is not the entire picture. For example, Vitousek *et al.* (1997) argue that introduced species can cause considerable change in the native community. To that extent it depends what is meant by change and what factors are acceptable to cause it. If we are just saying that any remodelling of a pre-existing community is change then we can have numerous examples from the global species changes of Low (2002) to the simplification of ecosystems due to the spread of agriculture (Tilman *et al.* 2001). Bowman (2001) complicates the picture still further by reminding us that conservation is an anthropocentric activity and that to make it acceptable we need to produce narratives of conservation to make the story more acceptable to the public (and, presumably, more likely to be accepted and funded!). To this mix we can add the fact that there is increasing support for any change effects to be global in extent but local in effect, i.e. patchiness (see, for example Cameron *et al.* 1997 and Huntley 1995).

Thus the final verdict is that, as presently constructed, conservation is not able to deal with the changes forecast. This is because those changes are not fully understood in terms of the spatial extent, magnitude and type of change. Ideas focusing on areas of high biodiversity (hotspots or crossroads – e.g. Spector 2002) might succeed because they have such a range of species under protection. At the same time, there is a body of work which suggests that it is possible to turn this situation around. There is a range of tools being tested which can help. For example, Wolff and Mendo (2000) working on fish population argue that it is possible to make some predictions from fisheries research. Cameron *et al.* (1997) put forward geographic information systems (GIS) as a suitable tool mainly because of the ability of GIS to handle large and diverse data sets. Huntley (1995) adds probability studies to the scene which suggests that it might be possible to model populations at the same time as climate change models. Bowman (2001) takes a more pragmatic approach by noting that a good historical analysis of the ecosystem/area, etc. could yield good results.

He suggests that this new field of historical ecology could be used to test ideas against the background of the two very different disciplines from which it is derived. Hannah *et al.* (2002) take this one stage further by outlining the advantages of taking a multidisciplinary approach to conservation.

Whatever the route chosen, we do need to do something and in this respect, biogeography is ideally situated to make a significant input!

15.6 Summary

Despite the considerable amount of public support for conservation it remains more of a human ideal than an ecological necessity. Since the majority of conservation involves keeping an area at a given stage in succession, it could be argued that it is even anti-nature! Leaving aside the semantics of the debate, it is obvious that conservation is growing as a form of land use and that it has a significant effect upon our actions. Conservation is not a fixed item: with the growth of the subject as an academic discipline we now know much more about it than even 15 years ago. The result of this is an ever-changing set of parameters with regard to what constitutes good conservation practice. Two aspects of the natural environment must be set against this. Firstly, we have the effect of conservation in the space/timescale continuum. Currently, conservation is relatively small scale (i.e. local and a matter of decades). This can be explained mainly by the fact that the discipline is so recent. However, as it spreads so we can expect it to become regional in perspective and attempt to plan for longer time periods. At this level the second factor, environmental change, becomes apparent. The central part of this chapter is that environmental change is becoming increasingly rapid especially due to the magnitude of human action (e.g. climate change). The central question for conservation then becomes not how to save a few species but how to respond to the many environmental changes that would appear to threaten far greater ecological disruption than we have experienced previously. As currently constituted, conservation would be unlikely to make a significant contribution, but recent work in this field holds out the hope that, with new tools and understanding, it can adapt.

APPLICATIONS – USING BIOGEOGRAPHY

Using ideas from this chapter in real-life situations

Environmental change is one of the key areas for biogeography. Although much of our work is aimed at gathering data we now have enough over a long enough timescale to be able to make some accurate comments and predictions about change. As might be expected, there are a range of global organisations looking into this. For example, the US Geological Survey's Earthshots programme uses banks of satellite images to compare scenes over 20 or more years. The set of Rondonia in the Brazilian Amazon region is particularly effective at highlighting change (http://edc.usgs.gov/earthshots/slow/Rondonia/Rondonia). The UK's Environmental Change Network (http://www.ecn.ac.uk/) aims to provide a local database on topics such as climate change and water quality. These and many other institutes focus on the way in which we are changing the earth and what some of the consequences could be. Although change is a natural process it is the rate of human-induced change that is causing many people to be concerned. The role of biogeography here is clear. By providing background information and helping to interpret data we can make accurate measurements of change over time which in turn allows predictions to be based more fully on reasonable assumptions.

Review questions

1. With the aid of relevant data show how the growth of the human population in a named area is related to the growth/decline of species and species protection.

2. Nature conservation is anti-nature. Discuss.

3. To what extent is it reasonable for conservationists to listen to the conservation opinions of the general public?

4. Is conservation stewardship or political expediency?

5. Describe the history of nature conservation in a named region.

6. Is conservation ecologically selective? Does it matter?

7. Why is there so little agreement about the contents of key conservation texts?

8. Have people yet reached a regional scale impact? What parameters would you use to evaluate this?

9. What is better – single species or multi-species conservation? Why?

10. Compare the change in protected areas in the UK with that in the rest of the world.

11. Do we need to worry about rapid environmental change?

12. For any one of the papers noted in Table 15.2, carry out a full academic critique including philosophical framework, methodology, results and discussion.

13. Using the information about Yakushima, discuss the question 'Should human beings be allowed anywhere near nature reserves, once the reserve has been established?'

14. How does endemism arise? How useful is it?

15. Investigate the criteria used by the IUCN for designation of important conservation sites. Why are such specialised criteria necessary?

Selected readings

There are a range of good texts here depending on your perspective. For a basic grounding, Caughley and Gunn (1996), although a little dated, is still good value. For the UK political/planning view Cullingworth and Nadin (2002) is the standard work, while Sheail (1998) is one of the best ecological historians. For more detailed work Magurran and May (1999) is a much cited text.

References

Agustí J, Rook L and Andrews P. (eds) 1999. *The Evolution of Neogene Terrestrial Ecosystems in Europe*. Cambridge University Press.

Andersen SO and Sarma KM. 2002. *Protecting the Ozone Layer – the United Nations History*. Earthscan.

Anon. 1992. *World Heritage List Nomination Yakushima (Yaku-Island)*. Environment Agency, Agency for Cultural Affairs, Forestry Agency, Government of Japan. 28 pp.

Anon. 1995. *Yakushima World Heritage Area Management Plan*. Environment Agency, Forestry Agency, Agency for Cultural Affairs, Government of Japan. 10 pp.

Begon M, Harper JL and Townsend CR. 1996. *Ecology*, 3rd edn. Blackwell Science.

Bennion H and Appleby P. 1999. An assessment of recent environmental change in Llangorse Lake using palaeolimnology. *Aquatic Conservation*, **9**(4): 361–75.

Bolen EG and Robinson WL. 1999. *Wildlife Ecology and Management*. 4th edn. Prentice Hall.

Bonan G. 2002. *Ecological Climatology*. Cambridge University Press.

Bowler JM, Johnston H, Olley JM, Prescott JR, Roberts RG, Shawcross W and Spooner NA. 2003. New ages for human occupation and climatic change at Lake Mungo, Australia. *Nature*, **421**: 837–40.

Bowman DMJS. 2001. Future eating and country keeping: what role has environmental history in the management of biodiversity? *Journal of Biogeography*, **28**(5): 549–64.

Bright PW and Morris PA. 1996. Why are dormice so rare? A case study in conservation biology. *Mammal Review*, **26**(4): 157–87.

Cameron GN, Seamon JO and Scheel D. 1997. Environmental change and mammalian richness: impact on preserve design and management in East Texas. *Texas Journal of Science*, **49**(3 suppl.): 155–80.

Caughley G and Gunn A. 1996. *Conservation Biology in Theory and Practice*. Blackwell Science.

Chamberlain DE and Fuller RJ. 2001. Contrasting patterns of change in the distribution and abundance of farmland birds in relation to farming system of lowland Britain. *Global Ecology and Biogeography Letters*, **10**(4): 399–409.

Chyo M. 1989. The estimation of tree numbers of Sugi, *Cryptomeria japonica* in Yakushima, Japan. *Science Bulletin of the Faculty of Agriculture Kyushu University*, **43**: 1–2 (abstract).

Cullingworth B and Nadin V. 2002. *Town and Country Planning in the UK*, 13th edn. Routledge.

Dincauze DF. 2000. *Environmental Archaeology: Principles and Practice*. Cambridge University Press.

Eley HAC, Lawes MJ and Piper SE. 1999. The influence of climate change on the distribution on indigenous forest in KwaZulu-Natal, South Africa. *Journal of Biogeography*, **26**(3): 596–617.

Ernst WG. (ed.) 2000. *Earth Systems: Processes and Issues*. Cambridge University Press.

Flowers RJ. 2001. Change, sustainability and aquatic ecosystem resilience in North African wetland lakes during the 20th C: an introduction to integrated biodiversity studies within the CASSARINA project. *Aquatic Ecology*, **35**(3–4): 261–80

Given DR. 1994. *Principles and Practice of Plant Conservation*. Chapman & Hall.

Goldsmith FB and Warren A. (eds) 1993. *Conservation in Progress*. Wiley.

Gubbay S. (ed.) 1995. *Marine Protected Areas*. Chapman & Hall.

Hannah L, Midgley GF and Millar D. 2002. Climate change-integrated conservation strategies. *Global Ecology and Biogeography*, **11**(6): 485–95.

Heywood VH and Watson RT. (eds) 1995. *Global Biodiversity Assessment*. Cambridge University Press.

Hogg ID, Eadie JM and De Lafontaine Y. 1998. Atmospheric change and diversity of aquatic invertebrates: are we missing the boat? *Environmental Monitoring and Assessment*, **49**(2–3): 291–301.

Hunter ML Jr. 1996. *Fundamentals of Conservation Biology*. Blackwell.

Huntley B. 1995. Plant species response to climate change: implications for the conservation of European birds. *Ibis*, **137**(suppl.1): S127–S138.

IPCC. 2001. *Climate Change 2001: Synthesis Report*. Cambridge University Press.

Jablonski D and Sepkoski JJ Jr. 1996. Paleobiology, community ecology and scales of ecological pattern. *Ecology*, **77**(5): 1367–78.

Kushlan JA. 1993. Colonial waterbirds as bioindicators of environmental change. *Colonial Waterbirds*, **16**(2): 223–51.

Levy D. 2002. Extreme climate variance sped extinction of local butterfly populations, researchers say. Stanford Report (accessed as http://news-service.stanford.edu/news/may15/butterfly-515.html).

Liu J, Daily GC, Ehrlich PR and Luck GW. 2003. Effects of household dynamics on resource consumption and biodiversity. *Nature*, **421**(6922): 530–3.

Low T. 2002. *The New Nature*. Viking.

Lowe JJ and Walker MJC. 1996. *Reconstructing Quaternary Environments*, 2nd edn. Prentice Hall.

Lucas PHC. 1992. *Protected Areas – a Guide for Policy-makers and Planners*. Chapman & Hall.

Magurran AE and May RM. 1999. *Evolution of Biological Diversity*. Oxford University Press.

Mannion AM. 1997. *Global Environmental Change*, 2nd edn. Longman.

Marren P. 2002. *Nature Conservation*. New Naturalist.

Meffe GK, Carroll CR. 1997. *Principles of Conservation Biology*, 2nd edn. Sinauer.

Nebel BJ and Wright RT. 2000. *Environmental Science*, 7th edn. Prentice Hall.

Numata, M. 1986. *The Natural Characteristics of Yaku Island*. Reprinted from *Memoirs Shukutoku University*, No. 20, pp. 15–20.

Park C. 1997. *The Environment: Principles and Applications*. Routledge.

Parmesan C and Yohe G. 2003. A globally coherent fingerprint of climate change impacts across natural systems. *Nature*, **421**: 37–42.

Pullin AS. 2002. *Conservation Biology*. Cambridge University Press.

Realé D, McAdam AG and Boutin S. 2003. Genetic and plastic responses of a northern mammal to climate change. *Proc. R. Soc. Lond. B*, **270**(1515): 591–6.

Roan S. 1990. *Ozone Crisis*. Wiley.

Root T, Price JT, Hall KR, Schneider SH, Rosenzweig C and Pounds JA. 2003. Fingerprints of global warming on wild animals and plants. *Nature*, **421**: 57–60.

Sepkoski JJ Jr. 1999. Rates of speciation in the fossil record. In Magurran AE and May RM. (eds) *Evolution of Biological Diversity*. Oxford University Press.

Setterfield SA. 2003. Seedling establishment in an Australian tropical savanna: effects of seed supply, soil disturbance and fire. *J. Appl. Ecol.*, **39**: 949–59.

Sheail J. 1998. *Nature Conservation in Britain: The Formative Years*. The Stationery Office.

Slate JE and Stevenson RJ. 2000. Recent and abrupt environmental change in the Florida Everglades indicated from siliceous microfossils. *Wetlands*, **20**(2): 346–56.

Spector S. 2002. Biogeographic crossroads as priority areas for biodiversity conservation. *Conservation Biology*, **16**(6): 1480–7.

Sutherland M and Britton D. 1980. *National Parks of Japan*. Kodansha International Ltd, Japan, 148 pp.

Tagawa H and Yoda K 1985. *A Case Study in the Biosphere on Yakushima Island*. Report on special Grant-in-Aid Environmental Sciences by the Ministry of Education, Science and Culture, 17 pp.

Tegler B, Sharp M and Johnson MA. 2001. Ecological monitoring and assessment network's proposed core monitoring variables: early warning of environmental change. *Environmental Monitoring and Assessment*, **67**(1–2): 29–56 (also http://eqb-dqe.cciw.ca/eman/).

Tilman D, Fargione J, Wolff B, D'Antonio C, Dobson A, Howarth R, Schindler D, Schlesinger WH, Simberloff D and Swackhamer D. 2001. Forecasting agriculturally driven global environmental change. *Science*, **292**(5515): 281–4.

UNEP. 2002. *Global Environmental Outlook 3*. Earthscan.

Vitousek PM. 1994. Beyond global warming: ecology and global change. *Ecology*, **75**(7): 1861–76.

Vitousek PM, D'Antonio CM, Loope LL, Rejmanek M and Westbrooks R. 1997. Introduced species: a significant component of human-caused global change. *New Zealand Journal for Ecology*, **21**(1): 1–16.

Warren A and French JR. (eds) 2001. *Habitat Conservation*. Wiley.

Warren A and Goldsmith FB. (eds) 1974. *Conservation in Practice*. Wiley

Warren A and Goldsmith FB. (eds) 1983. *Conservation in Perspective*. Wiley

Watson R *et al.* 2001. *Climate Change 2001: Synthesis Report*. Cambridge University Press.

Wolff M and Mendo J. 2000. Management of the Peruvian bay scallop (*Argopecten purpuratus*) metapopulation with regard to environmental change. *Aquatic Conservation*, **10**(2): 117–26.

CHAPTER 16

Looking forward – prospects for biogeography

Key points

- Subjects are dynamic and in an increasingly competitive world they need to be examined critically for justification of their area of study;
- Biogeography has been a part of study for thousands of years but only recently has it become a discrete academic topic;
- Research shows that its importance is growing especially since the 1990s;
- Biogeography is a dynamic subject area with many topics of relevance to current conditions and a growth in significance in public, business and political dimensions.

Figure 16.1 The modern idea of biogeography. Australia – a view from Taronga Zoo across Sydney Harbour to the City beyond: semi-natural habitat for introduced and endangered species to human landscape in a nation with one of the highest rates of species loss

16.1 Introduction

Only a few years ago it would have been unusual to try to justify a subject: its existence was proof enough. Today, we have an increasing diversity of subjects competing for a finite supply of money, research areas, students, staff and publicity (or even reactions to nature – see Figure 16.1). It is important to be able to put a case for the retention or even expansion of biogeography. In practical terms, as biogeographers, we could argue that it is no bad thing: after all, if we cannot give a good reason why we are studying a subject then one might question why we are! There is a certain vigour in a subject which can critique its own area. Competition for students and a limited fund of money mean that we can no longer rely on 'being there' to continue work. There is evidence in the UK, Australia and elsewhere that departments are merging and some areas of study are being lost. Since we are biogeographers it follows we need to support the subject (it also gives us insight into other subjects' behaviour who would presumably fight as hard for their piece of academic turf!). Fortunately, the need for biogeography has never been greater – all we need to do is to see how it has developed recently and how it can be used in the future.

One of the paradoxes of modern biogeography is that it is an ancient subject (the ancient Greeks had several 'biogeographers' among their academics even if they did not consider themselves as such) which has only recently taken off as a legitimate study in universities (and is still waiting to make a similar impact in the school syllabuses). Such is the recent strength in the subject that one leading UK academic could refer to the 'Cinderella' subject in her first edition (late 1960s), as a strong university subject (second edition late 1970s) and as a broadening discipline even reaching schools in the third edition (Tivy 1993).

The aim of this chapter is to take this argument further and to demonstrate the utility of biogeography in a number of areas – educational, political and social: to show that it has a place and that it can provide us with ideas and insights not readily available elsewhere.

16.2 The role of biogeography in higher education

In common with Tivy (1993), Brown and Lomolino (1998) note the antiquity of the subject and contrast this with its rapid recent rise in importance in universities. Although it is a very crude measure, a simple bibliographic search does give some indication of the amount of material published. One of the more popular abstracting services, Cambridge Life Sciences, was interrogated about the number of papers collected with the word 'biogeography' in the title (see Figure 16.2). Leaving aside those papers published prior to 1979 (and there were relatively few of them) we can see that there were just over 600 in the first five-year period. This increased modestly in the next five-year span but then almost doubled by the end of

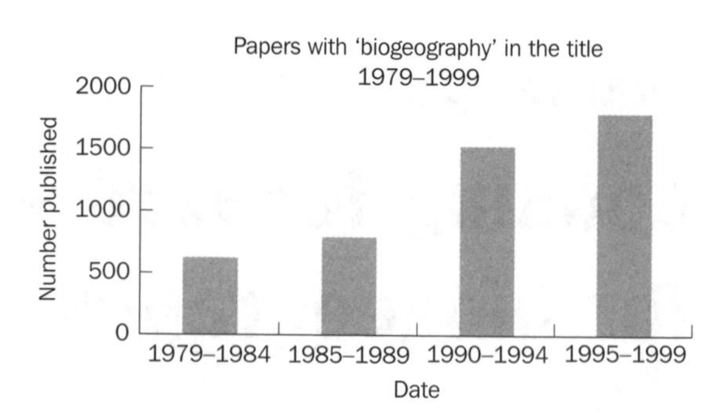

Figure 16.2 Analysis of academic papers

1994. The last time period shows a good rise but nothing to match the early 1990s. There are two questions we can ask from here: how good are the data and what does this 1990s rise mean?

The quality of the data is problematic. Such searches are valuable only if the authors all decide to use the keyword. Since it is not compulsory, some titles might slip past without the true nature of the work being recognised (a subject like biogeography covers a range of disciplines outside geography and so multidisciplinary work is common – others may not choose to use this keyword). However, set against these limitations we can at least argue that the magnitude of the responses gives some indication of relative interest in the topic area. It remains to consider why there is a marked rise in the 1990s (not a decade, many will recall, with an abiding interest in biogeography or any ecological topic). Although research in this area is poor we can at least point to some possibilities. Brown and Lomolino (1998) argue that biogeography is not an experimental science but a comparative, observational one. Biogeography is **the prime** synthesising discipline – it uses the ideas of other subjects to produce a holistic overview of organismal distribution. In the last 10–15 years there has been a rise in the use of impact statements, conservation projects, etc. These require biogeographic data: the rise we see in papers could thus mirror the rise in demand for data and the interpretation of data. Whatever the mechanism, there is a clear interest in the spatiotemporal distribution of organisms (which is after all what biogeography is). Alongside the rise of papers there is the rise of specialist journals. One of the first was the *Journal of Biogeography* which started in 1973. Today, this has expanded considerably.

What of biogeography as a subject area? Alongside the rise in published papers we see the rise of biogeography courses, departments and academics (Figure 16.3). Although there is no guaranteed link between increased personnel and increased papers, it is obvious that the more researchers one has, the more interest and research can be generated. Again, using a bibliographic database it is possible to demonstrate that the interest in the future development of biogeography is strong and follows a series of key

Figure 16.3 Biogeography and academic interest. From a medieval hunting technique and pasttime to a modern study in raptor ecology. This image of a long-eared owl is part of a display of falconry at an agricultural show in southern England

areas. Prime among these is the use of biogeography in producing a biological inventory. Ertter (2000) argues that far from being completed, there is a considerable need for a systematic (and systemic) analysis of plants in the USA. She considers that there are serious gaps, both taxonomic and distributional, that need to be filled if a serious effort is to be made in presenting a detailed biodiversity analysis of the USA. Similar themes are echoed by Compagno and Cook (1995) and Garnock-Jones and Breitwieser (1998). Frankenberg (1999) takes a similar perspective but here the aim is to use systematics to look at the development and future of an area. Potential pressure from housing, tourism, etc. in a fragile area of Israel can lead to damage: a biogeographical review can aid understanding and wise use. Wise use suggests conservation and it is here that the third focus can be seen. Pryde (1997) outlines an interesting case in New Zealand. Here, biogeography is being used to determine the impact of introduced species and also to gauge the effectiveness for conservation where these introduced species are selectively removed from small islands – thus creating an outdoor laboratory. Keddy (1999) describes a similar case only for the new field of restoration ecology. Like many of the examples mentioned here there is more than one perspective. Keddy integrates a practical biogeographical application (plant distribution), ecology (assembly rules) and social science (management) to suggest that restoration ecology needs intelligent application of existing ideas, not creation of new ideas and data (although there are several compelling cases where more data would be useful including those noted by Tikhomirov (1999), Irish (1994) and Hu (1993) among others). Fabricius and Burger (1997) highlight the relatively new field of indigenous biogeography where 'native' knowledge can be used to supplement or replace that gained from the Western scientific tradition. Even though these papers represent only a fraction of the work being done it does demonstrate forcefully the utility and value of biogeography.

16.3 The public perspective of biogeography

This can be a trickier case given that the public do not necessarily see biogeography as a discrete subject. However, interest in natural history (which could be seen as synonymous especially in its early days) has had a strong following for over 150 years. In the second half of the nineteenth century there were numerous UK Acts and organisations which demonstrate the interest in wildlife at this time. A Bird Protection Act in 1869 soon led to the founding of nature reserves and thus to the founding of what we know as the Royal Society for the Protection of Birds in 1889. The National Trust, long known for its interest in built heritage, was set up in 1895 and started by obtaining land for conservation (Nature Conservancy Council 1984). Given that these organisations have memberships of more than 2 million each it is obvious that there is support for their ideals.

Perhaps one of the greater impetuses for biogeography came from the foundation work of the British Broadcasting Corporation and its Wildlife Unit. As soon as television became an affordable medium in the 1950s, the BBC put out schools broadcasts complete with booklets. Such was the success of this that the BBC still produce work on this theme through television and multimedia. See Box 16.1.

16.4 Biogeography in commerce and business

Biogeography is the foundation of business in that the earliest forms of trade were for biological products, i.e. food (species), that needed to be gathered from specific locations (habitats). Along with biotechnology (bread and ale making) this represented the cornerstone of the early industrial/urban complex (see Figure 16.4). The impact of plant and animal distributions on human populations has long been accepted (Ucko and Dimbleby 1969); in fact it has been seen as the cornerstone of the development of the various civilisations through time.

Today, the situation has changed dramatically. Through increased use of technology fewer people are subject to direct environmental effects and so the use of biogeographical knowledge could be subject to decline. That this is far from the case can be seen through the increasing use of environmental impact assessment (EIA):

BOX 16.1

The public and biogeography

When discussing public interest it is important to note the contribution of certain authors. Although few will be explicitly biogeographical there is an ecological theme running through them that can interest the lay person. One of the first modern texts is Carson's *Silent Spring* (Carson 1962). This alerted people to the real problems of pesticide use in forested areas. Following on the heels of this work, the student unrest of the late 1960s and the activism focused on global and environmental issues paved the way for the first UN conference at Stockholm in 1972. Ward and Dubos's (1972) book *Only One Earth* was a key work in attracting public attention. In the UK this could be seen as the start of widespread public interest in the environment in general. From that time senior school syllabuses included courses in environmental studies and science where ecology and biogeography were key components. Subsequent UN conferences at Rio (1992) and Johannesburg (2002) have shown the enormous growth in interest (if not in solutions).

Figure 16.4 Commercial biogeography. Crayfish tanks in Western Australia

> EIA was conceived in the late 1960s in response to the groundswell of ecocentric concerns which began to challenge the technocentric view of the continued development of industry and the ability of science and managerial ingenuity to intervene to create economic growth and overcome environmental problems. (Petts 1999)

The aim of EIA is to create a systematic structured overview of the impacts of development upon the environment. Although the definition of EIA varies from country to country, the idea of pre-building assessment of impacts is central to its philosophy. Information is gained from a range of sources and the likely impacts of each one are assessed. It should therefore be possible to see beforehand what might happen and to suggest ways around these difficulties. The range of information is broad but does include, in the UK at least, reference to fauna, flora and landscape which are of direct interest to biogeography (Petts 1999). If EIA relates to a particular project then its close relative, strategic environmental assessment (SEA) has a broader focus, looking at plans, policies and programmes (Therivel and Paridario 1996). Both EIA and SEA require ecological assessments which put biogeography at the forefront.

Understanding potential problems is not the only use of biogeographical studies in business. In 1987, the United Nations started discussions which led to the Convention on Biological Diversity – CBD (CBD 2001). Although seen primarily as a way of controlling the loss of species, the CBD is also concerned about the economic value of species. This covers items such as agricultural plant genetic resources, plant collections, the use of plants in pharmacy, etc. Since the benefits to the pharmaceutical industry could be considerable it follows that plant and animal distributions are crucial data. Finally, biogeography is becoming used increasingly in risk assessment. Here, changes in vegetation cover, for example, can have significant impacts on rural production which can in turn impact upon agricultural and insurance companies (Kasperson and Kasperson 2001).

16.5 Biogeography and politics

Biogeography has been a major part of politics for centuries but, like its academic counterpart, this has only recently been made explicit (see Figure 16.5). Knowledge of natural biological resources has driven tribes, states and nations in the past no less than it will in the future (Diamond 1999). Hippocrates and Aristotle were concerned about the impact of the environment (Glassner 1996) and whereas the concern might be more its effect on them, today we worry about our effect on it!

Although biogeography is not taken 'as is' by governments the knowledge derived from its study is vital if we are to manage our biological resources. Concern over the environment has grown especially since the Stockholm Conference in 1972. Around this time, several nations set up crucial Acts dealing with environmental matters of which the USA's Environmental Policy Act 1970 was one of the first and most influential. Since that time it has become a major political battlefield with support from some presidents (e.g. Carter – Council for Environmental Quality 1982) and others with evidence that they are less well disposed towards it.

Figure 16.5 Biogeography and politics. The Law of the Sea is one of the earliest global agreements (which is still being decided!). Oceans are also subject to considerable loss of species with little sign of renewable resources agreements. Here, at part of the Twelve Apostles, a sea stack formation off the south Australian coast, we have the pressure of conservation areas to add to whale migration routes and overfishing pressures

Elsewhere there is concern over a range of biological resource issues. The United Nations Law of the Sea Conferences started in the late 1950s (Glassner 1996) and since that time have covered a range of issues which include wildlife distribution, abundance and use (key biogeographic issues). Biodiversity is another key issue for which biogeography can provide vital data. Irrigation, ocean fisheries, waste, wetlands, wildlife, feral animals and water are just some of the 'ecopolitical wars' that Glassner considers are of sufficient merit to be considered as geopolitical issues. Plate 16 shows peat digging on a wetland site. Areas such as this are extremely vulnerable to human activity of any sort and the loss of biodiversity consequent upon even moderate amounts of interference is irreplaceable. Wise management is the key, but does it happen?

Although this can seem like a losing battle there are encouraging signs. A range of international conferences supported by the UN has raised awareness of environmental issues. Major organisations like the United Nations Environment Programme have been established. Global warming (with its concomitant impact on wildlife distribution) is one of the major issues of the late twentieth and early twenty-first centuries, along with the impacts of transboundary pollution. The message is clear: there is an increasing recognition of, and demand for, biogeographical information.

16.6 The biogeography of the future

What are we looking at here? The future of biogeography which is looking quite optimistic as subject specialties combine, or the biogeography of the future which looks towards depletion of the basic organismic stock? It would be far too easy to state that we are at a pivotal time but there are certainly problems and prospects to be faced. The impact of human action upon the biosphere has altered radically the distribution of plants and animals. We tend to think of the extinctions but there is also the massive movement of introduced species to be considered (e.g. Low 1999). The net result is a biosphere map quite unlike that of a century ago. Of course since the distribution of plants and animals is always changing (as are the organisms themselves through evolution) it would be fair to challenge whether this modern change is of importance. At the same time, we are seeing an increased interest in the idea of biogeography in its broadest sense. Rarely has interest in 'nature' been more explicit in both public and private arenas. This might not translate into concern or action but it does raise the profile of the subject. As we start the new century it might be well to take stock of our progress and see how we want subject study and the subject itself to develop.

16.7 Summary

No subject has the 'right' to exist – its proponents need to be able to justify its value. If this is done in academic and public areas then it is likely to attract both attention and funding which is an integral part of subject development in the twenty-first century. Rather than arguing for biogeography just because it exists, it is possible to demonstrate clearly where the subject can contribute to a range of activities: educational, commercial and political. As our impact on the planet increases then there is every possibility that our need for biogeographical ideas and related skills will increase.

APPLICATIONS – USING BIOGEOGRAPHY

Using ideas from this chapter in real-life situations

This entire chapter has been given over to considering where this subject might be heading. From the range of applications boxes seen in other chapters it follows that the range of ideas available to us is considerable. It is difficult to select any one area as being more vibrant because, despite funding problems in higher education globally, there are still opportunities for research. On a personal level we have used biogeography in a range of situations from education to archaeology and ecology to impact statements. One topic which cuts across this and must be seen to be one of the more exciting prospects is GIS or geographic information systems. This database/map hybrid has the potential to provide unrivalled amounts of information which can be interrogated in a number of ways. Those of our readers in university courses would do well to look into this. Although global change is continuing, it is the critically important unifying power of biogeography which can help make sense of it.

Review questions

1. Repeat the database exercise noted in Figure 16.2 using different databases and different keywords. How do the results differ and why? Look at the 'help' section found in most databases. How do their search strategies change?

2. To what extent does popular media coverage affect the development of biogeography?

3. Describe the operation of environmental impact statements in your country. How does this practice differ from other countries?

4. Justify the role of the UN in global biogeographical initiatives like the CBD.

5. Describe the role played by biogeography in the development of the National Parks system in your country.

Selected readings

There are few texts which cover this topic alone. Brown and Lomolino (1998) have one of the largest sections in an introductory text. The CBD Secretariat (2001) give a good example of the complexity of this subject area, while the Council for Environmental Quality (1982) have produced one of the best texts on the way politics and biogeography intersect.

References

Brown J and Lomolino M. 1998. *Biogeography*. Sinauer.

Carson R. 1962. *Silent Spring*. Penguin.

CBD Secretariat. 2001. *Handbook of the Convention of Biological Diversity*. Earthscan.

Compagno LJV and Cook SF. 1995. The exploitation and conservation of freshwater elasmobranchs: status of taxa and prospects for the future. In Oetinger and Zorzi (eds) *The Biology of Freshwater Elasmobranchs*, Vol. VII. *Journal of Aquariculture and Aquatic Sciences*, **7**: 62–90.

Council for Environmental Quality. 1982. *The Global 2000 Report to the President*. Penguin.

Diamond J. 1999. *Guns, Germs and Steel*. Norton.

Ertter B. 2000. Our undiscovered heritage: past and future prospects for species-level botanical inventory. *Madrono*, **47**(4), October–December: 237–52.

Fabricius C. and Burger M. 1997. Comparison between a nature reserve and adjacent communal land in xeric succulent thicket: an indigenous plant user's perspective. *South African Journal of Science*, **93**(6): 259–62.

Frankenberg E. 1999. Will the biogeographical bridge continue to exist? *Israel Journal of Zoology*, **45**(1): 65–74.

Garnock-Jones PJ and Breitwieser I. 1998. New Zealand floras and systematic botany: progress and prospects. *Australian Systematic Botany*, **11**(2), May 28: 175–84.

Glassner MI. 1996. *Political Geography*, 2nd edn. Wiley.

Hu S. 1993. On general aspects of the Antarctica plants and vegetation and their prospects in botanical research. *Acta Botanica Sinica*, **35**(11): 868–76.

Irish J. 1994. Systematics and biogeography of southern African Lepismatidae (Thysanura): current progress and future prospects. *Acta Zoologica Fennica*, **195**: 81–6.

Kasperson J and Kasperson R. 2001. *Global Risk Assessment*. Earthscan.

Keddy P. 1999. Wetland restoration: the potential for assembly rules in the service of conservation. *Wetlands*, **19**(4), Dec: 716–32.

Low T. 1999. *Feral Future*. Viking.

Nature Conservancy Council. 1984. *Nature Conservation in Britain*. NCC.

Petts J. (ed.) 1999. *Handbook of Environmental Impact Assessment*, 2 vols. Blackwell.

Pryde P. 1997. Creating offshore island sanctuaries for endangered species: the New Zealand experience. *Natural Areas Journal*, **17**(3): 248–54.

Therivel R and Paridario MR. (eds.) 1996. *The Practice of Strategic Environmental Assessment*. Earthscan.

Tikhomirov V. 1999. Regional problems in plant chorology of European Russia. *Acta Botanica Fennica*, **162**, Jan. 15: 99–102.

Tivy J. 1993. *Biogeography*. Longman.

Ucko PJ and Dimbleby GW. 1969. *The Domestication and Exploitation of Plants and Animals*. Duckworth.

Ward B and Dubos R. 1972. *Only One Earth – care and maintenance of a small planet*. Penguin.

INDEX

Notes

1. Page numbers in **bold** indicate chapters; those in *italics* indicate illustrations.
2. There are entries for most countries and regions but United Kingdom is omitted as references are too numerous.